Health and Safety Law

Health and Safety Law

Fifth edition

Jeremy Stranks
MSc, FCIEH, FIOSH, RSP

Harlow, England • London • New York • Boston • San Francisco • Toronto
Sydney • Tokyo • Singapore • Hong Kong • Seoul • Taipei • New Delhi
Cape Town • Madrid • Mexico City • Amsterdam • Munich • Paris • Milan

PEARSON EDUCATION LIMITED

Edinburgh Gate
Harlow CM20 2JE
Tel: +44 (0)1279 623623
Fax: +44 (0)1279 431059
Website: www.pearsoned.co.uk

First published 1994
Second edition 1996
Third edition 1999
Fourth edition 2001
Fifth edition published in Great Britain 2005

© Jeremy Stranks 1994, 2005

The right of Jeremy Stranks to be identified as author
of this work has been asserted by him in accordance
with the Copyright, Designs and Patents Act 1988.

ISBN–10: 0 131 97646 X
ISBN–13: 978 0 131 97646 7

British Library Cataloguing in Publication Data
A CIP catalogue record for this book can be obtained from the British Library.

Transferred to Digital Print on Demand 2012

Typeset by 70 in 10pt Palatino
Printed and bound by CPI Group (UK) Ltd, Croydon, CR0 4YY

The Publisher's policy is to use paper manufactured from sustainable forests.

Contents

❺ **The principal cases 351**

Acknowledgements

We are grateful to Her Majesty's Stationery Office for use of Crown copy material. Crown copyright material is reproduced with the permission of the Controller of HMSO and the Queen's Printer for Scotland.

In some instances we may have been unable to trace the owners of copyright material and would appreciate any information that would enable us to do so.

Preface

Complying with the extensive requirements of health and safety law is, for many employers, something of a problem. To this extent, this edition of *Health and Safety Law* has been extended to provide recommendations for employers and their advisers with regard to documentation requirements, management systems, Statements of Health and Safety Policy, safety monitoring activities and the risk assessment process. The requirements for the appointment of competent persons and for providing information, instruction, training and supervision, together with health surveillance in certain cases, are also covered.

Health and safety law is in a continuing state of change. For example, the Health and Safety (Miscellaneous Amendments) Regulations 2002 have amended a number of Regulations and much of the legislation dealing with the transport of dangerous goods has been revoked. Recent legislation including that relating to lifts, asbestos, pressure systems, transportable pressure vessels, fire and rescue services, dangerous substances and explosive atmospheres has been included. It is regrettable, at this stage, that government proposals relating to 'corporate manslaughter' have still not been implemented.

The legal situation relating to stress-induced injury has now been clarified, to some extent, and a series of 'practical propositions' for courts dealing with claims for 'psychiatric injury' have been incorporated.

I hope that all who use this book will find it helpful, in particular those studying for NEBOSH examinations. Legal practitioners may also find this book useful.

Jeremy Stranks
2005

List of abbreviations

AC	Appeal cases
ACAS	Advisory, Conciliation and Arbitration Service
ACOP	Approved Code of Practice
ADR	The European Agreement concerning the International Carriage of Goods by Road
BATNEEC	Best available technology not entailing excessive costs
BPEO	Best practicable environmental option
BS	British Standard
CAC	Central Arbitration Committee
CBI	Confederation of British Industry
CDMR	Construction (Design and Management) Regulations
CEN	European Committee for Standardisation
CENELEC	European Committee for Electrotechnical Standardisation
CHIP 3	Chemicals (Hazard Information and Packaging for Supply) Regulations 2002
CHP	Combined Heat and Power
CJA	Criminal Justice Act 1991
COMAH	Control of Major Accident Hazards Regulations 1999
CORGI	Council for the Registration of Gas Installers
COSHHR	Control of Substances Hazardous to Health Regulations 2002
COTIF	The Convention concerning International Carriage by Rail
DDA	Disability Discrimination Act 1995
DSE	Display screen equipment
DWI	Disposable weekly income
EA	Environment Agency
EC	European Community
ECJ	European Court of Justice
EEA	European Environment Agency
EFTA	European Free Trade Association
EMA	Employment Medical Adviser
EMAS	Employment Medical Advisory Service
EPA	Environmental Protection Act 1990
ERA	Employment Rights Act 1996
EU	European Union
FA	Factories Act 1961
FPWR	Fire Precautions (Workplace) Regulations 1997
FRA	Fire and Rescue Authority
FSSPSA	Fire Safety and Safety of Places of Sport Act 1987

GMO	Genetically modified organism
HFL	Highly flammable liquid
HMIP	Her Majesty's Inspectorate of Pollution
HMSO	Her Majesty's Stationery Office
HSC	Health and Safety Commission
HSE	Health and Safety Executive
HSWA	Health and Safety at Work, etc. Act 1974
IPC	Integrated Pollution Control
IRLR	Industrial Relations Law Reports
IRR	Ionising Radiations Regulations 1999
LA	Local authority
LEV	Local exhaust ventilation
LOLER	Lifting Operations and Lifting Equipment Regulations 1998
LPG	Liquefied petroleum gas
LT	Law Times Reports
MAPP	Major Accident Prevention Policy
MEL	Maximum Exposure Limit
MHOR	Manual Handling Operations Regulations 1992
MHSWR	Management of Health and Safety at Work Regulations 1999
NEC	Network Emergency Co-ordinator
NPA	National Parks Authority
NRA	National Rivers Authority
NSNA	Noise and Statutory Nuisances Act 1993
OES	Occupational Exposure Standard
OLA	Occupiers Liability Acts 1957 and 1984
OSRPA	Offices, Shops and Railway Premises Act 1963
PPE	Personal protective equipment
PPEWR	Personal Protective Equipment at Work Regulations 1992
Ps.V	Product of pressure and volume
PUWER	Provision and Use of Work Equipment Regulations 1998
Reg	Regulation
REPAC	Regional Environmental Protection Advisory Committee
RID	Regulations Concerning the International Carriage of Dangerous Goods by Rail
RIDDOR	Reporting of Injuries, Diseases and Dangerous Occurrences Regulations 1995
RoSPA	Royal Society for the Prevention of Accidents
RPA	Radiation Protection Adviser
RPE	Respiratory protective equipment
RSA	Radioactive Substances Act 1960
RSI	Repetitive strain injury
SC	Session Cases (Scotland)
SEPA	Scottish Environment Protection Agency
SI	Statutory Instrument
SLT	Scots Law Times

SRSCR	Safety Representatives and Safety Committees Regulations 1977
SSGA	Safety of Sports Grounds Act 1975
SSP	Statutory sick pay
SWL	Safe working load
SWP	Safe working pressure
TPR	Transportable pressure receptacle
TPV	Transportable pressure vessel
UCTA	Unfair Contract Terms Act 1977
VDU	Visual display unit
VWF	Vibration-induced white finger
WCA	Waste Collection Authority
WDA	Waste Disposal Authority
WHSWR	Workplace (Health, Safety and Welfare) Regulations 1992
WRA	Waste Regulation Authority
YTS	Youth Training Scheme

Elements of health and safety law

SOURCES OF LAW

Health and safety law is derived from a number of sources. The three principal sources of English law are legislation, or statute law, common law, from which most case law is derived, and contract law.

Statute law

Statute law consists of Acts of Parliament, such as the Health and Safety at Work, etc. Act 1974 (HSWA), which is an enabling Act, together with a large number of statutory instruments (SIs) made under the Acts of Parliament. Statutory instruments are generally known as 'subordinate legislation', or, 'delegated legislation'. Subordinate legislation relating to workplace health and safety consists of statutory instruments proposed by the Health and Safety Commission (HSC), after consultation with industry and local authorities and other relevant bodies, to the Secretary of State for Employment and, subsequently, laid before Parliament. Statutory instruments take the form of regulations, such as the Management of Health and Safety at Work Regulations 1999 (MHSWR), the Control of Substances Hazardous to Health Regulations 2002 (COSHH 4) and Electricity at Work Regulations 1989. Since 1974, when the HSWA was introduced, all regulations concerning workplace health, safety and welfare have been passed in furtherance of HSWA.

Common law

Common law is an area of law that has developed since the eleventh century and is based on the decisions of the courts whereby *precedents* are established. Common law is, thus, the body of case law that is universally, or commonly, applied as a result of the judgments of the courts. Each judgment contains the judge's enunciation of the facts, a statement of the law applying to the case and their *ratio decidendi*, or legal reasoning, for the finding that has been arrived at. These various judgments are recorded in the series of *Law*

Reports and have developed into the body of decided case law we now have and continue to develop. Common law is, then, accumulated case law, recorded for the most part in law reports, underpinned by the doctrine of precedent, under which a court is bound to follow earlier decisions of courts of its own level and those of superior courts.

Judicial precedent

Judicial precedent is defined as 'a decision of a tribunal to which some authority is attached'. Precedents not only influence the development of law but are, in themselves, one of the material sources of the law.

To say that a case is a 'binding decision of precedent' means that the principle on which the decision was made will be binding in subsequent cases which are founded on similar facts.

A precedent may be authoritative or persuasive.

Authoritative precedents

These are decisions which judges are bound to follow. There is no choice in the matter; for instance, a lower court is bound by a previous decision of a higher court.

Persuasive precedents

These are decisions which are not binding on a court but to which a judge will attach some importance; for instance, decisions given by the superior courts in Commonwealth countries will be treated with respect in the English High Court.

The hierarchy of the courts

The doctrine force of precedent is based on the principle that the decisions of superior courts bind inferior courts. Thus:

1. The High Court is bound by a decision of the Court of Appeal, and the Court of Appeal is bound by a decision of the House of Lords, these courts being in the same hierarchy.
2. A High Court judge is not bound by a previous decision of the High Court. Such a decision enjoys only persuasive authority.
3. In civil matters, the Civil Division of the Court of Appeal will bind all inferior courts. It is bound by its own decisions and by those of the House of Lords.

   In criminal matters, the Criminal Division of the Court of Appeal will bind all inferior courts.
4. County Courts, Magistrates' Courts and Administrative Courts are bound, without question, by decisions of the superior courts.

Contract law

A contract is an agreement between two parties and, to be legally binding, must have certain basic features.

First, it must be certain in its wording. A contract consists of an *offer* made by one party that must be accepted unconditionally by the other. Second, there must be *consideration* that flows from one party to another, that is, the legal ingredient that changes an informal agreement into a legally binding contract, such as goods for money, work for pay. Third, there must be *intention*, whereby the parties concerned must intend tó enter into this legally binding agreement. Finally, both parties must have the legal *capacity* to make such a contract. Legally, capacity usually means that parties concerned be sane, sober and over the age of 18.

Contractors

A contractor is one who is engaged to perform a certain task without direction from the person employing him. A distinction must be made between an independent contractor engaged to perform a specific task and a servant who can be directed in terms of how to undertake a task.

The basic test of whether a person is an independent contractor is one of control over the undertaking of the work. Independent contractors working in another person's workplace are, in the main, responsible for their torts and those of their servants, i.e. employees. An occupier or employer must be careful in selecting a reliable independent contractor, that is, one who is competent to perform the task required.

Contractors are responsible for employing skilled and competent workers who can be relied upon to do the work in an efficient and safe manner. Moreover, an occupier cannot escape liability by putting out a job to a contractor which he, the occupier, has a statutory duty to perform.

Unfair Contract Terms Act 1977

This Act states that a person cannot by reference to any contract term or to a notice given to persons generally exclude or restrict his liability for death or personal injury resulting from negligence.

It is unlawful to contract out of, or to seek to modify, liability for personal injury or death caused by negligence. As far as negligent damage of property is concerned, the contract term must be fair and reasonable in the light of the circumstances at the time liability arose.

Criminal and civil law

A *crime* is an offence against the state. In such a case, a person who commits a crime, which is a breach of criminal law, is *prosecuted* and brought before a court. The burden of proving a criminal charge *beyond reasonable doubt* rests with the prosecution. If found guilty of the crime, the court will impose some form of *punishment*, such as a fine or jail sentence, or both. While it is not the function of a criminal court to compensate, compensation may be ordered by a court to be given to an individual to cover personal injury and damage to property. However, a compensation order is not possible for the dependants of a deceased person in consequence of that person's death. Criminal cases are heard in the magistrates' courts and in the Crown Court. The more serious cases, which pass to the Crown Court, are heard before a judge and jury.

A *civil action*, on the other hand, generally involves individuals. In such actions, a claimant *sues* a defendant for a remedy (or remedies) that is beneficial to the claimant. In most cases this is *damages*, a form of financial compensation. (Many accident claims, for instance, pass through the civil courts, with damages being awarded to the injured person.) In a substantial number of cases, the claimant will agree to settle out of court. The civil courts are the County Courts and the High Court. A civil case must be proved on the *balance of probabilities*, which is a lesser standard than that of 'beyond reasonable doubt' required in a criminal case.

Criminal and civil liability

Criminal liability refers to the responsibilities under statute and the penalties that can be imposed by criminal courts, such as fines, imprisonment and remedial orders. The criminal courts in question are the Magistrates' Courts, which handle the bulk of health and safety offences as these tend to be the less serious offences, and the Crown Courts, which deal with the more serious ones (see *The court hierarchy in England and Wales* later in this chapter). Appeals can be made to the High Court and beyond, to the Court of Appeal, and, assuming leave is given, to the House of Lords. Subject to certain criteria, a case may pass to the European Court of Justice.

Criminal law is based on a system of enforcement. Its statutory provisions are enforced by the State's enforcement agencies, such as the police. Health and safety law is enforced mainly by the Health and Safety Executive (HSE), local authorities and fire authorities. In contrast to civil liability in tort, criminal liability cannot be insured against by employers, directors or employees. However, it is lawful to insure against court costs and the legal expenses of counsel in connection with a prosecution for health and safety offences.

Civil liability refers to the 'penalty' that can be imposed by a civil court. The civil courts are the County Court, High Court, Court of Appeal (civil division) or House of Lords. In general terms, the first two courts are courts of first instance, that is, they try cases for the first time, whereas the latter two courts are courts of appeal. Civil liability consists of awards of damages for injury, disease and/or death at work:

- in circumstances disclosing a breach of common law and/or statutory duty (normally negligence (see Chapter 2) on the part of an employer/factory occupier)
- arising out of and in the course of employment.

The damages that can be recovered following an injury-causing accident, contracting of an occupational disease or death fall into two categories:

- *general damages* – these relate to losses incurred after the hearing of the action, namely actual and probable (not merely 'possible') loss of future earnings following the accident)
- *special damages* – these relate to quantifiable losses incurred before the hearing of the case, and mainly consist of:
  - medical expenses incurred before the hearing of the case
  - loss of earnings incurred before the hearing of the case.

In the case of fatal injury, compensation for death negligently caused is payable under the Fatal Accidents Act 1976. Moreover, a fixed lump sum is payable under the Administration of Justices Act 1982 in respect of bereavement.

As far as assessment of future probable earnings under the heading of 'general damages' is concerned, such future earnings are assessed net (that is, as if tax had been deducted). Moreover, where, as is likely, an injured employee is in receipt of social security benefits, the Social Security (Consequential Provisions) Act 1975 provides that half of the value of certain social security benefits, including particularly sickness benefit and disablement benefit, must be deducted from the subsequently awarded damages.

THE COURT HIERARCHY IN ENGLAND AND WALES

There are two distinct systems, the courts dealing with criminal or civil actions respectively. Some courts have both criminal and civil jurisdiction, however.

The Magistrates' Courts

This is the lowest of the courts and deals mainly with criminal matters. Magistrates determine and sentence for many of the less serious offences. They also hold preliminary examinations into other offences to see if the prosecution can show a prima-facie case, as a result of which the accused may be committed for trial at a higher court.

The Crown Courts

Serious criminal charges (and some not so serious, but where the accused has the right to jury trial) are heard on indictment in the Crown Court. This court also hears appeals from magistrates' courts.

The County Courts

Each of these courts deal with civil matters only.

The High Court of Justice

More important civil matters – because of the sums involved or their legal complexity – will start in the High Court of Justice. The High Court has three divisions:

- Queen's Bench, which deals with contract and torts
- Chancery, which deals with matters relating to, for instance, land, wills, partnerships and companies
- Family.

In addition, the Queen's Bench division hears appeals on matters of law:

- from the magistrates' courts and from the Crown Court on the procedure called 'case stated'
- from some tribunals, for example the finding of an industrial tribunal on an enforcement notice under the HSWA.

The Queen's Bench division also has some supervisory functions over the lower courts and tribunals if they exceed their powers or fail to undertake their functions properly, or at all.

The High Court, the Crown Courts and the Court of Appeal are known collectively as the Supreme Court of Judicature.

The Court of Appeal

The Court of Appeal has two divisions:

- the civil division, which hears appeals from the County Courts and the High Court
- the criminal division, which hears appeals from the Crown Courts.

The House of Lords

Further appeal (in practice, this is on important matters of law only) is made to the House of Lords from the Court of Appeal and, in restricted circumstances, from the High Court.

European Court of Justice

This is the supreme law court, whose decisions on the interpretation of European Community law are sacrosanct. Such decisions are enforceable through the network of courts and tribunals in all member states.

Cases can only be brought before this court by organisations or individuals representing organisations.

For a summary of the positions of these courts in the judicial hierarchy and which hear criminal and which civil cases, see Figure 1.1.

COURT PROCEDURES

Civil actions

In a civil action a claimant sues a defendant. Most civil actions involve an action in tort, commonly negligence and/or breach of statutory duty, and a claim for damages in respect of personal injury, death or damage to property.

The first stage of a civil action is the issue of a *Writ of Summons*, which must be acknowledged by the defendant following service by returning a completed form of acknowledgement of service to the issuing court office, i.e. the Central Office of the High Court or the District Registry. Where a defendant fails to acknowledge service of a writ within the specified time limit, judgment may be entered against him.

Where an issue goes for trial, the counsel representing the claimant commences by outlining the allegations against the defendant, as indicated in the *Statement of Claim* previously delivered to the defendant. Witnesses may be called, including designated expert witnesses. In most cases, a civil action will be heard by a judge sitting alone.

A defence counsel has a choice of several submissions, namely:

(a) 'No case to answer' – in this case, following a reply from the claimant's counsel, the judge gives his verdict; or

(b) 'No evidence to be called' – the claimant's counsel addresses the court, followed by the defence counsel; or

(c) 'Defence with evidence' – the defendant's counsel submits the case for the defence, calling witnesses as appropriate.

The judge will then find for or against the claimant. If successful, the claimant is awarded damages. Costs are generally paid by the losing party. In certain cases a claimant may be required to pay his own costs, particularly where a court feels such an action should never have been brought.

Standard procedure in the County Court is covered by the Courts and Legal Services Act 1990. County Court actions generally involve actions where damages not exceeding £50,000 are involved. In these actions a summons, which sets out the details of the parties to the action and the nature and amount of the claim, is requested by the claimant. The Registrar, an officer of the court, prepares the summons and serves it on the defendant, which the defendant or his solicitor must accept.

The Registrar frequently carries out a pre-trial review with the objective of identifying the key issues of the claim and any points of contention. The actual trial is taken by a County Court judge. Where the claim is admitted by the defendant, or where the defendant does not present himself to answer the claim, judgment is made in favour of the claimant. In case of dispute, the trial proceeds on lines similar to those of the High Court.

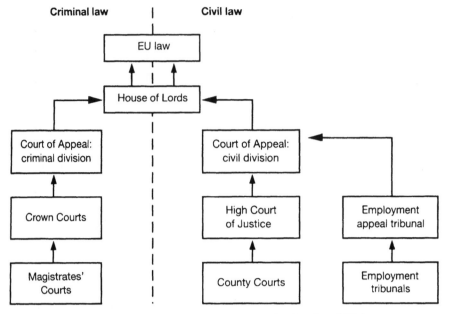

● FIG 1.1 The court structure and hierarchy in England and Wales

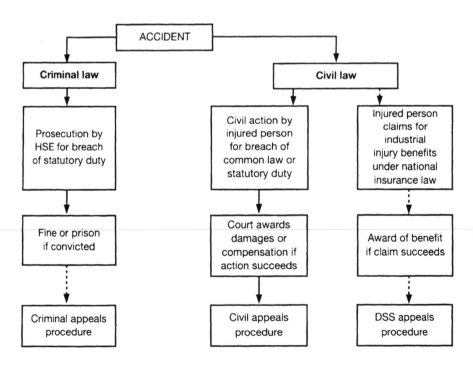

● FIG 1.2 The routes that could be taken through the legal system following an accident at work

Criminal prosecution

Any breach of the HSWA or the relevant statutory provisions is a criminal offence and proceedings are taken in a criminal court.

Offences may be *summary offences, indictable offences* or *triable either way*.

Summary offences

Such offences can only be dealt with by a court of summary jurisdiction, namely a Magistrates' Court. These include a range of minor offences, such as a driving offence, drunken and disorderly behaviour and failure to comply with an Improvement Notice. As such, Magistrates' Courts are the court of first instance, and have the power to refer criminal cases to a higher court, namely the Crown Court.

Summary offences include:

(a) contravening a requirement under section 14 of HSWA (power of the HSC to direct investigations and inquiries);

(b) contravening a requirement of an inspector; and

(c) preventing a person from appearing before an inspector, or from answering questions asked by an inspector.

Indictable offences

A Magistrates' Court can decide whether there is sufficient evidence to commit an offender to a Crown Court for trial. Where such evidence exists, namely a *prima facie* case, magistrates can instigate *committal proceedings* through issue of an *indictment*.

Offences triable only on indictment are:

(a) breaching any of the relevant statutory provisions through by, perhaps, undertaking a process without a licence from the HSE or breaching a term, condition or restriction attached to such a licence;

(b) acquiring or attempting to acquire, possessing or using an explosive article or substance.

Triable either way

Certain offences permit the accused to be heard in either a Magistrates' Court or the Crown Court, depending upon the choice made by the accused and the severity of the offence.

Many offences are triable either way, including:

(a) breach of duties under sections 2 to 7 of the HSWA;

(b) breach of section 8; i.e. intentionally or recklessly interfering with anything provided in the interests of health, safety and welfare;

(c) breach of section 9; i.e. levying charges on employees in respect of anything done or provided in pursuance of any specific requirement of the relevant statutory provisions;

(d) breaching regulations made under HSWA;

(e) obstructing an inspector intentionally;

(f) intentionally or recklessly making a false statement;

(g) intentionally making a false entry in a book, register, etc., required to be maintained;

(h) impersonating an inspector; and

(i) breach of any requirement or prohibition imposed by an Improvement Notice or Prohibition Notice.

Criminal court procedure

Procedures in the criminal courts take the following forms:

- a defendant, one who is accused of a crime, can be brought before the court, either by:

    (a) a summons, namely a written order, signed by a magistrate, ordering that person to appear before a certain court on a certain date at a specified time to answer the accusation, which is the principal means; or

    (b) a warrant, namely a written authority issued by a magistrate, addressed to a constable directing him to carry out some specified act, namely to arrest the person named in the warrant, and to bring that person before the court.

- a summons is served by post or by hand, usually by a police constable
- proceedings for Health and Safety offences can be commenced only by an inspector or with the consent of the Director of Public Prosecutions
- offences under the HSWA can be prosecuted thus:

    triable only summarily – Magistrates' Court
    triable either way – Magistrates' or Crown Court
    triable only on indictment – Crown Court

- prosecution procedures takes the following steps:

    (a) information laid before the magistrate or clerk to the justice; examining magistrate decides whether there is a charge to be answered;
    magistrate decides whether the charge will be tried summarily or on indictment;

    (b) if tried summarily, the decision is explained to defendant who consents or otherwise to the summary trial;

    (c) if tried on indictment, Magistrates' Court proceedings are known as committal proceedings.

- bail may be granted to the accused pending trial, depending upon the severity of the offence, and the accused can appeal to a judge in chambers where bail is refused
- not all witnesses need attend a Magistrates' Court hearing, as written

statement may be accepted as evidence

- two sides are involved – prosecution and defence
- in Magistrates' Courts the procedure is governed by the Magistrates' Court Act 1980, whereby the charge is read out and the defendant asked to state whether guilty or not guilty
  (a) where the defendant denies the charge, witnesses may be called and the evidence of witnesses challenged through the process of cross-examination;
  (b) after the cross-examination stage, the magistrates make a decision upon conviction, the defendant will be sentenced.
- in the Crown Court, a jury decide on guilt or otherwise, and a judge decides on the sentence
  (a) Crown Court procedure starts with an arraignment, namely the calling of the accused, reading of the indictment and calling on the accused to plead;
  (b) where a not-guilty plea is entered, a jury will be sworn in;
  (c) procedure takes the form of reading the prosecution charge and giving of evidence, followed by defence submission of 'no case to answer';
  (d) where 'no case to answer' is accepted by the judge, the jury are directed to find not guilty; where this submission is rejected, the trial continues, with defendant and witnesses giving evidence;
  (e) on completion of evidence and cross-examination stage, judge sums up and the jury considers a verdict;
  (f) a judge may accept a majority verdict of not less than 10:2;
  (g) where a jury cannot agree a verdict, the judge may make a direction to return a not-guilty verdict;
  (h) where jury return a guilty verdict, the judge must decide on sentence to be imposed;
  (i) on a guilty verdict, the judge can hear evidence of both good and bad character which he can take into consideration or not;
  (j) the judge may impose a sentence or discharge the accused; i.e. fine and/or imprisonment, suspended sentence, conditional discharge, absolute discharge.

Appeal procedure – criminal courts

Where convicted of a summary offence there is right of appeal to a Crown Court against:

(a) conviction, sentence or both, if the accused pleaded guilty

(b) sentence only if he pleaded guilty

- the prosecution has no right of appeal
- appeals from the Crown Court are heard by the Court of Appeal, Criminal Division

- both prosecution and defendant may appeal on a point of law by way of 'case stated' to the High Court; the lower court states a case for the High Court to adjudicate on.

TRIBUNALS

Employment tribunals were first established under the Industrial Training Act 1964 to deal with appeals against industrial training levies by employers. They now cover a wide range of other industrial matters, including industrial relations issues, unfair dismissal, equal pay and sex discrimination.

Composition

Each tribunal consists of a legally qualified chairperson appointed by the Lord Chancellor and two lay members – one from management and one from a trade union – selected from panels maintained by the Department of Employment following nominations from employers' organisations and trade unions.

Decisions

When all three members of a tribunal are sitting, the majority view prevails.

Complaints relating to health and safety issues

Employment tribunals deal with the following employment/health and safety issues:

- appeals against improvement and prohibition notices
- time off for the training of safety representatives (Safety Representatives and Safety Committees Regulations 1977 (SRSCR), regulation 11(1)(a))
- failure of an employer to pay a safety representative for time off for undertaking their functions and training (SRSCR regulation 11(1)(b))
- failure of an employer to make a medical suspension payment (Employment Protection (Consolidation) Act 1978, section 22)
- dismissal, actual or constructive, following a breach of health and safety law, regulation and/or term of employment contract.

CONTRACT OF EMPLOYMENT

This is fundamentally a contract between employer and employee and features the following elements:

- an offer of a job which is accepted

- terms of employment, which are express (stated verbally or in writing), implied (by custom and practice), and may be incorporated (through collective agreements)
- under the Employment Protection (Consolidation) Act 1978 (as amended by the Employment Acts 1980 and 1982), a contract must incorporate a 'written statement of particulars' specifying parties to the contract, date of employment, rate of payment, hours of work, benefits, job title, length of notice and other conditions
- implied terms can include presumed intentions of the parties, common law duties and custom and practice
- incorporated terms generally apply collectively to groups or the entire workforce, such as collective agreements, Wages Council agreements, company rules and policy documents
- statutory restraints must be considered by an employer, such as pay, as laid down in the Equal Pay Act 1970 (as amended by the Employment Protection Act (Amendment) Regulations 2003), and benefits
- provision for variation of contract is generally incorporated in a contract of employment.

DISMISSAL

Unfair dismissal

This is concerned with whether the decision to dismiss can be justified by the reasons given for the decision and the manner in which the dismissal is handled.

Dismissal occurs when:

- the employer terminates the employee's contract, with or without notice;
- the employee terminates the contract by resigning because of the employer's behaviour (*constructive dismissal*); or
- the employer fails to renew a fixed term contract of two years or more when it expires.

A fair dismissal is one which satisfies a two-stage test of fairness.

1. It can be specified that the reason for dismissal is one of the following:

    - capability or qualifications of the employee;
    - conduct of the employee;
    - redundancy;
    - where continued employment would contravene some other statute or regulation; or
    - 'some other substantial reason'.

2. The employer acted reasonably in the circumstances.

Where any of the above criteria have not been met by an employer, a tribunal may confirm 'unfair dismissal'.

MISCONDUCT

Misconduct implies some forms of behaviour which are deemed unacceptable and which could result in disciplinary action.

Misconduct is an action which is deemed a breach of rules but not sufficiently serious to merit instant or summary dismissal, and would result in a formal warning, e.g. non-observance of safety procedures, persistent lateness.

Gross misconduct, on the other hand, is an act which is so serious as to amount to the employee 'smashing the employment contract', with the result that dismissal without notice is deemed to be appropriate, e.g. physical assault, fraudulent practices, theft of company property.

Company rules should specify aspects of behaviour which constitute both forms of misconduct.

ACAS

The *Advisory, Conciliation and Arbitration Service (ACAS)* was established by the Employment Protection Act 1975 (EPA).

The functions of ACAS are:

(a) to advise employers and trade unions on any matter concerned with employment policies and industrial relations;

(b) to conciliate in matters such as maternity leave, trade union membership and activities, dismissals and redundancies; and

(c) to arbitrate, at the request of both parties.

ACAS has a general duty to promote good industrial relations and to encourage the extensions of collective bargaining.

ACAS is composed of a full-time chair and nine members – three representing employers, three representing employees and three independent members.

Conciliation is undertaken through a conciliation officer who endeavours to reach a settlement before the case goes to tribunal.

Arbitration, on the other hand, with the consent of all parties, follows unsuccessful conciliation and can be through an independent arbitrator or the Central Arbitration Committee.

ACAS gives advice in the following areas to employers, employees and their representative bodies:

● organisation of workers for collective bargaining purposes;
● recognition of trade unions by employers;
● negotiating machinery and joint consultation;
● disputes and grievance procedures;
● communications between employers and employees;
● facilities for trade union officials;

- procedures relating to termination of employment;
- disciplinary matters; and
- manpower planning, labour turnover and absenteeism recruitment, retention, promotion and vocational training payment systems, equal pay and job evaluation.

The Central Arbitration Committee (CAC) was established by the EPA 1975 and consists of a chair, deputy chair, a panel of persons representing employers and employees appointed by the Secretary of State after consultation with ACAS.

- CAC deals with matters referred to it by ACAS, in particular, complaints regarding employers who refuse to disclose information by law, matters relating to payment of wages, amendments to collective agreements, pay structures or wages orders, and disputes within a statutory joint industrial council.

THE EUROPEAN DIMENSION

The European Union (EU)

The EU is administered by the following four bodies.

- *The Commission* – The Commission performs a civil service-like function in the EC and is headed by a body of commissioners from the various Member States. It is empowered to take action against any Member State not complying with EU legislation. The Commission is also empowered to make proposals for future EU legislation to the Council of Ministers (see Figure 1.3). These normally take the form of *Directives*.
- *The Council of Ministers* – This is the final decision-making body in the EU, that is, the EU Government. The governments of Member States are represented on this body.
- *The European Parliament* – This consists of members (MEPs) who represent constituencies in their own Member States. The Parliament's principal role is one of discussion of EU proposals, and it must be consulted on all proposed legislation.
- *The European Court of Justice (ECJ)* – This is the supreme law court, the decisions of which regarding the interpretation of EU law are sacrosanct. EU law is enforced through a network of courts and tribunals in all Member States. Cases can only be brought before this Court by organisations or individuals representing organisations.

EU standards

Under the Treaty of Rome, the Council of the European Commission can issue Directives. Directives usually provide for 'harmonisation' of the laws

● **FIG 1.3 The legislative process of the EU**

of the member states, including those covering occupational health and safety. The process commences when representatives of the member states meet to agree the content of draft Directives. Once agreed, drafts are presented to the European Parliament for ratification. Directives impose a duty on each member state to:

- make legislation to conform to the Directive
- enforce such legislation.

Directives are legally binding on all governments, which must introduce national legislation or use certain administrative procedures. *Framework Directives* set out overall objectives, which are subsequently dealt with individually by *Daughter Directives*.

European Directives

UK health and safety law is now largely driven by European Directives. This process commenced in 1992 with the Framework Directive and the five Daughter Directives.

Framework Directive

The Council Directive on 'the introduction of measures to encourage improvements in the safety and health of workers at work' was implemented in UK as the Management of Health and Safety at Work Regulations 1992.

Daughter Directives

1. The Council Directive on 'the introduction of measures to encourage improvements in the safety and health of workers at the workplace' was implemented in the UK as the Workplace (Health, Safety and Welfare) Regulations 1992.
2. The Council Directive concerning 'the minimum safety and health requirements for the use by workers of machines, equipment and installations' was implemented in the UK as Provision and Use of Work Equipment Regulations 1992.
3. The Council Directive on 'the approximation of the laws of Member States relating to personal protective equipment' was implemented in the UK as the Personal Protective Equipment at Work Regulations 1992.
4. The Council Directive on 'the minimum health and safety requirements for handling heavy loads where there is a risk of back injury to workers' was implemented in the UK as Manual Handling Operations Regulations 1992.
5. The Council Directive 'concerning the minimum health and safety requirements for work with visual display units' was implemented in the UK as the Health and Safety (Display Screen Equipment) Regulations 1992.

The Single European Act

This Act fundamentally identified the need to eliminate technical barriers to trade, one of which was the varying legal standards for health and safety requirements throughout the Community. It introduced the *new approach* to technical harmonisation and standards, the development of the *essential safety requirements* philosophy and of harmonisation Directives that establish these essential safety requirements. The Act further recognised the need to encourage improvements in the working environment.

Agreement on European standards

The 'new approach' depends heavily on the availability of European standards. These are usually prepared by the European standards bodies – which are the European Committee for Standardisation (CEN) and the European Committee for Electrotechnical Standardisation (CENELEC) – as a result of mandates being agreed with the European Commission. CEN and CENELEC are based in Brussels and bring together the national standards bodies of the Community and European Free Trade Association (EFTA). The British Standards Institute is the UK member.

Setting standards in CEN and CENELEC is essentially a process of consensus among the national members (see Figure 1.4). Once an acceptable text has been developed in the relevant committee, it is circulated to all national members for comment. The committee then reviews the comments and the draft standard is circulated to national members for adoption by weighted majority voting.

Alternatively, particularly if the proposal is to adopt an existing international standard as a European standard, a questionnaire procedure can be used to secure endorsement for a text. This can avoid the need to convene any meeting.

All national members have agreed to adopt the resulting European standards as their national standards (in the case of the United Kingdom, as British Standards) and to withdraw any existing and conflicting national standards.

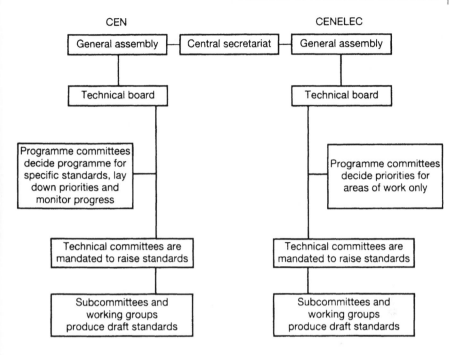

● **FIG 1.4 The procedure followed for developing harmonised standards in Europe**

Civil liability

INTRODUCTION

The two main areas of civil liability at common law are *contract* and *tort*. This chapter examines the duties that the law of tort has placed on employers. Contract law has generally not been concerned with liability for injuries, disease or death.

AN EMPLOYER'S LIABILITY IN TORT

A tort is a 'civil wrong'. The three principal torts are those of *negligence, trespass* and *nuisance*. The rule of common law is that everyone owes a duty to everyone else to take *reasonable care* so as not to cause them foreseeable injury. Where an employee suffers injury or disease at work, he may be in a position to sue his employer within the tort of negligence or breach of statutory duty.

The common duty of care

The position at common law is that employers must take reasonable care to protect their employees from the risks of foreseeable injury, disease or death at work. The effect of this requirement is that if an employer knows of a health and safety risk to employees, or ought, in the light of knowledge current at that time, to have known of the existence of a hazard, he will be liable if an employee is injured or killed or suffers illness as a result of the risk, or if the employer failed to take reasonable care to avoid this happening.

An employer's duties under common law were identified in general terms by the House of Lords in *Wilsons & Clyde Coal Co. Ltd* v. *English* (1938) AC 57 2 AER 628. The common law requires that all employers provide and maintain:

- a safe place of work with safe means of access and egress
- safe appliances and equipment and plant for doing the work
- a safe system for doing the work
- competent and safety-conscious personnel.

These duties apply even though an employee may be working away on a third party's premises, or where an employee has been hired out to another employer, but where the control of the task he is performing still lies with the permanent employer. The test of whether or not an employee has been temporarily 'employed' by another employer is one of 'control'.

Vicarious liability

The doctrine of vicarious liability is based on the fact that if an employee, while acting in the course of his employment, negligently injures another employee or the employee of a contractor working on the premises, or even a member of the public, the *employer*, rather than the employee, will be liable for that injury. As most accidents at work are caused in this way, rather than as a result of personal negligence on the part of the employer, vicarious liability is the ground on which most civil claims for injury-type accidents are won.

Vicarious liability rests on the employer simply as a result of the fact that he is the employer and is deemed to have ultimate control over the employee in what is known as a master and servant relationship. This liability *must* be insured against under the Employers' Liability (Compulsory Insurance) Act 1969, and, indeed, employers cannot contract out of this liability as it is prohibited by the Law Reform (Personal Injuries) Act 1948 and the Unfair Contract Terms Act 1977.

The key to liability is that the accident causing the injury, disease or death arises, first, out of and, second, in the course of employment. This does not, however, normally include the time travelling to and from work (though it would if the mode of transport for such travelling was within the employer's control or was provided by them or in arrangement with them).

Notwithstanding vicarious liability, the employee can be sued instead of, or as well as, an employer where the employee has been negligent (see *Lister v. Romford Ice & Cold Storage Co. Ltd* (1957) and Chapter 5).

THE TORT OF NEGLIGENCE

'Negligence' can be defined as 'careless conduct injuring another'. The duties of employers at common law listed above are part of the general law of negligence and, as such, are specific aspects of the duty to take reasonable care. *Negligence* has been defined at common law as:

- the existence of a duty of care owed by the defendant to the claimant
- breach of that duty
- damage, loss or injury resulting from or caused by that breach (see *Lochgelly Iron & Coal Co. Ltd* v. *M'Mullan* (1934) AC 1)

These three circumstances must be established by an injured employee before they will be entitled to bring a civil claim for damages. (Cases illustrating these three aspects of negligence are covered in Chapter 5.)

Employers' defences

There are two defences available to an employer sued for a breach of duties at common law, namely:

- voluntary assumption of risk (*volenti non fit injuria*)
- contributory negligence.

The first of these is a complete defence and means that no damages will be payable. The second is a partial defence and means that the injured employee's damages will be reduced to the extent to which they are adjudged to be blamed for their injuries.

Volenti non fit injuria

The English translation of this term is 'to one who is willing, no harm is done'. It applies to a situation where an employee, being fully aware of the risks that he is taking in not complying with safety instructions, duties and so on, after being exhorted and supervised, and having received training and instruction in the dangers involved in not following safety procedures and statutory duties, suffers injury, disease and/or death as a result of not complying with them. Here, it is open to an employer to argue that the employee, in effect, agreed to run the risk of the injury, disease and/or death involved in such action. If this defence is successfully pleaded, the employer will be required to pay no damages.

This defence has not generally been successful in cases brought by injured employees, as there is a presumption that employment and the dangers sometimes inherent in it are not voluntarily accepted by workers, but, rather, that employment is an economic necessity. This has been the position since the decision of *Smith* v. *Baker & Sons* (1891) AC 325, where the appellant, who was employed to drill rocks, was injured by stones falling from a crane operated by a fellow employee. Although the appellant knew of the risk he was running from the falling stones, it was held by the House of Lords that he had not agreed to run the risk of being injured.

An exception to this general position occurred in the case of *ICI Ltd* v. *Shatwell* (1964) 2 AER 999. Here, two brothers, the respondents, were employees of the appellant, and were employed as skilled shotfirers in rock blasting, for which they were highly paid. A statutory duty was placed on the two employees personally to take specific safety precautions when shotfiring was about to commence. The two employees had been thoroughly briefed regarding the dangers of the work and the risks involved, for example, premature explosion. They knew of the statutory prohibition in question. Nevertheless, they decided to test although a cable was too short to reach the shelter, rather than wait a few minutes until a workmate could go to get another cable. One brother handed the other brother two wires and the latter applied them to the galvanometer terminals. An explosion resulted, injuring both employees. On an appeal to the House of Lords, it was held that the employer was *not* liable for the following reasons:

- the employer was not in breach of statutory duty, that is, the breach of statutory duty was committed by the employees
- the two employees were highly skilled operators, part of a well-paid team, who were well aware of the risks.

The 'double-barrelled' action

Because an injured employee is entitled to sue his employer for damages for injury resulting from a breach of both a duty at common law and a statutory duty, this has led to the emergence of the 'double-barrelled' action against employers. In such cases, an injured employee sues separately, though simultaneously, for breach of both duties on the part of the employer. This development can be traced back to the decision in *Kilgollan* v. *Cooke & Co. Ltd* (1956) 1 AER 294, a case that involved the fencing requirements of the FA 1937.

It should be noted, however, that section 47 of the HSWA provides that:

- a breach of any of the general duties in sections 2–8 will *not* give rise to civil liability, but
- breach of any duty contained in regulations made under the Act *will* give rise to civil liability, unless the regulations state otherwise.

Limitations of actions

Generally, civil actions may not be brought after the expiration of six years from the date of the tort (*Limitations Act 1939*). However, in the case of personal injuries, this period stands at three years (*Limitations Act 1975*) from the date that the claimant was aware that he had grounds for an action. Special provisions were introduced to cover people who were unaware that they had contracted occupational diseases, such as asbestosis. On this basis, they could not commence an action in the same way as someone who was aware of an injury or physical condition.

The *Limitations Act 1980* replaced the 1975 version, with the limitation periods remaining at six years and three years respectively. In the case of an accident at work involving injury, the limitation period is three years either from the date of the accident or from the deceased's personal representative's knowledge of the accident, whichever is the later. Where an injured person dies within three years, the period recommences from the date of death or of his personal representative's knowledge.

Personal injury is defined as including any disease and any impairment of a person's physical or mental condition. Under the Act a court has discretion to allow time-barred actions according to certain criteria. This is particularly appropriate in cases of occupational disease with a long latent period, such as asbestosis, or in cases of noise-induced hearing loss.

Access to information

Where an employee is suing an employer he will need access to certain information.

Under section 28 of HSWA, as amended by the Employment Protection Act 1975, anyone involved in civil proceedings may obtain a written statement of relevant factual information obtained in the exercise of his powers from an inspector.

Under section 27 of HSWA, the HSC can require any person to provide specific information concerning health and safety matters, either to the HSC or to an enforcing authority.

Disclosure of information

Disclosure is directly connected with the legal concept of 'discovery and inspection' of documents. Discovery of documents is the procedure whereby a party discloses, to a court or to any another party, the relevant documents in the action that he has, or has had, in his possession, custody or power.

Documentary evidence plays an important part in nearly all civil cases. In an industrial injuries action, for example, the employers are likely to have internal accident reports, machinery maintenance records, records of complaints, etc., which it is very much in the claimant's interests to see, while on the other hand the claimant may have documents relating to his medical condition, or to his earnings since the accident, or state benefits which he has received, all of which may be highly relevant to the qualification of his claim by the defendants.

Disclosure for purposes of civil liability proceedings

In civil cases, discovery may take place in two ways, under the Rules.

1. Discovery without order (Automatic Discovery)
   Parties must, in any action commenced by writ, make discovery by exchanging lists within 14 days of the close of the pleadings. The lists are in prescribed form, i.e.

   (a) relevant documents, listed numerically, which the party has in his possession, custody or power and which he does not object to produce;

   (b) relevant documents which he objects to produce; and

   (c) relevant documents which have been, but at the date of service of the list, are not, in the possession, custody or power of the party in question.

   A party serving a list of documents must also serve on his opponent a notice to inspect the documents in the list (other than those which he objects to producing) setting out a time within 7 days and the place where the documents in (a) above may be inspected.

   A right to inspect includes the right to take copies of the documents in question.
2. Discovery by order
   In certain cases an order for discovery will be required. The first case is

where the action is one to which the automatic discovery rule does not apply or where any party has failed to comply with that rule. In this case an Application for an Order for Discovery must be made, should discovery be required. The second case arises where a party is dissatisfied with his opponent's list. He may then apply for discovery of documents.

In either case an application must be made to the court for an order for discovery of the documents in question. The application must be accompanied by an affidavit setting out the grounds for the deponent's belief and identifying the documents of which discovery is required.

Irrespective of the above provisions, any party may, at any time, require inspection of any document referred to in any other party's pleadings or affidavits.

Where any party fails to comply with an order for discovery, the judge may order his statement of claim to be struck out, if he is a claimant, or his defence to be struck out and judgments entered against him, if a defendant. In a case of extreme and wilful disobedience the offender may be committed for contempt of court.

The above procedures are detailed in the Administration of Justice Act 1970, whereby orders for discovery and inspection of documents against proposed parties may be made before the commencement of proceedings.

Under section 34 of the Supreme Court Act 1981 there is, in addition, in actions for damages or death or personal injuries, power to order a person who is not a party to the proceedings, and who appears to be likely to have or has had in his possession, custody or power any documents which are relevant to an issue arising out of the action, to disclose whether these documents are in his possession, custody or power and to produce those documents which he has to the applicant. Such a power may be used, for example, by defendants to obtain a sight of a claimant's hospital notes, or by claimants to obtain copies of manufacturers' instructions for using and maintaining machines, reports by employers to the Health and Safety Executive and documents of that type.

Documents privileged from production

The only circumstances in which a party may refuse to produce a document is where the document is one of a class which the law recognises as privileged. The privilege which attaches to documents must be carefully distinguished from privilege in the law of defamation. Thus a document which attaches qualified privilege for the purposes of defamation, such as the minutes of a company board meeting, is not necessarily privileged from discovery. The former is a rule of law, the latter a rule of evidence or procedure.

A party who wishes to claim privileges for a document must include it under those relevant documents which he objects to produce, together with a statement in the body of the list of the ground upon which he claims privilege for it. The principal classes of document which the law recognises as being privileged from production are documents which relate solely to the deponent's own case, incriminating documents, documents attracting legal professional privilege and documents whose production would be injurious to the public interest.

Disclosure of information by enforcement officers

Section 28 of HSWA requires that no person shall disclose any information obtained by him as a result of the exercise of any power conferred by sections 14 or 20 (including, in particular, any information with respect to any trade secret obtained by him in any premises entered by him by virtue of any such power) except:

- for the purposes of his functions
- for the purposes of any legal proceedings, investigation or inquiry, for the purpose of a report of any such proceedings or inquiry or of a special report made by virtue of section 14
- with the relevant consent.

Information must not normally be disclosed except with the consent of the person providing it.

Disclosure may be made in certain cases:

- for the purposes of any legal proceedings, investigation or inquiry held at the request of the HSC
- with the relevant consent
- for providing employees or their representatives with health- and safety-related information.

INSURANCE

The Employers' Liability (Compulsory Insurance) Act 1969 requires employers to insure against claims by employees suffering personal injury, damage or loss. The following points must be considered:

- the certificate of insurance must be conspicuously displayed at the work-place
- it is standard practice to extend such a policy to provide insurance against public liability
- an employer must disclose all information to an insurer for it to be valid
- the policy must be approved by virtue of the Employers' Liability (Compulsory Insurance) General Regulations 1999 and both policy and certificate must be made available to an enforcement officer
- the policy must state that any person under a contract of service or apprenticeship who sustains injury, disease or death caused during the period of insurance and arising out of the course of employment will be covered for any legal liability on the part of the employer to pay compensation.
- under the Employers' Liability (Defective Equipment) Act 1969 the employer is deemed liable where injury is caused by defective equipment provided by the employer for use in his business and the defect is attributable (wholly or in part) to the fault of a third party.

The Employers' Liability (Compulsory Insurance) Regulations 1999 consolidate with amendments former Regulations made under the 1969 Act and supplement the requirements of the 1969 Act relating to the compulsory insurance of risks relating to employees.

Under Regulation 2 certain conditions are prohibited in policies of insurance, namely any condition which provides (in whatever terms) that no liability (either generally or in respect of a particular claim) shall arise under the policy, or that any liability so arising shall cease, if:

- some specified thing is done or omitted to be done after the happening of the event giving rise to a claim under the policy
- the policyholder does not take reasonable care to protect his employees against the risk of bodily injury or disease in the course of their employment
- the policyholder fails to comply with the requirements of any enactment for the protection of employees against the risk of bodily injury or disease in the course of their employment
- the policyholder does not keep specified records or fails to provide the insurer with or make available to him information from such records.

Regulation 3 sets the limit of the sum to be insured at not less than £5,000,000. Regulation 4 and Schedule 1 place obligations on authorised insurers as to the issue of certificates including the form of the certificates. Certificates must also be kept by employers. Certificates must be displayed at each place of business where he employs employees of the class or description to which the certificate relates (regulation 5).

Where an employer is required to produce a certificate of insurance by form of a written notice from an inspector he must produce same or send it to any person specified in the notice (regulation 6). Similarly, an employer shall, during the currency of the insurance, permit the policy of insurance, or a copy of it, to be inspected:

- at such reasonable time as the inspector may require
- at such place of business of the employer (which, in the case of an employer who is a company, may include its registered office) as the inspector may require.

Schedule 1 states the form to be taken by a Certificate of Employers' Liability Insurance.

Policy Number
1. Name of policyholder
2. Date of commencement of insurance policy
3. Date of expiry of insurance policy.

We hereby certify that subject to paragraph 2:

1. The Policy to which this certificate relates satisfies the relevant law applicable in Great Britain; and

2. (a) the minimum amount of cover provided by this policy is no less than £5,000,000; or
   (b) the cover provided under this policy relates to claims in excess of £X but not exceeding £Y.

Signed on behalf of:

(Authorised Insurer) _____

Signature _____

Employers' Liability (Compulsory insurance) Regulations 1999

These regulations consolidate with amendments of former regulations made under the 1969 Act and supplement the requirements of the Act relating to compulsory insurance of risks relating to employees.

Prohibited conditions

Under regulation 2 certain conditions are prohibited in policies of insurance, namely any condition which provides (in whatever terms) that no liability (either generally or in respect of a particular claim) shall arise under the policy, or that any liability so arising shall cease, if

- some specified thing is done or omitted to be done after the happening of the event giving rise to a claim under the policy;
- the policyholder does not take reasonable care to protect his employees against the risk of bodily injury or disease in the course of their employment;
- the policyholder fails to comply with the requirements of any enactment for the protection of employees against the risk of bodily injury or disease in the course of their employment; or
- the policyholder does not keep specified records or fails to provide the insurer with or make available to him information from such records.

Limit of the sum to be insured

Regulation 3 sets the limit of the sum to be insured at not less than £5,000,000.

Obligations on insurers

Regulation 4 and Schedule 1 place obligations on authorised insurers as to the issue of certificates, including the form of the certificates.

Obligations on employers

Certificates must be kept by insurers. They must be displayed at each place of business where an employer employs employees of the class or description to which the certificate relates (regulation 5).

Where an employer is required to produce a certificate of insurance by form of a written notice from an inspector, he must produce same or send it to any person specified in the notice (regulation 6).

An employer shall, during the currency of an insurance, permit the policy of insurance, or a copy of it, to be inspected:

- at such reasonable time as the inspector may require; and
- at such place of business of the employer (which, in the case of an employer who is a company, may include its registered office) as the inspector may require.

Certificate of employers' liability insurance

Schedule 1 states the form to be taken by a Certificate of Employers' Liability Insurance thus:

Policy Number
1. Name of policyholder
2. Date of commencement of insurance policy
3. Date of expiry of insurance policy

We hereby certify that subject to paragraph 2:

1. The Policy to which this certificate relates satisfies the relevant law applicable in Great Britain; and
2. (a) the minimum amount of cover provided by this policy is no less than £5,000,000; or
   (b) the cover provided under this policy relates to claims in excess of £X but not exceeding £Y.

Signed on behalf of:

(Authorised Insurer) _____

Signature _____

No-fault liability

Under normal circumstances claims for injury are negotiated between solicitors acting on behalf of an employer's insurance company and an injured employee's trade union. In most cases the injured employee has, in effect, to prove that he was owed a duty of care and that the employer had been negligent for a claim to succeed.

Under a no-fault liability system the employer would automatically be liable for injury to his employees and the insurance companies would compensate according to a set scale of damages. The employer would not admit fault as to the injury and, therefore, would not be strictly liable.

The advantage of a no-fault liability system is that there would be a quick settlement of the claim without the need to prove negligence, etc. The principal disadvantage is that the system could be open to abuse and could lead to more expensive insurance claims.

Fatal accidents at work

Under the *Fatal Accidents Act 1976* dependants of a person killed at work may claim compensation for financial loss suffered by them as a result of the death.

Where death is caused by any wrongful act, neglect or default which is such as would (if death had not ensued) have entitled the person injured to maintain an action and recover damages, the person who would have been liable if death had not ensued, shall be liable for damages (Administration of Justice Act 1982, amending the Fatal Accidents Act 1976).

Under the *Administration of Justice Act 1982*, a lump sum is also payable to dependants.

Subsequent remarriage or the prospect of remarriage of a dependant must not be taken into account in assessing fatal damages.

THE TORT OF NUISANCE

Halsbury's Laws of England states that 'nuisances are divisible into common law and statutory nuisances'. 'A common law nuisance is one which apart from statute, violates the principles which the common law lays down for the protection of the public and of individuals in the exercise and enjoyment of their rights.' A 'nuisance' has also been defined as 'an act not warranted by law, or an omission to discharge a legal duty, which act or omission obstructs or causes inconvenience or damage to the public in the exercise of rights common to all Her Majesty's subjects'.

Nuisances at common law may be private nuisances or public nuisances. A *private nuisance* can take many forms, but in all cases, they constitute some form of act, or failure to act, on the part of an individual or group that results in obstruction, inconvenience or damage to another individual or group. An action for private nuisance lies where there has been interference with the enjoyment of land. Such interference need not be intentional, however. Private nuisances are associated with, for example, the keeping of animals – say, dogs barking – the emission of smoke and gases from the burning of refuse and the longstanding problem of noise nuisance from neighbours. Such interference must be sufficiently significant and unreasonable, that is without thought for a neighbour or other person who could be affected, for such a case to be successful.

Actions in respect of private nuisances may be brought by a person injured by the nuisance and they may make a claim for damages and/or obtain an injunction. In appropriate cases, they may abate the nuisance themselves, for example, by lopping overhanging trees.

Public nuisances, on the other hand, have a direct effect on the public at large. Typical examples include obstruction of a public right of way or footpath, failure to make secure an excavation in a public car park and hosing down of brickwork whereby spray falls on members of the public in close proximity.

In cases of public nuisances, the person primarily responsible for the initiation of any action is the Attorney General, who has complete discretion in the matter. The Attorney General may, however, permit an individual or a local authority to use their name to bring a *relator* action if they are satisfied that it is justified.

However, a private individual may bring an action in their own name if they have suffered some direct and substantial injury, different in kind from that suffered by the rest of the public, or if the act is also a private nuisance against their land.

THE TORT OF TRESPASS

Trespass implies the intentional invasion of somebody's person, land or goods. An action for trespass involves a civil claim for damages resulting from false imprisonment, unlawful entry on to the land of another, assault and battery.

Defamation

Defamation implies 'a statement exposing a man to hatred, ridicule or contempt or causing him to be shunned or avoided by right-thinking members of society'.

Defamation in the permanent form is termed libel (written) and in the transitory form slander (spoken).

In a civil claim, the claimant must prove the statement was defamatory, referred to him, was published and damage was suffered.

A defendant has recourse to the following defences, namely that the statement was justified, fair comment, made under privilege, by way of apology and/or in offer of amends.

AN OCCUPIER'S LIABILITY

While health and safety law is largely concerned with the relationships between employers and employees, consideration must also be given to the duties of those people and organisations, such as local authorities and companies, who occupy land and premises. Their land and premises are visited by people for a variety of purposes, such as to undertake work, provide goods and services, settle accounts and so on. The HSWA, section 4, requires those people *in control* of premises to take reasonable care towards such others, and failure to comply with this duty can lead to prosecution and a fine on conviction.

Furthermore, anyone who is injured while visiting or working on land or premises may be in a position to sue the occupier for damages, even though that injured person may not be their employee. Lord Gardner in the case of

Commissioner for Railways v. *McDermott* (1967) 1 AC 169 explained the position thus: 'occupation of premises is a ground of liability and is not a ground of exemption from liability. It is a ground of liability because it gives some control over and knowledge of the state of the premises, and it is natural and right that the occupier should have some degree of responsibility for the safety of persons entering his premises with his permission . . . there is "proximity" between the occupier and such persons and they are his "neighbours". Thus arises a duty of care . . . .'

The Occupier's Liability Acts 1957 and 1984

Occupier's liability is a branch of civil law concerned with the duties of occupiers of premises to all those who may enter on to those premises. The legislation covering this area of civil liability is the Occupier's Liability Act (OLA) 1957 and, specifically in the case of trespassers, the OLA 1984.

The OLA 1957

Previously, common law had given some form of protection to lawful visitors to premises that was based on proof of negligence against the occupier. The level of protection was related to whether or not the visitor was an invitee or licensee. This distinction is now obsolete and there is now one, common duty on the part of an occupier under the OLA – namely *a common duty of care* – to all lawful visitors. This common duty of care is defined as 'a duty to take such care as in all the circumstances of the case is reasonable to see that the visitor will be reasonably safe in using the premises for the purposes for which he is invited or permitted by the occupier to be there'.

Section 1 of the OLA defines the duty owed by the occupiers of premises to all persons lawfully on the premises in respect of 'dangers due to the state of the premises or to things done or omitted to be done on them'. Such liability is not confined to buildings and has been held to include, for instance, that of the main contractors retaining general control over a tunnel being constructed (see *Bunker* v. *Charles Brand & Son Ltd* (1969) 2 AER 59).

The OLA regulates the nature of the duty imposed in consequence of a person's occupation of premises (section 1(2)). The duties are not personal duties, but, rather, are based on the occupation of premises and extend to 'a person occupying, or having control over, any fixed or moveable structure, including any vessel, vehicle or aircraft' (section 1(3)).

Visitors

Visitors to premises are classed as both invitees and licensees. Protection is afforded to *all* lawful visitors, whether they enter for the occupier's benefit, such as customers or clients, or for their own benefit, for instance, a police officer, though not to persons exercising a public or private right of way over premises (section 2(6)).

Warning notices

Under the OLA 1957, occupiers have a duty to erect notices warning visitors of any imminent danger, such as an uncovered pit or obstruction. However, section 2(4) states that a warning notice does not, in itself, absolve the occupier from liability, unless, in all the circumstances, it was sufficient to enable the visitor to be reasonably safe. Moreover, while an occupier, under the provisions of the OLA, could have excused his liability by a suitable, prominent and carefully worded notice, the chance of such avoidance is not permitted as a result of the Unfair Contract Terms Act 1977 (UCTA). This Act states that it is not permissible to exclude liability for death or injury due to negligence by a contract or by a notice, including a notice displayed in accordance with section 2(4) of the OLA.

Trespassers

A trespasser is defined in common law as a person who:

- goes on to premises without invitation or permission
- although invited or permitted to be on premises, goes to a part of the premises to which the invitation or permission does not extend
- remains on premises after the invitation or permission to be there has expired
- deposits goods on premises when not authorised to do so.

The OLA 1984

The common law was, for many years, biased against trespassers. However, the OLA 1984 took a more humane approach. Section 1 of this Act imposes a duty on an occupier in respect of trespassers, namely 'persons who may have lawful authority to be in the vicinity or not', who may be at risk of injury on the occupier's premises. This duty can be discharged by issuing some form of warning, such as the display of hazard warning notices, but such warnings must be very explicit. For example, it is insufficient to display a notice that merely states 'Eye hazard' where there may be a risk to visitors from welding activities. A suitable notice in such a circumstance might read,

RISK OF EYE INJURY FROM WELDING ACTIVITIES. NO PERSON IS ALLOWED TO ENTER THIS AREA UNLESS WEARING APPROVED EYE PROTECTION.

It is not good enough, however, to merely *display* such a notice. The requirements of such notices must be actively *enforced* by management. Thus, the old maxim 'A notice without more is no defence in law' applies in this case. Generally, the displaying of a notice, the clarity, legibility and explicitness of such a notice, and evidence of regular reminding of people of the message outlined in the notice, may count to a certain extent as part of the defence when sued for injury by a simple trespasser under the OLA 1984.

Individual risk taking

Under the OLA 1984, there is no duty on the part of occupiers to persons who willingly accept risks (section 2(5)). Also, the fact that an occupier has taken precautions to prevent persons going into their premises or on to their land where some form of danger exists, does not mean that the occupier has reason to believe that someone would be likely to come into the vicinity of the danger, thereby owing a duty to the trespasser under the OLA 1984 (section 1(4)).

Children

Children generally, from a legal viewpoint, have always been deemed to be less responsible than adults. The OLA 1957 is quite specific on this matter. Section 2(3)(a) requires an occupier to be prepared for children to be less careful than adults. Where, for instance, there is something or a situation on the premises that is a lure or an attraction to a child, such as an old motor car, a pond or scaffolding, this can constitute a 'trap' as far as a child is concerned. Should a child be injured as a result of this 'trap', the occupier could then be liable. Much will depend on the location of the premises, for instance, whether it is close to houses or a school or isolated, such as a farmyard deep in the countryside but, in all cases, occupiers must consider the potential for child trespassers and take appropriate precautions.

Structural defects

An occupier's liability for injury to a visitor resulting from structural defects, such as a defective roof or a hole in the road, only applies to those defects of which the occupier had prior knowledge or had reason to believe existed. This duty was extended to trespassers under section 1 of the OLA 1984.

Contractors and their employees

The relationship between occupiers and contractors has always been a tenuous one. Section 2(3)(b) of the OLA 1957 states that an occupier may expect that a person in exercising their calling, such as a bricklayer, painter or window cleaner, will appreciate and guard against any risks ordinarily incident to it, say, the risk of falling, so far as the occupier gives them leave to do so. This means that the risks associated with the system of work on a third party's premises are the responsibility of the contractor's employer, not the occupier. (It should be appreciated, however, that while the above may be the case in civil law, the situation of criminal law, namely the duties of an employer towards a non-employee under section 3 of the HSWA, is quite different – see *Section 3: General duties of employers and the self-employed to people other than their employees*, page 53.)

Where work is being done on premises by a contractor, the occupier is not liable if they:

● took care to select a competent contractor

- satisfied themselves that the work was being properly done by the contractor (OLA, section 2(4)(b)).

However, in many cases, an occupier may not be competent or knowledgeable enough to ascertain whether the work *is* being 'properly done'. For instance, an occupier may feel that an unsafe system of work adopted by a contractor's employee, such as cleaning windows to the fourth floor offices without using any form of protection, such as a suspended scaffold or safety line, is standard practice among window cleaners! In such cases, an occupier might need to be advised by a surveyor, architect or consultant health and safety specialist in order to be satisfied that the work is being done properly.

THE EMPLOYER'S LIABILITY (DEFECTIVE EQUIPMENT) ACT 1969

As a result of the Employer's Liability (Defective Equipment) Act 1969, employers have been strictly liable for injuries to employees caused by defective equipment, where the defect is wholly or in part the result of manufacture, that is by a third party.

This Act was passed largely to reverse the effect of the case of *Davie* v. *New Merton Board Mills Ltd* (1956) 1 AER 379. This case involved a worker who had been blinded in one eye when a drift that he was using broke and a piece flew out, hitting him in the eye. The employers argued that they had acquired the drifts from a well-known and reputable supplier. On this basis, the House of Lords ruled that the company had exercised reasonable care.

The Act provides that an injury suffered by an employee is to be attributable to negligence by the employer in the following situations:

- where an employee suffers personal injury (including death) in the course of employment in consequence of a defect in equipment
- the equipment was provided by the employer for use in the employer's business
- the defect is attributable, wholly or in part, to the fault of a third party, whether identified or not, such as a manufacturer, supplier, distributor, importer.

Employers, therefore, are liable for defects in manufacture and supply and should protect themselves against this by way of contractual indemnity against the manufacturer or importer.

DEFENCES AVAILABLE IN CIVIL LIABILITY CLAIMS

When presented with a civil claim, an occupier or employer may make a denial of liability based on the following:

- that the duty alleged to have been breached by the defendant was never owed in the first place
- that the nature of the duty was different to that pleaded by the defendant
- that the duty was complied with
- that the breach of duty did not lead to the damage in question
- that the claimant was guilty of contributory negligence that resulted wholly in the damage.

A further aspect of any defence could be the fact that the conduct of the claimant, which constituted contributory negligence in this case, caused and/or resulted in the damages they suffered and that a proportion of the damages should be reduced accordingly. Moreover, in certain situations, it may be possible to show that the accident was the fault of some other party. Where another party is blamed in the defence, the usual result is that they are joined as co-defendant by the claimant, and they sue both.

Res ipsa loquitur

This term means 'the thing speaks for itself'. It is a term that commonly arises in civil actions, implying that the defendant was negligent or careless, or that if they had taken proper precautions or ordered their work correctly, the damage would not have arisen. In other words, the negligence is self-evident. On this basis, it is for the defendant to prove the absence of fault rather than for the claimant to prove fault on the part of an employer or occupier.

In the case of *Ward* v. *Tesco Stores Ltd* (1976) 1 AER 219, while shopping the appellant slipped on some yogurt that had been spilled on the floor, was injured and sued the respondent for negligence. It was held that it was the duty of the respondents to keep the floors clean and free from spillages and that it was for the respondents to show that the accident had not arisen from a lack of reasonable care on their part. In the absence of a satisfactory explanation as to how the yogurt was spilled on the floor, the inference was that the spillage had occurred because the respondents had failed to exercise reasonable care.

Res ipsa loquitur applies, therefore, in the following circumstances:

- the thing or event causing the accident must be under the control of the employer, or under their management/control
- the injury-causing accident must be one that does not happen in the ordinary course of events
- the accident would not have happened if management had been exercising reasonable care
- absence of explanation, on the part of the defendant, to indicate that they had used reasonable care.

A defendant can set aside the presumption against them by:

- proving that reasonable care *was* taken
- providing an alternative explanation for the accident that is equally probable and does not involve negligence on their part
- providing a complete analysis of the facts, that is, by stating the specific facts of the case and inviting full consideration of liability.

2

Statute law

A QUICK GUIDE TO STATUTES, REGULATIONS, APPROVED CODES OF PRACTICE AND GUIDANCE NOTES

Statutes

These, as we have seen, are Acts of Parliament, such as the Factories Act 1961, the Offices, Shops and Railway Premises Act 1963 and the Health and Safety at Work Act 1974.

Regulations

Most statutes give the Minister(s) power to make regulations (subordinate or delegated legislation) without referring the matter to Parliament. The regulations may be drafted by the HSE and submitted through the HSC to the Secretary of State or Minister, such as the Control of Substances Hazardous to Health Regulations 2002, the Noise at Work Regulations 1989, the Safety Representatives and Safety Committees Regulations 1977 and the Management of Health and Safety at Work Regulations 1999.

There is a general requirement for the HSC and HSE to keep interested parties informed of, and adequately advised on, such matters.

Approved Codes of Practice (ACOPs)

The need to provide elaboration on the implementation of regulations is recognised in section 16 of the HSWA, which gives the HSC power to prepare and approve codes of practice on matters contained not only in regulations, but in sections 2 to 7 of the Act. Before approving a Code, the HSE, acting for the HSC, must consult with any interested body.

An ACOP is a quasi-legal document and, although non-compliance does not constitute a breach, if the contravention of the Act or regulations is alleged, the fact that the ACOP was not followed would be accepted in court as evidence of failure to do all that was reasonably practicable. A defence would be to prove that works of equivalent nature had been carried out or something equally good or better had been done.

Examples of ACOPs are Control of Asbestos at Work, Control of Substances Hazardous to Health, Protection of Persons against Ionising Radiation Arising From Any Work Activity and Safety of Pressure Systems.

HSE Guidance Notes

The HSE issues Guidance Notes in some cases to supplement the information in ACOPs. Such Guidance Notes have no legal status and are purely of an advisory nature.

HSE Guidance Notes fall into six categories:

- general safety (GS)
- chemical safety (CS)
- environmental hygiene (EH)
- medical series (MS)
- plant and machinery (PM)
- health and safety (Guidance) (HS(G)).

Examples of HSE Guidance Notes are:

EH40   Occupational Exposure Limits

MS20   Pre-employment Health Screening

HS(G)37 Introduction to Local Exhaust Ventilation

PM41   Application of Photoelectric Safety Systems to Machinery

GS20   Fire Precautions in Pressurised Workings.

THE PARLIAMENTARY PROCESS

Statutes

One of the main functions of Parliament is the enactment of *statutes*, or, *Acts of Parliament*.

Most statutes commence their lives as *bills* and most government-introduced bills start the Parliamentary process in the House of Commons (though some bills of a non-controversial nature may start in the House of Lords). The process in the House of Commons commences with a formal *first reading*. This is followed by a *second reading*, at which stage, there is discussion of the general principles and the bill's main purpose. Following the second reading, the bill goes to the *committee stage* for detailed consideration by an appointed committee, comprising both Members of Parliament and specialists. After due consideration, the committee reports back to the House with recommendations for amendments. Such amendments are considered by the House and they, in turn, may make amendments at this stage and/or return the bill to the committee for further consideration. After this *report stage*, the bill receives a *third reading* where only verbal alterations are made.

The bill is then passed to the House of Lords where it goes through a similar process. The House of Lords either passes the bill or amends it. In the case of an amendment it is returned to the House of Commons for further consideration.

After a bill has been passed by both Houses, it receives the *Royal Assent*, which is always granted, and then becomes an *Act of Parliament*.

Regulations

A statute generally confers power on a Minister or Secretary of State to make *statutory instruments*, which may indicate more detailed rules or requirements for implementing the overall objectives of the statute. Most statutory instruments take the form of *regulations* and come within the area of law known as *delegated or subordinate legislation*.

The HSWA is an *enabling Act*. This means that, in the case of health and safety regulations, the Secretary of State for Employment has powers conferred under the HSWA (section 15, Schedule 3 and section 80) to make regulations on a wide variety of issues, but within the general objectives and aims of the parent Act. Examples of regulations made under the HSWA are the:

- Electricity at Work Regulations 1989
- Health and Safety (First Aid) Regulations 1981
- Management of Health and Safety at Work Regulations 1999.

Regulations may:

- repeal or modify existing statutory provisions
- exclude or modify the provisions of sections 2–9 of the HSWA or existing provisions in relation to any specified class
- make a specified authority, such as a local authority, responsible for enforcement
- impose approval requirements
- refer to specified documents to operate as references
- give exemptions
- specify the class of person who may be guilty of an offence, for example, employers
- apply to a particular case only.

THE PRINCIPAL STATUTES IN FORCE BEFORE THE HEALTH AND SAFETY AT WORK, ETC. ACT 1974

Prior to the HSWA, the principal health and safety legislation was contained in the Factories Act 1961, the Offices, Shops and Railway Premises Act 1963 and the Mines and Quarries Act 1954. Much of this legislation, including regulations made under these Acts, was repealed by subsequent statutes and regulations made under the HSWA, for instance the Fire Precautions Act 1971 and the Pressure Equipment Regulations 1999, the Workplace (Health, Safety and Welfare) Regulations 1992, the Provision and Use of Work Equipment Regulations 1998, and the Personal Protective Equipment at Work Regulations 1992 respectively (see also Chapter 4).

THE FACTORIES ACT 1961

This Act is a *consolidating Act* as it brings together earlier separate pieces of legislation relating to health, safety and welfare in factories. It incorporates both general and specific provisions relating to working conditions, machinery safety, structural safety and the welfare of people working in factories.

Whilst much of the Factories Act 1961 (FA) has been repealed or revoked by the Workplace (Health, Safety and Welfare) Regulations 1992 (WHSWR) and the Provision and Use of Work Equipment Regulations 1998 (PUWER), the following remaining provisions are of importance.

- *Section 20: Cleaning of machinery by young people* This section requires that no young person shall clean any part of a prime mover or of any transmission machinery while the prime mover or transmission machinery is in motion, and shall not clean any part of any machine if the cleaning thereof would expose the young person to risk of injury from any moving part either of that machine or of any adjacent machinery.

- *Section 21: Training and supervision of young people working at dangerous machines* This section prohibits young people from working at any prescribed dangerous machine unless they have been fully instructed as to the dangers arising in connection with it and the precautions to be observed, and:

  - have received sufficient training in work at the machine
  - are under adequate supervision by a person who has thorough knowledge and experience of the machine.

- *Section 30: Dangerous fumes and lack of oxygen* This section is principally concerned with the hazards that may exist in confined spaces and the precautions necessary, such as the operation of a permit to work a system. A 'confined space' is defined as including any chamber, tank, vat, pit, pipe, flue or similar confined space and the following precautions must be taken:

  - where there is no adequate means of egress from the confined space, a manhole must be provided of a minimum size
  - a responsible person must certify that the space is safe to enter for a specified length of time without the need for breathing apparatus
  - where the confined space is not certified as being safe, no one must enter or remain in the confined space unless they are wearing breathing

apparatus, have authority to enter from a responsible person and, where practicable, are wearing belts with a rope securely attached and a person, keeping watch outside and capable of pulling them out, is holding the free end of the rope

- there must be sufficient trained people to use the breathing apparatus during the operation
- there is provided and kept readily available a sufficient supply of approved breathing apparatus, of belts and ropes, and of suitable reviving apparatus and oxygen, and the apparatus, belts and ropes shall be maintained and thoroughly examined at least once a month or at such other intervals as may be prescribed, by a competent person, and a report of every such examination, signed by the person making the examination and containing the prescribed particulars, shall be kept available for inspection
- a sufficient number of people employed shall be trained and practised in the use of the apparatus mentioned above and in reviving those who have stopped breathing
- no person shall enter or remain in any confined space in which the proportion of oxygen in the air is liable to have been substantially reduced unless either:

  (i)   they are wearing suitable breathing apparatus or
  (ii) the space has been and remains adequately ventilated and a responsible person has tested and certified it as safe for entry without breathing apparatus

- no work shall be permitted in any boiler furnace or boiler flue unless it has been sufficiently cooled by ventilation or otherwise to make work safe for the persons employed.

- *Section 31: Precautions with respect to explosive or inflammable dust, gas, vapour or substance* Where, in connection with any grinding, sieving or other process giving rise to dust, there may escape dust of such a character and to such an extent as to be liable to explode on ignition, all practicable steps shall be taken to prevent such an explosion by enclosure of the plant used in the process, and by removal or prevention of accumulation of dust that may escape in spite of the enclosure, and by exclusion or effective enclosure of possible sources of ignition.

  Where there is present in any plant used in any such process as aforesaid dust of such character and to such an extent as to be liable to explode on ignition, then, unless the plant is so constructed as to withstand the pressure likely to be produced by any such explosion, all practicable steps shall be taken to restrict the spread and effects of such an explosion by the provision in connection with the plant, of chokes, baffles and vents, or other equally effective appliances.

  Where any part of a plant contains any explosive or inflammable gas or vapour under pressure greater than atmospheric pressure, that part shall not be opened, except in accordance with the following provisions:

- before the fastening of any joint of any pipe connected with the part of the plant or the fastening of the cover of any opening into the plant is loosened, any flow of the gas or vapour into the part or into any such pipe shall be effectively stopped by a stop-valve or otherwise
- before any such fastening is removed, all practicable steps shall be taken to reduce the pressure of the gas or vapour in the pipe or part of the plant to atmospheric pressure

and if any such fastening has been loosened or removed, no explosive or inflammable gas or vapour shall be allowed to enter the pipe or part of the plant until the fastening has been secured or, as the case may be, securely replaced, but nothing in this subsection applies to a plant installed in the open air.

No plant tank or vessel that contains, or has contained, any explosive or flammable substance shall be subjected:

- to any welding, brazing or soldering operation
- to any cutting operation that involves the application of heat
- to any operation involving the application of heat for the purpose of taking apart or removing the plant, tank or vessel or any part of it until all practicable steps have been taken to remove the substance and any fumes arising from it, or to render them non-explosive or non-inflammable. If any plant, tank or vessel *has* been subjected to any such operation, no explosive or inflammable substance shall be allowed to enter the plant, tank or vessel until the metal has cooled sufficiently to prevent any risk of igniting the substance.

- *Section 34: Steam boilers (restrictions on entry)* No person shall enter or be in any steam boiler that is one of a range of two or more steam boilers unless:

  - all inlets through which steam or hot water might otherwise enter the boiler from any other part of the range are disconnected from that part
  - all valves or taps controlling the entry of steam or hot water are closed and securely locked, and, where the boiler has a blow-off pipe in common with one or more other boilers or delivering into a common blow-off vessel or sump, the blow-off valve or tap on each boiler is so constructed that it can only be opened by a key that cannot be removed until the valve or tap is closed and is the only key in use for that set of blow-off valves or taps.

- *Section 63: Removal of dust or fumes* In every factory in which, in connection with any process carried on, there is given off any dust or fume or other impurity of such a character and to such an extent as to be likely to be injurious or offensive to the people employed, or any substantial quantity of dust of any kind, all practicable measures shall be taken to protect these people against inhalation of the dust or fume or other impurity and to prevent its accumulating in any workroom and, in particular, where the nature of the process makes it practicable, exhaust appliances shall be provided and maintained, as near as possible to the point of origin of the dust or fume or other impurity, so as to prevent it entering the air of any workroom.

No stationary internal combustion engine shall be used unless:

- provision is made for conducting the exhaust gases from the engine into the open air
- the engine (except when used for the purpose of being tested) is so partitioned from any workroom or part of a workroom in which people are employed, other than those attending to the engine, so as to prevent any injurious fumes from the engine entering the air of the room or part of the room.

EMPLOYMENT RIGHTS ACT 1996

Part V of the Employment Rights Act deals with 'protection from suffering detriment in employment'. Section 44 covers the right not to suffer detriment in 'health and safety cases'.

Health and safety cases

An employee has the right not to be subjected to any detriment by any act, or any deliberate failure to act, by his employer done on the ground that:

(a) having been designated by the employer to carry out activities in connection with preventing or reducing risks to health and safety at work, the employee carried out (or proposed to carry out) any such activities;

(b) being a representative of workers on matters relating to health and safety at work or a member of a safety committee:

    (i) in accordance with arrangements established under or by virtue of any enactment; or

    (ii) by reason of being acknowledged as such by the employer,

the employee performed (or proposed to perform) any functions as such a representative or a member of such a committee;

(c) being an employee at a place where:

    (i) there was no such representative or safety committee; or

    (ii) there was such a representative or safety committee but it was not reasonably practicable for the employee to raise the matter by those means,

he brought to his employer's attention, by reasonable means, circumstances connected with his work which he reasonably believed were harmful or potentially harmful to health or safety;

(d) in circumstances of danger which the employee reasonably believed to be serious and imminent and which he could not reasonably have been expected to avert, he left (or proposed to leave) or (while the danger persisted) refused to return to his place of work or any dangerous part of his place of work; or

(e) in circumstances of danger which the employee reasonably believed to be serious and imminent, he took (or proposed to take) appropriate steps to protect himself or other persons from thedanger.

For the purposes of sub-section (e) above, whether steps which an employee took (or proposed to take) were appropriate is to be judged by reference to all the circumstances including, in particular, his knowledge and the facilities and advice available to him at the time.

An employee is not to be regarded as having been subjected to any detriment on the ground specified in subsection (e) above if the employer shows that it was (or would have been) so negligent for the employee to take the steps which he took (or proposed to take) that a reasonable employer might have treated him as the employer did.

CHILDREN AND YOUNG PEOPLE

A 'child' is a person under compulsory school-leaving age, which, at present, is under 16 years, while a 'young person' is someone who has ceased to be a 'child', but who is not yet 18 years old.

Health and Safety (Training for Employment) Regulations 1990

These regulations outline the requirements for protection of employees under the relevant statutory provisions with respect to the provision of 'relevant training'.

Relevant training means work experience provided pursuant to a training course or programme, or training for employment, or both, except if:

(a) the immediate provider of the work experience or training for employment is an educational establishment and it is provided on a course run by the establishment; or

(b) received under a contract of employment.

The regulations extend the definition of 'work' under the HSWA to include relevant training (as defined).

Similarly, a person provided with relevant training shall be treated as being the employee of the person whose undertaking (whether carried on by him for profit or not) is for the time being the immediate provider to that person of the training.

Restrictions and prohibitions on the employment of young people

Under the FA 1961, specific requirements are laid down as regards:

• *conditions and hours of work*, that is, they should work a maximum of 9 hours per day, 48 hours in any week, or for young people under 16 years old, 44 hours a week

- *periods of employment* must not exceed 11 hours in one day, not begin before 7 am, nor end later than 8 pm, or 1 pm on Saturdays
- *continuous employment*, that is, a young person must not be employed for more than four and a half hours without an interval of at least half an hour for a meal break, and there must be a minimum ten-minute rest break per four and a half hours work period
- *meals and rest breaks*, that is, young people must not work during meal or rest periods
- *holidays*, that is, young people must not work on Sundays or public holidays as a rule, but subject to certain provisions
- *shift work*, that is, arrangements must be made with the HSE for all young people employed in factories to work on a shift system and then only within certain defined limits
- *work involving machinery*, that is, as mentioned above, no young person must work with specified machines unless fully instructed in the dangers of, and precautions to be taken with, them
- *restrictions on employment in specific processes*, for example, blasting operations.

The Offices, Shops and Railway Premises Act 1963

The WHSWR 1992 and PUWER 1992 revoked substantial parts of the Offices, Shops and Railway Premises Act 1963 (OSRPA). The following remaining provisions are of importance.

- *Section 18: Exposure of young persons to danger in cleaning machinery* No young person, that is someone under the age of 18, must clean any machinery if, by so doing, they expose themselves to risk of injury from it or any adjacent machinery
- *Section 19: Training and supervision for working at dangerous machines* No one shall work at any machine that is prescribed as dangerous, unless they have been fully instructed of the dangers arising in connection with it and the precautions to be observed and either have sufficient training in work at the machine or are under the supervision of a person who has thorough knowledge and experience of the machine
- *Section 49: Notification* Before work commences in offices or shops, the employer must notify the enforcing authority (local authority) on Form OSR1.

FIRE SAFETY LEGISLATION

The Fire Precautions Act (FPA) 1971, as amended by the Fire Safety and Safety of Places of Sport Act (FSSPSA) 1987, applies to all premises actually in use – industrial, commercial or public – and is enforced by the various fire authorities. Where premises incorporate intrinsically hazardous substances, such as the storage of flammable substances and explosives, control is exercised by the HSE. Fire safety legislation is principally concerned with ensuring the provision of:

- means of escape in the event of fire
- the means for fighting fire

and this is implemented by means of a process of fire certification.

The FPA required that a fire certificate be issued for certain classes of factory and commercial premises, based on the concept of *designated use*. The FSSPSA, in effect, deregulated many premises that formerly required certification under the FPA, including:

- factories, offices and shops where:
  - more than 20 people are employed at any one time
  - more than ten people are employed at any one time elsewhere than on the ground floor
  - buildings containing two or more factory and/or office premises, where the aggregate of people employed in all of them at any one time is more than 20
  - buildings containing two or more factory and/or office premises, where the aggregate of people employed at any one time in all of them, elsewhere than on the ground floor, is more than ten
  - factories where explosive or highly flammable materials are stored, or used in or under the premises, unless, in the opinion of the fire authority, there was no serious risk to employees from fire

- hotels and boarding houses where there is sleeping accommodation provided for guest or staff:
  - for six or more people
  - at basement level
  - above the first floor.

The effect of this deregulation was to exempt certain low-risk premises from the certification requirements. Thus, the FSSPSA empowers local fire authorities to grant exemptions in respect of certain medium-risk and low-risk premises that were previously 'designated use' premises. The exemption certificate must, however, specify the maximum number of persons who can safely be in or on the premises at any one time. Furthermore, the exemption can be withdrawn by the fire authority without prior inspection and by service of notice of withdrawal, where the degree of risk associated with the premises increases.

Fire Precautions Act 1971

- *Section 5*: *Application for fire certificates* Premises will require a fire certificate where they have not been granted an exemption under the

FSSPSA. An application for a fire certificate should normally be made to the local fire authority on Form FP1, by the occupier in the case of factories, offices and shops. In certain cases, however, it must be made by the owner, thus:

- premises consisting of part of a building, all parts of which are owned by the same people, that is, multioccupied buildings in single ownership
- premises consisting of part of a building, the different parts being owned by different people, that is, multioccupied buildings with plural ownership.

An application must specify the premises concerned, the use to be covered and, if required by the fire authority, be supported within a specified time by plans of the premises. Before a fire certificate is issued, inspection of the premises is essential under the FPA. Where, following an inspection, the fire authority does not consider the premises safe against fire, they will normally require the occupier to carry out improvements before issuing a certificate.

- *Section 6: Fire certificates* A fire certificate specifies:
  - the particular use or uses of the premises that it covers
  - the means of escape in case of fire, indicated in a plan of the premises
  - the means for securing that the means of escape can be safely and effectively used at all relevant times
  - the means for fighting fire for use by persons in the premises
  - the means for giving warnings in the case of fire
  - particulars as to explosive or highly flammable liquids stored and used on the premises.

The fire certificate may also impose requirements relating to:

- the maintenance of the means of escape and keeping it free from obstruction
- the maintenance of other fire precautions outlined in the certificate
- the training of employees as to action in the event of fire and the maintenance of suitable records of such training
- limiting the number of people who, at any one time, may be on the premises
- any other relevant fire precautions.

A fire certificate, or a copy of same, must be kept in the building concerned.

- *Section 8: Inspection of premises* So long as a fire certificate is in force for any premises, the fire authority may cause inspections to be made from time to time to ascertain whether the requirements of the certificate are being maintained or are becoming inadequate.

  Before carrying out any structural alterations or any material internal alterations to premises requiring a fire certificate, it is necessary, first, to notify the fire authority, or in the case of 'special premises', the HSE, of the proposed changes. Similar requirements apply in cases of proposed alterations to equipment or furniture on the premises. Following notification, the premises will be inspected before the alteration work can go ahead. Failure to follow this procedure is an offence under the FPA.

Fire Safety and Safety of Places of Sport Act 1987

Legal requirements relating to the fire certification of premises, as contained in the FPA, were modified by the FSSPSA. One of the main general purposes of the FSSPSA was to concentrate effort on premises where there is a high risk of fire, at the same time reducing the need for certification in the case of low-risk premises. The FSSPSA is, therefore, a deregulating measure. It empowers fire authorities to grant exemptions in respect of low-risk premises that were previously 'designated use' premises under the FPA and required a fire certificate. An occupier does not have to formally apply for such exemption. It may be granted either on application for a fire certificate or while a fire certificate is in force. Generally, however, premises would *not* be exempted unless they had been inspected within the previous 12 months. The consequence of this is that if an exemption is granted:

- on application for a fire certificate, it disposes of the application
- while the certificate is in force, the certificate no longer has any effect.

The exemption certificate *must*, however, specify the maximum number of people who can safely be in or on the premises at any one time. Depending on the relative degree of fire risk, the exemption can be withdrawn by the fire authority without prior inspection, on notice being given to the occupier of such a withdrawal.

The following sections of the FSSPSA are important.

- *Section 5: Means of escape in case of fire* Although low- or medium-risk premises may be exempt from certification under the FSSPSA, occupiers must still provide a means of escape and fire fighting equipment. 'Escape' is defined in the Act as 'escape from premises to some place of safety beyond the building, which constitutes or comprises the premises, and any area enclosed by it or within it; accordingly, conditions or requirements can be imposed as respects any place or thing by means of which a person escapes from premises to a place of safety'.
- *Section 8: Alterations to exempted premises* An occupier who intends to carry out any material alterations to a premises while an exemption order is in force must first notify the fire authority of the proposed alterations. Failure to notify is an offence under the Act.

Notification is particularly necessary in cases where:

- the proposed extension of, and/or structural alteration to premises may affect the means of escape
- any alteration to the interior of the premises, in furniture or equipment, may affect the means of escape
- the proposed storage of explosive or highly flammable material in or on the premises in quantities/aggregate quantities is greater than that

specified by the current certificate
- it is proposed that a greater number of people be on the premises than specified by the certificate.

Enforcement provisions of the FSSPSA

- *Section 9: Improvement notices* Where a fire authority is of the opinion that an occupier has not fulfilled their duty with regard to the provision of:

  - means of escape in case of fire
  - means for fighting fire

  the authority may serve on the occupier an *improvement notice*, detailing the steps that should be taken in the way of improvements, alterations and other measures to remedy this breach of the Act. The occupier must normally undertake remedial work within 21 days, unless they submit an appeal against the notice. Such an appeal must be lodged within 21 days from the date of service of the notice, and it has the effect of suspending the operation of the notice. Where such an appeal fails, the occupier must undertake the remedial work specified in the improvement notice. Failure to do so can result in a fine on summary conviction and, on conviction, an indictment, an indefinite fine or imprisonment for up to two years, or both.
- *Section 10: Prohibition notices* Where there is considered to be a *serious* risk of injury to employees and visitors from fire on the premises, the fire authority can serve a *prohibition notice* on the occupier, requiring that remedial work be carried out in the interests of fire safety or, alternatively, have the premises closed down. The power to serve prohibition notices applies to all former 'designated use' premises, but not places of public religious worship or single private dwellings.
  A prohibition notice may be served on premises:

  - providing sleeping accommodation, such as hotels
  - providing treatment/care, such as nursing homes
  - for the purposes of entertainment, recreation or instruction, or for a club, society or association
  - for teaching, training or research
  - providing access to members of the public, whether for payment or otherwise
  - which are places of work.

A prohibition notice is most likely to be served where means of escape are inadequate or non-existent or where there is a need to improve means of escape. As with a prohibition notice under the HSWA, such a notice can be served with immediate effect or deferred. An appeal does not suspend the operation of a prohibition notice.

Places of sport

The FSSPSA extends the existing Safety of Sports Grounds Act (SSGA) 1975, which only applied to sports stadia, that is, sports grounds where the accommodation for spectators wholly or substantially surrounds the activity taking place, to all forms of sports ground.

The legal situation relating to sports grounds under the FSSPSA is as follows:

- a *general safety certificate* is required for any sports ground
- the SSGA is extended to any sports ground that the Secretary of State considers appropriate
- the validity of safety certificates no longer requires the provision at a sports ground of a police presence, unless consent has been given by a chief constable or chief police officer
- there is provision for the service of *prohibition notices* in the case of serious risk of injury to spectators, prohibiting or restricting the admission of spectators in general or on specified occasions.

Prohibition notices

A prohibition notice may specify steps that must be taken to reduce risk, particularly from a fire, to a reasonable level, including structural alterations (irrespective of whether or not this may contravene the terms of a safety certificate for the ground issued by the local authority, or for any stand at the ground; see the next section, **Safety certificates**). Where a prohibition notice requires the provision of a police force, such requirements cannot be specified without the consent of the chief constable or chief police officer.

Under section 23 of the FSSPSA, a prohibition notice may be served on any of the following people:

- the holder of a general safety certificate
- the holder of a *specific safety certificate*, that is, a safety certificate for a specific sporting activity or occasion
- where no safety certificate is in operation, the management of the sports ground
- in the case of a specific sporting activity for which no safety certificate is in operation, the organisers of the activity
- where a general safety certificate is in operation for a stand at a ground, the holder of this certificate
- where a specific safety certificate is in operation for a stand, the holder of the certificate.

Under section 25 of the FSSPSA, sports grounds *must* be inspected at least *once* a year.

Safety certificates

Where a sports ground provides covered accommodation in stands for spectators, a safety certificate, issued by the local authority, is required for each stand providing covered accommodation for 500 or more spectators, that is, a *regulated stand*. In certain cases, safety certificates may be required for stands accommodating smaller numbers.

Records

Section 27 of the Act gives the local authority power to require the keeping of the following records in the case of stands at sports grounds:

- the number of spectators in covered accommodation
- procedures relating to the maintenance of safety in the stand.

Sports grounds with regulated stands must be inspected periodically.

Offences under the FSSPSA

The principal criminal offences under the Act are committed in respect of regulated stands when:

- spectators are admitted to a regulated stand at a sports ground on an occasion when a safety certificate should be, but is not, in operation
- any term or condition of a safety certificate for a regulated stand at a sports ground is contravened.

Both the management of the sports ground and, in the second case, the holder of the certificate, are guilty of an offence (section 36).

Under section 36 of the FSSPSA, a number of defences are available, namely:

- that either:
  - the spectators were admitted when no safety certificate was in operation
  - the contravention of the safety certificate occurred without their consent
- that they took all reasonable precautions and exercised all due care to avoid the offence being committed either by themselves or by people under their control.

Where a person or corporate body is charged with an offence under section 36, that is they had no safety certificate for a regulated stand, the defendant may plead that they did not know that the stand had been designated as being a regulated stand.

THE HEALTH AND SAFETY AT WORK, ETC. ACT 1974

This Act covers all people at work, except domestic workers in private employment, whether they be employers, employees or the self-employed. It is aimed at people and their activities, rather than premises and processes.

The legislation includes provisions for both the protection of people at work and the prevention of risks of the health and safety of the general public that may arise from work activities.

The objectives of HSWA

These are:

- to secure the health, safety and welfare of all people at work

- to protect others from the risks arising from workplace activities
- to control the obtaining, keeping and use of explosive or highly flammable substances
- to control emissions into the atmosphere of noxious or offensive substances.

Specific duties

- *Section 2: General duties of employers to their employees* It is the duty of every employer, so far as is reasonably practicable, to ensure the health, safety and welfare at work of all their employees. More particularly, this includes:

  - the provision and maintenance of plant and systems of work that are, so far as is reasonably practicable, safe and without risk to health
  - arrangements for ensuring, so far as is reasonably practicable, safety and absence of risk to health in connection with the use, handling, storage and transport of articles and substances
  - the provision of such information, instruction training and supervision as is necessary to ensure, so far as is reasonably practicable, the health and safety at work of employees
  - so far as is reasonably practicable as regards any place of work under the employer's control, the maintenance of it in a condition that is safe and without risk to health and the provision and maintenance of means of access to and egress from it that are safe and without such risk
  - the provision and maintenance of a working environment for their employees that is, so far as is reasonably practicable, safe, without risk to health, and adequate as regards facilities and arrangements for their welfare at work.

  Employers must prepare and, as often as is necessary, revise a written Statement of Health and Safety Policy, and bring the Statement and any revision of it to the notice of all their employees.

  Every employer must also consult appointed safety representatives with a view to making and maintaining arrangements that will enable them and their employees to co-operate effectively in promoting and developing measures to ensure the health and safety at work of the employees, and in checking the effectiveness of such measures.
- *Section 3: General duties of employers and the self-employed to people other than their employees* Every employer must conduct their undertaking in such a way as to ensure, so far as is reasonably practicable, that people not in their employ who may be affected by the business are not exposed to risks to *their* health or safety. (Similar duties are imposed on the self-employed.)

  Every employer and self-employed person must give to others (those not in their employ) who may be affected by the way in which they conduct their business the prescribed information about such aspects of the way in which they work that might affect their health and safety.

- *Section 4: General duties of people concerned with premises to people other than their employees* This section has the effect of imposing duties in relation to those who:

  - are not their employees, but
  - use non-domestic premises made available to them as a place of work.

  Every person who has, to any extent, control of premises must ensure, so far as is reasonably practicable, that the premises, all means of access to it or egress from it, and any plant or substances on the premises or provided for use there, is or are safe and without risk to health.

- *Section 5: The general duty of people in control of certain premises in relation to harmful emissions into the atmosphere* Any person having control of any premises of a class prescribed for the purposes of section 1(1)(d) must use the best practicable means for preventing the emission into the atmosphere from the premises of noxious or offensive substances and for rendering harmless and inoffensive such substances as may be emitted.

- *Section 6: General duties of manufacturers and so on regarding articles and substances for use at work* Any person who designs, manufactures, imports or supplies any article for use at work:

  - must ensure, so far as is reasonably practicable, that the article is so designed and constructed as to be safe and without risk to health when properly used
  - must carry out or arrange for the carrying out of such testing and examination as may be necessary to comply with the above duty
  - must provide adequate information about the use for which it is designed and has been tested to ensure that, when put to that use, it will be safe and without risk to health.

  Any person who undertakes the design or manufacture of any article for use at work must carry out or arrange for the carrying out of any necessary research with a view to the discovery and, so far as is reasonably practicable, the elimination or minimising of any risk to health or safety that the design or article may pose.

  Any person who erects or installs any article for use at work must ensure, so far as is reasonably practicable, that nothing about the way it is erected or installed makes it unsafe or a risk to health when properly used.

  Any person who manufactures, imports or supplies any substance for use at work:

  - must ensure, so far as is reasonably practicable, that the substance is safe and without risk to health when properly used
  - must carry out or arrange for the carrying out of such testing and examination as may be necessary
  - must take such steps as are necessary to ensure adequate information about the results of any relevant tests is available in connection with the use of the substance at work.

- *Section 7: General duties of employees at work* It is the duty of every employee while at work:
  - to take reasonable care for the health and safety of themselves and of others who may be affected by their acts or omissions at work
  - as regards any duty or requirement imposed on their employer, to co-operate with them so far as is necessary to enable that duty or requirement to be performed or complied with.
- *Section 8: The duty not to interfere with or misuse things supplied pursuant to certain provisions* No one shall intentionally or recklessly interfere with or misuse anything provided in the interests of health, safety or welfare in order to satisfy any of the relevant statutory provisions.
- *Section 9: The duty not to charge employees for things done or provided pursuant to certain specific requirements* No employer shall levy or permit to be levied on any employee of theirs any charge in respect of anything done or provided in order to comply with any specific requirement of the relevant statutory provisions.

CORPORATE LIABILITY, CORPORATE MANSLAUGHTER AND CORPORATE KILLING

Corporate liability

Under the HSWA directors, managers, company secretaries and similar officers of the body corporate have both general and specific duties. Breaches of these duties can result in individuals being prosecuted.

Offences committed by companies (section 37(1))

Where a breach of one of the relevant statutory provisions on the part of a body corporate is proved to have been committed with the consent or connivance of, or to have been attributable to any neglect on the part of, any director, manager, secretary or other similar officer of the body corporate or a person who was purporting to act in any such capacity, he as well as the body corporate shall be guilty of that offence and shall be liable to be proceeded against and punished accordingly.

Breach of this section has the following outcomes:

1. Where an offence is committed through neglect by a board of directors, the company itself can be prosecuted as well as the directors individually who may have been to blame.
2. Where an individual functional director is guilty of an offence, he can be prosecuted as well as the company.
3. A company can be prosecuted even though the act or omission was committed by a junior official or executive or even a visitor to the company.

Generally, most prosecutions under section 37(1) would be limited to that body of persons, i.e. the board of directors and individual functional directors, as well as senior managers.

Offences committed by other corporate persons (section 36)

Section 36 makes provision for dealing with offences committed by corporate officials, e.g. personnel managers, health and safety specialists, training officers, etc. Thus:

Where the commission by any person of an offence under any of the relevant statutory provisions is due to the *act* or *default* of some other person, that other person shall be guilty of the offence, and a person may be charged with and convicted of the offence by virtue of this subsection whether or not proceedings are taken against the first mentioned person.

Corporate manslaughter

Manslaughter is of two kinds, that is, voluntary and involuntary. The former, which is essentially murder but reduced in severity owing to, say, diminished responsibility, is not relevant to health and safety. Involuntary manslaughter extends to all unlawful homicides where there is no malice aforethought or intent to kill.

There are two forms of involuntary manslaughter, that is, *constructive* manslaughter and *reckless* manslaughter. The former applies to situations where death results from an act unlawful at common law or by statute, amounting to more than mere negligence. Reckless manslaughter or gross negligence arises where death is caused by a reckless act or omission, and a person acts recklessly 'without having given any thought to the possibility of there being any such risk or, having recognised that there was some risk involved, has none the less gone on to take it' (*R*. v. *Caldwell* (1981) 1 AER 961).

Corporate killing

An offence of 'corporate killing' to make it easier to punish companies whose blameworthy conduct causes the death of employees or members of the public was proposed by the Law Commission in a report to Parliament on 5 March 1996.

The Commission recommended that corporations should be liable to an unlimited fine and that judges should have power to order them to remedy the cause of the death.

This situation arose following a number of disasters for which no successful prosecutions for manslaughter were brought, even though corporate bodies were found to be at fault. Examples include the King's Cross fire, the Piper Alpha disaster and the Clapham rail crash. The main reason for the lack of successful prosecutions was that, under the present law, corporate manslaughter charges can only be brought where the corporation has acted through the 'controlling mind' of one of its agents, said the Commission. In practice it was often not possible to identify one person who has been the 'controlling mind'. As a result, there had been only four prosecutions for corporate manslaughter and one conviction. In that case, the firm had been a 'one man company', so it was easy to identify the 'controlling mind'.

The Law Commission said that it saw no reason why companies should continue to be effectively exempt from the law of manslaughter. The report also recommended that the present offence of involuntary manslaughter should be replaced with two new offences of reckless killing and killing by gross carelessness.

THE SOCIAL SECURITY ACT 1975

Under this legislation:

- employees must notify their employer of any accident resulting in personal injury in respect of which benefit may be payable (notification may be given by a third party if the employee is incapacitated)
- employees must enter the appropriate particulars of all accidents in an accident book (Form BI 510), which may be done by another person if the employee is incapacitated; such an entry is deemed to satisfy the requirements given in the first point above
- employers must investigate all accidents of which notice is given by employees and any variations between the findings of this investigation and the particulars given in the notification must be recorded
- employers must, on request, furnish the Department of Social Security with such information as may be required relating to accidents in respect of which benefit may be payable, such as Forms 2508 and 2508A
- employers must provide and keep readily available an accident book in an approved form in which the appropriate details of all accidents can be recorded (Form BI 510) and such books, when completed, should be retained for three years after the date of the last entry
- for the purposes of the above, the appropriate particulars should include:

  - name and address of the injured person
  - date and time of the accident
  - the place where the accident happened
  - the cause and nature of the injury
  - the name and address of any third party giving the notice.

THE CONSUMER PROTECTION ACT 1987

This Act implements in the UK the provisions of the EC Directive of 25 July 1985 on the approximation of the laws, regulations and administrative provisions of the member states concerning liability for defective products.

The Act also:

- consolidates, with amendments, the Consumer Safety Act 1978 and the Consumer Safety (Amendment) Act 1986
- makes provision with respect to the giving of price indications
- amends Part I of the HSWA and sections 31 and 80 of the Explosives Act 1975
- repeals the Trade Descriptions Act 1972 and the Fabrics (Misdescription) Act 1913.

Part I, which deals with product liability, is significant from a health and safety viewpoint.

Part I: Product liability

Some important definitions

Within the context of Part I, a number of definitions are of significance. For example, *producer* in relation to a product means the following:

- the person who manufactured it
- in the case of a substance that has not been manufactured, but has been won or abstracted, the person who won or abstracted it
- in the case of a product that has not been manufactured, won or abstracted, for which the essential characteristics are attributable to an industrial process having been carried out, such as in relation to agricultural produce, the person who carried out that process.

Product means any goods or electricity and (subject to the proviso below) includes products that are used in another product, whether by virtue of being a component part or raw material for it or otherwise. The proviso is that a person who supplies any product in which products are used whether by virtue of being component parts or raw materials or otherwise, shall not be treated by reason only of their supply of that product, but as supplying any of the products so comprised.

Liability for defective products (section 2)

Where any damage is caused wholly or partly as a result of there being a defect in a product, all of the following people shall be liable for the damage:

- the producer of the product
- any person who, by putting their name on the product or using a trade mark or other distinguishing mark in relation to the product, has held themselves out to be the producer of the product
- any person who has imported the product into a Member State from a place other than one of the Member States in order, in the course of a business of theirs to supply it to another.

Where any damage is caused wholly or partly as a result of a defect in a product, any person who supplied the product (whether to the person who suffered damage, to the producer of any product in which the product in question is a part or to any other person) shall be liable for the damage if:

- the person who suffered the damage requests the supplier to identify one or more of the people (whether still in existence or not) detailed in the list above in relation to the product
- this request is made within a reasonable period after the damage occurs and at a time when it is not reasonably practicable for the person making the request to identify all these people
- the supplier fails within a reasonable period after receiving the request, either to comply with the request or to identify the person who supplied the product to them.

The meaning of 'defect' (section 3)

Subject to the following provisions, there is a defect in a product if the safety of the product is not of the kind one would generally be entitled to expect. For those purposes *safety*, in relation to a product, shall include safety with respect to products used in that product and safety in the context of any risks of damage to property, as well as in the context of risks of death or personal injury.

In determining what people are generally entitled to expect in relation to a product, all the circumstances shall be taken into account, including:

- the manner in which, and the purposes for which, the product has been marketed, its get-up, the use of any mark in relation to the product and any instructions for, or warnings with respect to doing or refraining from doing anything with or in relation to the product
- what might reasonably be expected to be done with or in relation to the product
- the time when the product was supplied by its producer to another

and nothing in this section requires a defect to be inferred, on its own, from the fact that the safety of a product that is supplied after that time is greater than the safety of the product in question.

Defences (section 4)

In any civil proceedings against any person ('the person proceeded against') in respect of a defect in a product, it shall be a defence if they can show that:

- the defect is attributable to compliance with any requirement imposed by or under any enactment or with any Community obligation
- that the person proceeded against did not at any time supply the product to another
- that the following conditions are satisfied:
  - that the only supply of the product to another by the person proceeded against was otherwise than in the course of a business of that person's
  - that section 2 above does not apply to that person or applies to them by virtue only of things done otherwise than with a view to profit

- that the defect did not exist in the product at the relevant time
- that the *state of scientific or technical knowledge* at the relevant time was not such that a producer of products of the kind in question might have been expected to have discovered the defect in the products while they were under their control ('state of the art' defence)
- that the defect:
  - constituted a defect in a product ('the subsequent product') of which the product in question was a part
  - was wholly attributable either to the design of the subsequent products or to compliance by the producer of the product in question with instructions given to them by the producer of the subsequent product.

Damage giving rise to liability (section 5)

Subject to the following provisions, by the word *damage* is meant death or personal injury or any loss of or damage to any property (including land).

A person shall *not* be liable under section 2 for any defect in a product, for the loss or damage to the product itself or for the loss or damage to the whole or part of any product if it has been supplied as part of another product.

A person shall not be liable under section 2 for any loss of or damage to any property that, at the time it is lost or damaged, is not:

- ordinarily intended for private use, occupation or consumption
- intended by the person suffering the loss or damage mainly for their own private use, occupation or consumption.

No damages shall be awarded to any person under Part I regarding any loss of or damage to any property if the amount that would be so awarded apart from this subsection and any liability for interest, does not exceed £275.

In determining, for the purposes of this Part, *who* has suffered any loss or damage to property and *when* any such loss occurred, the loss or damage shall be regarded as having occurred at the earliest time at which a person with an interest in the property had *knowledge* of the material facts regarding any loss or damage.

For the purposes of the above subsection, the *material facts* regarding any loss of or damage to any property are those that relate to the kind of loss or damage that would lead a reasonable person with an interest in the property to consider the matter sufficiently serious as to justify their instituting proceedings for damages against a defendant who did not dispute liability and was able to satisfy a judgment.

By a person's *knowledge* is meant knowledge that the person might reasonably have been expected to acquire:

- from the facts observable or ascertainable by this person
- from facts ascertainable by the person with the help of appropriate expert advice that it is reasonable to seek

but a person shall not be taken (by virtue of this subsection) to have knowledge of a fact ascertainable by them only with the help of expert advice *unless* they have failed to take all reasonable steps to obtain (and, where appropriate, act on) that advice.

Prohibition on exclusion from liability (section 7)

The liability of a person by virtue of this Part to a person who has suffered damage caused wholly or partly as a result of a defect in a product, or to a dependant or relative of such a person, shall not be limited or excluded by any contract term, by any notice or by any other provisions. By 'notice' is meant notice in writing.

Part II: Consumer safety

Part II deals with a *general safety requirement*, that making it an offence to supply unsafe consumer goods. This provision supplements certain provisions of the Consumer Safety Act 1978 providing coverage where there are no specific safety regulations under that Act.

Note that although the general safety requirement of Part II is restricted to consumer goods, section 6 of the HSWA applies in respect of the safety of articles and substances used at work. Also, Part I of the Act is not restricted to consumer goods – it can apply to anything described as a product.

THE ENVIRONMENTAL PROTECTION ACT 1990 (EPA)

The EPA brought in fundamental changes regarding the control of pollution and the protection of the environment. It repealed completely the Alkali, etc. Works Regulations Act 1906 and the Public Health (Recurring Nuisances) Act 1969, and certain sections of the well-established environmental protection legislation, such as the Public Health Acts 1936 and 1961, Control of Pollution Act 1954 and the Clean Air Act 1956. The Act covers eight specific aspects, namely:

- integrated pollution control and air pollution control by local authorities (Part I)
- waste on land (Part II)
- statutory nuisance and clean air (Part III)
- litter (Part IV)
- amendments to the Radioactive Substances Act 1960 (Part V)
- genetically modified organisms (Part VI)
- nature conservation (Part VII)
- miscellaneous provisions, for instance relating to the control of stray dogs and stubble burning (Part VIII).

Part I: Integrated pollution control and air pollution control by local authorities

A number of terms are of significance in the interpretation of this Part of the EPA:

- *environment* consists of all, or any, of the mediums of the air, water and land (and the medium of air includes the air within buildings and other natural or man-made structures above or below ground)
- *pollution of the environment* means pollution of the environment due to the release (into any environmental medium), from any process, of substances that are capable of causing harm to human beings or any other living organisms supported by the environment
- *harm* means harm to the health of living organisms or other interference with the ecological systems of which they form part and, in the case of humans includes offence caused to any of their senses or their property (*harmless* has a corresponding meaning)
- *process* means any activities carried out in Great Britain, whether on premises or by means of mobile plant, that are capable of causing pollution of the environment, and *prescribed process* means a process prescribed under section 2(1)
- *authorisation* means an authorisation for a process (whether on premises or by means of mobile plant) granted under section 6; and a reference to the conditions of an authorisation is a reference to the conditions subject to which, at any time, the authorisation has effect
- a substance is *released* into any environmental medium whenever it is released directly into that medium, whether it is released into it within or outside Great Britain, and release includes:

  - in relation to air, any emission of the substance into the air
  - in relation to water, any entry (including any discharge) of the substance into water
  - in relation to land, any deposit, keeping or disposal of the substance in or on land.

Part 1 of the EPA identifies industrial processes that are scheduled for control either by Her Majesty's Inspectorate of Pollution (HMIP) or by local authorities (LAs). Industrial processes are split into a two-part schedule for the purposes of enforcement.

Category A processes are subject to *integrated pollution control* (IPC) by HMIP. IPC applies the principle of *best practicable environmental option* (BPEO), and Category A processes have controls applied to all waste streams. BPEO is not defined in the Act, but was considered at length by the Royal Commission on Environmental Protection, whose definition is as follows:

A BPEO is the outcome of a systematic consultative and decision making procedure which emphasises the protection and conservation of the environment across land, air and water. The BPEO procedure establishes, for a given set of objectives, the option that proves the most benefit or least damage to the environment as a whole, at acceptable cost, in the long term as well as in the short term.

IPC covers waste streams in to the air, water and land. The general approach is to minimise these waste streams to ensure that the BPEO for the process is applied. Authorisations and controls are applied to all emissions (section 7). Fundamentally, the approach hinges on a prior authorisation procedure. Authorisations for industrial processes can have stringent conditions applied to them and these conditions apply the concept of the use of *best available techniques not entailing excessive costs* (BATNEEC). The primary objective of this concept is that of minimising pollution, or predicted pollution, of the environment as a whole from an industrial process. In other words, having regard to the BPEO (section 7(2)) the BATNEEC concept specifies the objectives to be considered in the conditions of authorisation as:

- ensuring that, in carrying out a prescribed process, the best available techniques not entailing excessive cost will be used:
  - for preventing the release of substances prescribed for any environmental medium into that medium or, where this is not practicable by such means, for reducing the release of such substances to a minimum and for rendering harmless any such substances that are so released
  - for rendering harmless any other substances that might cause harm if released into any environmental medium
- compliance with any directions that the Secretary of State has given for the implementation of any obligations in the United Kingdom under the Community Treaties or international law relating to environmental protection
- compliance with any limits or requirements and achievements of any quality standards or quality objectives prescribed by the Secretary of State under any of the relevant enactments
- compliance with any requirements applicable to the granting of authorisations specified by or under a plan made by the Secretary of State.

Those processes identified as Category B industrial processes are subject to a similar prior authorisation procedure administered by the LA's environmental health department for discharges into the air only. Processes discharging substances in to water are regulated by the National Rivers Authority's (NRA) discharge consent procedure and discharges into land are controlled by the Waste Regulation Authority (WRA), using controls detailed in Part II of the Act. Liaison between central and local government inspectorates is maintained through HMIP.

Under Part I of the Act, enforcing authorities must establish and maintain public information registers that contain details of the authorisations given for, and conditions applied to, processes.

Enforcement arrangements

Section 13 of the EPA states that where the enforcing authority is of the opinion that the person carrying on a prescribed process under an authorisation is contravening any condition of that authorisation, or is likely to contravene any such conditions, the authority may serve an *enforcement notice*.

An enforcement notice shall:

- state that the authority is of the said opinion
- specify the matters constituting the contravention or the matters making it likely that the contravention will arise, as the case may be
- specify the steps that must be taken to remedy the contravention or to remedy the matters that make it likely that the contravention will arise, whichever is relevant in the circumstances
- specify the period within which these steps must be taken.

Section 14 of the EPA makes provision for the service of *prohibition notices*. If the enforcing authority is of the opinion, as respects the carrying out of a prescribed process under an authorisation, that to continue to carry it out or to do so in a particular manner, involves *an imminent risk of serious pollution of the environment*, the authority shall serve a prohibition notice on the person carrying out the process.

A prohibition notice may be served *whether or not* the manner of carrying out the process in question contravenes a condition of the authorisation and may relate to any aspects of the process, whether these are regulated by conditions of the authorisation or not.

A prohibition notice shall:

- state the authority's opinion
- specify the risk involved in the process
- specify the steps that must be taken to remove it and the period of time in which they must be taken
- direct that the authorisation shall, until the notice is withdrawn, wholly or to the extent specified in the notice, cease to authorise the carrying out of the process.

Also, where the direction applies to only part of the process, it may impose conditions to be observed in the carrying out of the part that is authorised to continue to be carried out.

Section 17 provides inspectors appointed under the EPA with considerable powers regarding premises on which a prescribed process is, or is believed to be, carried out and to those on which a prescribed process has been carried out, the condition of which is believed to pose a risk of serious pollution of the environment. Such powers include those of being able to:

- enter premises at any reasonable time if there is reason to believe that a prescribed process is or has been carried out that is or will give rise to a risk of serious pollution

- take a constable where obstruction on entry is anticipated, plus any necessary equipment or materials
- make such examinations and investigations as may be necessary
- direct that premises, or any part of the premises, remain undisturbed
- take measurements and photographs and make such recordings as are considered necessary
- take samples of articles and substances, and of air, water or land
- cause any article or substance that has caused, or is likely to cause, pollution to be dismantled or subjected to any process or test
- take possession of and detain any above article or substance for the purpose of examination, to ensure it is not tampered with before examination and is available for use as evidence
- require any person to answer questions as the inspector thinks fit and sign a declaration that their answers are the truth
- require the production of written or computerised records and take copies of these
- require that any person provide the facilities and assistance necessary to enable them to exercise their powers
- any other power conferred on them by the regulations.

3

Part II: Waste on land

The EPA imposes a duty of care on anyone who imports, carries, keeps, treats or disposes of waste. Such people must take all reasonable steps to ensure that the waste is collected, transported, treated and disposed of by licensed operators, that is, those issued with a *waste management licence.*

Public registers must be maintained by the authorities that detail the conditions of the licences issued and details of any enforcement actions taken. This Part should be read in conjunction with the system for IPC that is enforced by HMIP on prescribed Category A process industries.

Part III: Statutory nuisances and clean air

Section 79 deals with statutory nuisances and inspections. The following matters constitute *statutory nuisance:*

- any *premises* that are in such a state as to be prejudicial to health or cause a nuisance
- *smoke* emitted from premises that is prejudicial to health or causes a nuisance
- *fumes or gases* emitted from premises that are prejudicial to health or cause a nuisance
- any *dust, steam, smell or other effluent* arising on industrial, trade or business premises that are prejudicial to health or cause a nuisance
- any *accumulation or deposit* that is prejudicial to health or causes a nuisance
- any *animal* kept in a place or manner that is prejudicial to health or causes a nuisance

- *noise* emitted from premises that is prejudicial to health or causes a nuisance
- *any other matter* that is declared by any enactment to be a statutory nuisance.

It shall be the duty of every LA to ensure that its area is inspected from time to time to detect any statutory nuisances that ought to be dealt with under section 80 (Summary proceedings for statutory nuisances) and, where a complaint of a statutory nuisance is made to it by a person living within its area, to take such steps as are reasonably practicable to investigate the complaint. Where an LA is satisfied that a statutory nuisance exists, or is likely to occur or recur, that is in the area of the authority, the LA shall serve an *abatement notice*, imposing all or any of the following requirements:

- requiring the abatement of the nuisance or prohibiting or restricting its occurrence or recurrence
- requiring the execution of any works, and the taking of other steps, that may be necessary for any of these purposes

and the notice shall specify the time or times within which the requirements of the notice are to be complied with. Such a notice shall be served:

- except in the next two cases below, on the person responsible for the nuisance
- where the nuisance arises from any defect of a structural character, on the owner of the premises
- where the person responsible for the nuisance cannot be found or the nuisance has not yet occurred, on the owner or occupier of the premises.

A person served with an abatement notice may appeal to a Magistrates' Court within 21 days, beginning with the date on which they were served with the notice. If a person is served an abatement notice and, without reasonable cause, contravenes or fails to comply with any requirement or prohibition imposed by the notice, they shall be guilty of an offence.

Failure to comply with an abatement notice can result in the defendant being subjected to a fine not exceeding level 5 on the standard scale, together with a further fine of an amount equal to one-tenth of that level for each day the offence continues after the defendant has been convicted. A person who commits an offence on industrial, trade or business premises, however, shall be liable to a fine not exceeding £20,000 (see Table 3.1 for these penalties and others under the EPA). It is a defence to prove that *best practicable means* were used to prevent, or to counteract the effects of, the nuisance.

Section 81 gives the LA power, where an abatement notice has not been complied with, and whether or not they have already taken proceedings, to abate the nuisance themselves and do whatever may be necessary to execute the notice. Any expenses reasonably incurred may be recovered by them from the person whose actions of failure to act have caused the nuisance or be apportioned among several persons accordingly. This section also gives power to an LA, where proceedings for an offence would afford an inadequate remedy in the case of any statutory nuisance, to take proceedings in the High Court for the purpose of securing the abatement, prohibition or restriction of the nuisance, and the proceedings shall be maintainable notwithstanding the LA having suffered no damage from the nuisance.

A person who is aggrieved by the existence of a statutory nuisance need not necessarily pursue the matter through their local authority. Under section 82, a Magistrates' Court may act on a complaint made by any person on the ground that they are aggrieved by the existence of a statutory nuisance. If the court is satisfied that the alleged nuisance exists, or that, although abated, it is likely to recur on the same premises, the court shall make an order for either or both of the following:

- requiring the defendant to abate the nuisance (an abatement order) within a time specified in the order and to execute any works necessary to achieve this purpose
- prohibiting a recurrence of the nuisance (a prohibition order) and requiring the defendant, within a time specified in the order, to execute any works necessary to prevent the recurrence

and may also impose on the defendant a fine not exceeding level 5 on the standard scale.

If the court is satisfied that the alleged nuisance exists and is such that, in the opinion of the court, it renders the premises unfit for human habitation, an order (as above) may prohibit the use of the premises for human habitation until the premises are, to the satisfaction of the court, rendered fit for this purpose.

See Figure 3.1 for a summary of the procedure in such cases.

Part IV: Litter and so on

The EPA brought in new procedures regarding the control of litter. Section 87 created the offence of 'leaving litter' and section 88 brought in fixed penalty notices for leaving litter. Section 99 also gives LAs the power to deal with abandoned shopping and luggage trolleys.

Part V: Amendment to the Radioactive Substances Act 1960

Part V of the Act makes a number of amendments to the Radioactive Substances Act (RSA) 1960, the principal amendments being:

- provision for the appointment of inspectors
- provision for a scheme of fees and charges payable for registration and authorisation under the RSA
- new powers of enforcement, that is enforcement notices
- withdrawal of the exemption in favour of the UK Atomic Energy Authority from certain requirements of the RSA
- applications of the RSA to the Crown.

Table 3.1 Penalties under the Environmental Protection Act 1990

Part	Contravention	Penalties
Part I **Integrated pollution control**	Operating without an authorisation	Summary conviction – £20,000 fine
	Failing to comply with or contravening an enforcement or prohibition notice	Indictment: fine, 2 years imprisonment or both
	Failing to comply with a court order	Summary conviction – £20,000 fine
	Failing to supply required information	Indictment: fine, 2 years or both
Part III **Statutory nuisances**	Obstructing or preventing an inspector from carrying out their duties	£2000 fine
		On conviction for nuisance – £2000
		Daily penalty – £200
		On conviction for nuisance arising from industrial trade or business premises – £20,000
		Daily penalty – £2000
Part IV **Litter**	Leaving litter	On summary conviction – £1000
		Fixed penalty ticket scheme – £10

Part VI: Genetically modified organisms (GMOs)

The purpose of this Part is to prevent or minimise any damage to the environment that may arise from the escape or release from human control of GMOs. The following definitions are important here.

- the term *organism* means any acellular, unicellular or multicellular entity (in any form), other than humans or human embryos, and, unless the context otherwise requires, the term also includes any article or substance consisting of biological matter – *biological matter* meaning anything (other than an entity mentioned above) that consists of or includes:

- **FIG 3.1 Statutory nuisance procedure under the Environmental Protection Act 1990**

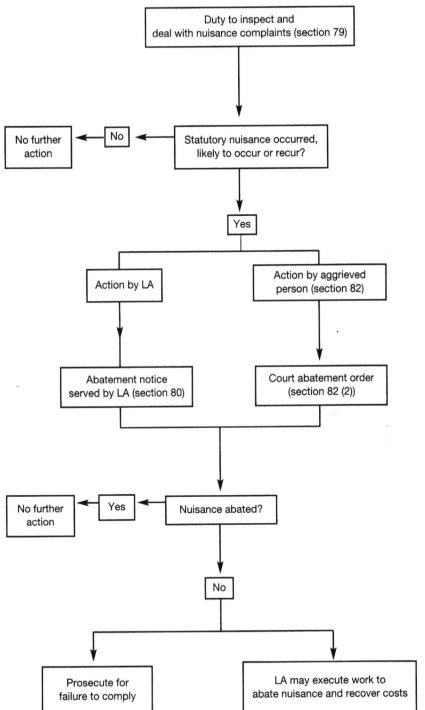

- tissue or cells (including gametes or propagules) or subcellular entities, of any kind, capable of replication or of transferring genetic material
- genes or other genetic material, in any form, that are so capable

and it is immaterial (in determining if something is or is not an organism or biological matter) whether it is the product of natural or artificial processes of reproduction and, in the case of biological matter, whether it has ever been part of a *whole* organism

- an organism is *genetically modified* if any of the genes or other genetic material in the organism:

  - have been modified by means of an artificial technique prescribed in regulations by the Secretary of State
  - are inherited or otherwise derived, through any number of replications, from genes or other genetic material (from any source) that were so modified

- the *techniques* that may be prescribed for the above purposes include:

  - any technique for the modification of any genes or other genetic material by the recombination, insertion or deletion of, or of any component parts of, that material from its previously occurring state
  - any other technique for modifying genes or other genetic material that, in the opinion of the Secretary of State, would produce organisms which should, for the purposes of this Part, be treated as having been genetically modified,

  but do not include techniques that involve no more than, or no more than the assistance of, naturally occurring processes of reproduction (including selective breeding techniques or *in vitro* fertilisation).

Section 108 of the Act requires that no person shall import or acquire, release or market any GMOs unless, before so doing, they:

- have carried out an assessment of the risks of damage to the environment that would be caused as a result of such an act
- in prescribed cases and circumstances, have given notice of this intention and prescribed information to the Secretary of State.

General duties relating to the importation, acquisition, keeping, release or marketing of GMOs are detailed in section 109. Section 110 empowers the Secretary of State to serve a prohibition notice on any person there is reason to believe is:

- proposing to import or acquit, release or market GMOs
- keeping any such organisms

if the Secretary of State is of the opinion that doing any such act to those organisms or continuing to keep them would involve a risk of causing damage to the environment.

A system of *consents*, however, is operated by the Secretary of State's office whereby a person importing or acquiring, releasing or marketing GMOs may be subject to certain limitations and conditions (sections 111 and 112).

Similar provisions relating to enforcement, offences and the powers of a court to make orders apply as with other Parts of the EPA.

Part VII: Nature conservation

This Part established Nature Conservancy Councils for England, Scotland and Wales and their functions, both generally and specifically.

Part VIII: Miscellaneous

The most significant of the miscellaneous provisions of this Part are:

- power to restrict the importation, use, supply or storage of substances or articles for the purpose of avoiding pollution or harm to human beings, animals or plants
- power to restrict the importation or exportation of waste for the purpose of preventing pollution or harm to human health or for conserving facilities or resources for dealing with waste
- power to make provision for obtaining information about substances that have the potential to cause pollution or harm to human health
- provision of public registers of potentially contaminated land
- amendments of the legislation of control of hazardous substances
- increase in the maximum penalties in respect of water pollution offences
- amendments to legislation as to marine deposits and the creation of public registers as to such deposits and marine incineration
- amendments to the provisions regarding oil pollution offences from ships
- provisions for the control of stray dogs
- provision regarding the burning of straw, stubble and other crop residues.

THE ENVIRONMENT ACT 1995

This Act brought in a number of new arrangements for dealing with environmental pollution, in particular the creation of the *Environment Agency for England and Wales*. From 1 April 1996 the Environment Act brought together HM Inspectorate of Pollution, the National Rivers Authority and local waste regulation authorities. Similar arrangements were made for Scotland with the creation of the *Scottish Environment Protection Agency*. Local air pollution control, formerly the responsibility of district and islands councils, comes under this Agency.

The Act contains detailed provisions for dealing with a range of environmental problems and issues.

Air quality

- Early legislation to establish a national strategy and framework of ambient air quality standards and targets for nine main pollutants.
- New powers for local authorities to review air quality in their districts and create *Air Quality Management Areas* where levels fall short of targets.

The Act provides the statutory framework for the new system. Much of the detailed local arrangements will be made under secondary legislation and guidance.

Contaminated land

The 'polluter pays' principle was reinforced, but there was also recognition that land-owners should be responsible for some aspects of the land, if the original polluters cannot be found. No new classes of liability were created.

The measures are based on the *suitable for use* approach, the removal of real environmental hazards without the imposition of unnecessary costs. This approach requires remedial action only where there is *significant harm* or pollution of controlled waters and where there are appropriate and cost-effective means available to take such action, taking into account the actual or intended use of the site.

Contaminated land is defined as any land which appears to be in such a condition that significant harm or pollution of controlled waters is being caused or is likely to be caused. *Harm* relates to the health of living organisms, or interference with the ecological systems of which they form part.

Pollution from abandoned mines

Statutory protection will be removed from the owners and operators of all mines abandoned after the end of 1999. This will allow the national agencies to deal with discharges through *consents*, as with other pollution discharges. Failure to comply with these consents will lead to prosecution. Mine operators will be required to give the agencies six months' notice of their intention to abandon a mine, thus allowing steps to be taken to prevent future mine water pollution.

National Parks Authorities (NPAs) in England and Wales

Local authority members will be drawn from parish, rather than county, councils, and will be appointed by the Secretary of State. NPAs will be required to foster the economic and social well-being of their local communities as well as protecting the natural beauty of the areas in their stewardship. Government and other public bodies must consider National Park purposes in carrying out their functions.

Waste strategy

National waste strategies would be drawn up by the new agencies for England and Wales and for Scotland, and national waste surveys would be carried out to inform these strategies. *Sustainable development* will be the cornerstone of the strategy, that is, making the best possible use of unavoidable waste and minimising the risk of pollution or harm to human health arising from waste disposal or recovery.

Producer responsibility

Regulations would be introduced to impose *producer responsibility* to increase the re-use, recovery or recycling of any product or material. The powers would be applicable to any waste stream. It is through regulations on packaging under these powers that the Government intends to fulfil its obligations to implement the recovery targets in the EC Directive on Packaging and Packaging Waste.

Minerals planning permissions

There would be an initial review and updating of old mineral permissions phased over six years. Future reviews of mineral permissions will be held every 15 years.

Water conservation

Water companies would be required to promote the efficient use of water by their customers. The Director General of Water Services will monitor this work and publish conservation performance league tables.

Water quality

The Environment Agency would have powers to require action to prevent water pollution and require polluters to clean up after pollution incidents.

Combined Heat and Power (CHP)

The Act amends the Electricity Act 1989 to enable CHP to compete effectively for support under the fossil fuel levy with other forms of non-fossil electricity generating schemes.

Nuisance provisions

The framework for the control of statutory nuisance contained in Part III of the Environmental Protection Act 1990 would be extended to Scotland.

THE EUROPEAN ENVIRONMENT AGENCY

This Agency was set up in 1994 following the adoption by the Council of Ministers in May 1990 of Council Regulation 1210/90. It is based in Copenhagen.

The functions of the Agency are:

(a) to provide Member States with objective, reliable and comparable information about the environment; and

(b) to ensure that the public is properly informed about the state of the environment.

The Management Board of the EEA is made up of one representative from each Member State, two representatives from the European Commission and a further two designated by the European Parliament.

The EEA is assisted by the *European Environment Information and Observation Network*. The main criticism of the EEA is that it has no role to play in the enforcement of environmental law.

THE ENVIRONMENT AGENCY

In November 1994 the Environment Bill was introduced before Parliament, its principal purpose being the creation of the *Environment Agency for England and Wales*. The creation of the agency was seen as an important means of developing a consistent and cohesive approach to environmental protection, which will also mean that the regulation and control of pollution is more readily understandable to those who are subject to the controls.

The agency brings together a number of functions formerly exercised by HMIP, NRA and the Waste Regulatory Authorities (WRAs).

In particular, the agency is responsible for the following functions:

(a) those formerly exercised by the NRA, which was abolished;

(b) waste management functions exercised by the WRAs;

(c) HMIP responsibilities under Part I of the EPA (Integrated Pollution Control);

(d) those relating to radioactive substances; and

(e) certain enforcement functions under the Health and Safety at Work etc. Act 1974 dealing with control of emissions.

The Environment Agency is assisted by *Regional Environment Protection Advisory Committees (REPACs)* that must be consulted when the Environment Agency decides how to undertake its functions in the respective regions. REPACs comprise mainly experts in environmental management.

SEX DISCRIMINATION

Sex Discrimination Act 1975

This Act, in its application to employment issues, makes discrimination unlawful in the following areas:

Unlawful discrimination against women or men

in respect of:

(a) arrangements for recruitment and selection;

(b) the terms on which employment is offered; or

(c) refusal or deliberate omission of an offer of employment on grounds of sex.

Similarly, discrimination in areas of training, transfer, promotion, or the provision of benefits, facilities or services relating to employment is unlawful.

Unlawful discrimination against married persons

In these situations a married person may be the subject of unlawful discrimination compared to a person of the same sex or of the opposite sex.

Unlawful indirect discrimination

'Indirect discrimination' includes discrimination which arises by the imposition of a requirement (other than sex or marital status) which, fundamentally, has the effect of excluding a major proportion of applicants of a particular sex or marital status.

Whilst direct discrimination might arise as a result of an advertisement seeking applications from men only, indirect discrimination results from imposing a term which might exclude more men than women; for instance, 'Applicants must be fit and not weigh more than 150 lb'.

Sex Discrimination Act 1986

This Act aimed to rectify a number of deficiencies in the 1975 Act. In particular this Act requires that employers do not discriminate against women in relation to retirement, and covering employment with regard to dismissal, promotion, training, provision of facilities and other benefits. Furthermore, the Act requires non-discriminatory retirement ages for men and women. This means there should be only one normal retirement age for each category of employee, regardless of sex. Where there is no *normal retirement age*, then all employees, regardless of sex, have the right to claim unfair dismissal up to the age of 65 years.

Employment tribunals

Any person who believes that he has been the subject of sex discrimination may raise a complaint with an employment tribunal. Cases are referred to the *Equal*

Opportunities Commission who may support the claim if they believe discrimination has occurred. On this basis, tribunals may award damages in these cases.

Both the above Acts were amended by the Sex Discrimination Act (Amendment) Regulations 2003.

THE RACE RELATIONS ACT 1976

This Act promotes the view that *race* is not a valid criterion on which employment discrimination should be based. Discrimination on the grounds of race covers colour, race, nationality and national or ethnic origins. 'Religion' and religious discrimination is not covered by the Act.

Principal provisions

Discrimination on the grounds of race, as defined, is unlawful in terms similar to those relating to sex discrimination. Indirect discrimination, such as discrimination on the basis of English language ability or national dress, could be unlawful.

In certain circumstances, racial discrimination is permitted, for instance in the case of genuine occupational qualifications, such as jobs providing services to ethnic groups, dramatic performances and modelling. Private households are excluded, unless the alleged discrimination takes the form of victimisation.

It is possible to discriminate in certain training activities, such as providing courses in a language not used by delegates.

Enforcement

This is through the *Commission for Racial Equality* and employment tribunals.

The Race Relations Act was amended by the Race Relations Act (Amendment) Regulations 2003.

THE SALE OF GOODS ACT 1979

This Act required that goods must be of merchantable quality, fit for normal use and give reasonable service for their price.

The seller must guarantee that a 'good title' passes to the buyer, i.e. the property is owned by the seller and that he has the legal right to sell.

In particular, goods must be 'fit for the purpose' the customer has asked the seller for. Where one of the above conditions is broken, the customer must return the goods at once, and the seller must refund the money.

Where the manufacturer is to blame, the seller can claim compensation from the manufacturer within a reasonable time.

THE SALE AND SUPPLY OF GOODS ACT 1994

This Act amended the Sale of Goods Act 1979 with regard to quality and fitness of goods. Thus:

- goods are of satisfactory quality if they meet the standard that a reasonable person would regard as satisfactory, taking account of the description of the goods, the price (if relevant) and other relevant circumstances;
- the quality of goods includes their state and condition and where appropriate fitness for all of the purposes for which goods of the kind in question are commonly supplied, i.e. appearance and finish, freedom from minor defects, safety and durability;
- 'satisfactory quality' does not extend to any matter making the quality of goods unsatisfactory:

    (a) which is specifically drawn to the buyer's attention before the contract is made;

    (b) where the buyer examines the goods before the contract is made, which that examination ought to reveal; and

    (c) in the case of a contract for sale by sample, which would have been apparent on a reasonable examination of the sample.

Both statutes were amended by the Sale and Supply of Goods to Consumers Regulations 2002.

BENEFITS FOR SICKNESS AND DISABILITY

A person who is incapable of work because of sickness or disability, either temporarily or permanently, may qualify for one of the following benefits:

- statutory sick pay; or
- incapacity benefit.

Statutory sick pay

Statutory sick pay (SSP) is paid by an employer to an employee who is:
16 years and over and below 65 years; and

- incapable of work because of sickness or disability; and
- earning at least as much as the national insurance lower earnings limit.

    SSP is not paid:

- if the person was getting incapacity benefit within the past 8 weeks (in some cases, 52 weeks); in this case the person should reclaim incapacity benefit;
- if the person is in the maternity allowance period.

SSP will not be paid for the first three days of being incapable of work but if the employee was getting SSP within the last 8 weeks it will be paid from the first day off work. It can be paid for up to 28 weeks. When it runs out the employee may be able to claim incapacity benefit. (See below.)

Incapacity benefit

Incapacity benefit is paid to a person who:

- is aged below 60 for a woman, 65 for a man; and
- is incapable of work because of sickness or disability; and
- is not entitled to SSP; and
- has paid sufficient national insurance contributions.

There are three types of incapacity benefit:

- *lower rate short-term incapacity benefit* is paid for the first 196 days and an adult dependant addition can be paid;
- *higher rate short-term incapacity benefit* is paid from the 197th day of sickness to the 364th day; an adult dependant addition can be paid;
- *long-term incapacity benefit* is paid from the 365th day of sickness; additions can be paid for an adult and child.

For the first 196 days the person will usually have to show that he is incapable of work by sending in medical certificates. After 196 days the person will usually have to pass the personal capability assessment to show he is incapable for work. Most people have to have a medical examination to decide if they pass the personal capability assessment. The assessment looks at the person's capability to perform a range of specified activities.

Extra costs of disability

There are two benefits which are payable to a person to help with the extra costs of disability, namely:

- *disability living allowance* paid to a person who has specific care needs and mobility needs; and
- *attendance allowances*, where a person has special care needs, such as assistance with bodily functions, needs supervision to stop himself being a danger to himself and/or is terminally ill.

Disablement benefit

A person who has a disability because he has had an accident at work or has a prescribed occupational disease may be able to qualify for disablement benefit. He may also be able to claim a benefit for not being able to work due to sickness and/or the extra costs of disability and/or an income-related benefit.

This benefit is paid if:

- the person was a paid employee at the time of the accident or when he contracted the disease; and
- his disability is assessed at 14% or more; people with pneumoconiosis and certain respiratory diseases can get benefit for a disablement assessment of less than 14%.

Disablement benefit can be paid for a fixed period or for life and the amount paid depends on the assessment of disablement.

Extra allowances

There are a number of extra allowances that may be paid in addition to or instead of disablement benefit.

- *Reduced earnings allowance* This can be paid to a person who has a disablement assessment of 1% or more, and has been unable to work, or is unable to work at his normal job or has worked less hours at his normal job. This allowance is only applicable to persons where the accident or onset of disease took place before 1 October 1990.
- *Retirement allowance* Where a person gives up regular employment after reaching 60 (women) or 65 (men) reduced earnings allowance becomes retirement allowance.
- *Constant attendance allowance* This allowance is made to people who are so disabled that they need constant care and attention and who are getting disablement benefit for 100% disablement assessment.
- *Exceptionally severe disablement allowance* This is paid to people who get constant attendance allowance at one of two higher rates and where the need for attendance is likely to be permanent.

DISABILITY DISCRIMINATION ACT 1995

Under the Act a person has a disability if he has a physical or mental impairment which has a 'substantial and long-term adverse effect' on his 'ability to carry out normal day-to-day activities'.

Physical or mental impairment – Guidance to the Act

The Guidance issued by the Secretary of State for Employment, which deals with the matters to be considered in relation to the discrimination of disability, refers to 'mental impairment' as not including any impairment from a mental illness unless that mental illness is 'clinically well-recognised'.

The Guidance outlines matters to be taken into account in determining:

- whether an impairment has a substantial adverse effect on a person's ability to carry out normal day-to-day activities; or

- whether such an impairment has a long-term effect.

The Guidance gives examples of:

- effects which it would be reasonable, in relation to particular activities, to regard for purposes of the Act as substantial adverse effects;
- effects which it would not be reasonable, in relation to particular activities, to regard for such purposes as substantial adverse effects;
- substantial adverse effects which it would be reasonable to regard, for such purposes, as long-term;
- substantial adverse effects which it would not be reasonable to regard, for such purposes, as long-term.

Substantial and long-term adverse effects

Indicators of potentially 'substantial effects' include:

- the time taken to carry out tasks
- the way in which tasks are carried out

Consideration must also be given to the cumulative effects of impairments which may not in themselves be substantial if taken individually. Moreover, environmental conditions, such as temperature and humidity, are relevant to the issue as to whether the adverse effect is substantial or not.

The extent to which an employee can reduce the effects of his condition by modifying his behaviour would be taken into account though the beneficial effects of treatment would be disregarded in assessing the severity of these effects.

A typical 'substantial adverse effect' is severe disfigurement, unless it is a disfigurement acquired by tattoo or body piercing.

'Long-term' means at least 12 months or the remainder of the person's life (if less than 12 months). Where a condition is recurrent, it is treated as continuing if it is likely to recur. A typical condition in this case is epilepsy.

Ability to carry out normal day-to-day activities

A person is defined as disabled only if his impairment affects his ability to carry out one of the following day-to-day activities:

- mobility
- manual dexterity
- physical co-ordination
- continence
- ability to lift, carry or otherwise move everyday objects
- speech, hearing, eyesight
- memory or ability to concentrate, learn or understand
- perception of risk or physical danger.

The Guidance gives examples of effects which would be classed as impairment. For example, a person who can walk no more than a mile without discomfort would have difficulty in establishing this as an impairment. On the other hand, inability to pick up objects of moderate weight with one hand would be classed as a substantial adverse effect, but an inability to move heavy objects without some form of mechanical aid would not be classed as such.

Disability and Discrimination (Meaning of Disability) Regulations 1996

These regulations list conditions which will not be treated as amounting to an impairment, including addictions to alcohol, nicotine or other substance unless originally caused by taking prescribed drugs.

Discrimination

The Disability Discrimination Act (DDA) makes it unlawful for an employer to discriminate against a disabled job applicant or worker in:

- selection for a job
- terms and conditions of employment
- promotion or transfer
- training
- employment benefits
- dismissal or any other detrimental treatment.

The DDA covers not just employees and workers but also the self-employed, temporary and contract workers. The employment sections of the Act do not apply to employers with less than 15 employees. Certain Crown employees including prison officers, operational fire fighters, people serving in the armed forces and police officers are currently excluded from the protection of the Act.

Discrimination against a disabled person

There are two elements to establish whether an employer has discriminated against a disabled person. An employer will have discriminated if:

- for a reason which relates to the disabled person's disability he treats him less favourably than he treats or would treat others to whom that reason does not or would not apply
- he cannot show that the treatment in question is justified.

'Justification' – the employer's defence

An employer who discriminates against a disabled person has a defence if he can establish that the discriminatory treatment is justified. The burden of proof rests with the employer.

Discrimination can only be justified if 'the reason for it is both material to the circumstances of the particular case and substantial'.

The duty to adjust

Employers must make reasonable adjustments where working arrangements and/or the physical features of premises cause a substantial disadvantage for a disabled person in comparison with persons who are not disabled.

The DDA gives examples of 'reasonable steps' which an employer could take including:

- altering working hours
- allowing time off for rehabilitation or treatment
- allocating some of the disabled person's duties to another person
- transferring the disabled person to another vacancy or another place of work
- giving or arranging training
- providing a reader or interpreter
- acquiring or modifying equipment or reference manuals
- adjusting the premises (see below).

An employer's failure to make a reasonable adjustment is a separate act of discrimination unless it can be justified.

Arrangements

An employer is under a duty to make a reasonable adjustment to:

- arrangements to determine who is offered employment, i.e. selection arrangements
- terms and conditions or other arrangements on which employment, promotion, transfer, training or other benefit is offered.

Physical features of the premises

Physical features which need consideration are:

- those arising from the design or construction of the building exits or access to buildings
- fixtures, fittings, furnishings, equipment or materials
- any other physical element or quality of land or the premises.

Typical examples of physical features causing disadvantage to a disabled person include poor levels of lighting that is too dim for a person with restricted vision, doors that are too narrow for a wheelchair and a start time for work which may cause problems for persons with restricted mobility.

In considering what is a 'reasonable step' for an employer to take reference may be made to the cost of adjustments, their effectiveness and the financial resources of the employer.

Employment tribunals

A complaint of discrimination can be taken to an employment tribunal and the procedures and remedies available are similar to those of the Sex Discrimination Act 1975 and the Race Relations Act 1976. In particular:

- there is a 3-month time limit for lodging application with a tribunal the tribunal has the power to:

- make a declaration about the rights of the disabled employee
- recommend that the employer takes reasonable action within a pre-scribed period of time to prevent or reduce any disadvantage to the dis-abled person, or to make a reasonable adjustment
- award compensation, on which interest is payable, to the disabled employee
- there is provision for a questionnaire procedure to gather information from employers.

Code of Practice

The DDA is accompanied by a code of practice for the elimination of disability discrimination. It does not impose legal obligations but can be used in evidence before a tribunal and must be taken into account. This Code of Practice pro-vides a wide range of examples as to what amounts to unlawful discrimination.

The health and safety implications of the DDA

The Act clearly strengthens the duty of care that an employer owes towards disabled employees, and outlines 'reasonable steps' to be taken and 'adjust-ments' to be made in order to reduce the disadvantages at work suffered by such persons.

In particular, the DDA makes special provisions with respect to the physical and environmental features of workplaces which commonly put disabled per-sons at some form of disadvantage compared with non-disabled employees.

Employers need, in many cases, to consider the ergonomic aspects of tasks undertaken by disabled persons, with respect to the time taken to complete the task and the way in which it is carried out. This may entail some form of study directed at modifying the system of work to take into account the par-ticular disability which could create, for example, an inability to lift certain items, a lack of physical co-ordination or an inability to remember a particu-lar procedure.

One of the outcomes of such a study would be the provision of training specifically directed at undertaking tasks safely, bearing in mind the partic-ular disability suffered.

Procedures for ensuring disabled employees are protected from work-place hazards may well need to be incorporated in some form of in-house documentation, including a Statement of Policy on the Safety of Disabled Employees.

THE TRADE UNION REFORM AND EMPLOYMENT RIGHTS ACT 1993

The objective of this legislation was to reform industrial relations and employment.

Section 28 gives employees the right not to be dismissed, regardless of hours of work, age or length of service for any of the following reasons:

(a) taking steps to protect themselves or others in dangerous circumstances;

(b) carrying out health and safety functions designated by the employer;

(c) quitting a dangerous part of the workplace if it is believed to be dangerous and refusing to return whilst danger is still present;

(d) bringing reasonable health and safety concerns to the employer's notice in the absence of a representative or relevant committee; or

(e) carrying out duties as an acknowledged (by the employer) health and safety representative or committee member.

Part II of the Act covers rights to maternity leave, granting pregnant employees 14 weeks' maternity leave irrespective of length of service or hours of work.

THE NOISE AND STATUTORY NUISANCES ACT 1993

The provisions in the EPA relating to noise nuisance were amended by the *Noise and Statutory Nuisances Act 1993 (NSNA)*. In particular, the NSNA incorporated a new statutory nuisance into the EPA at section 79(1)(ga).

Sections 79(1)(g) and 79(1)(ga) provide that noise can be a statutory nuisance in the following circumstances:

Section 79(1)(g) – noise that is prejudicial to health or a nuisance; and

Section 79(1)(ga) – noise that is prejudicial to health or a nuisance and is emitted from or caused by a vehicle, machinery or equipment in a street.

The EPA does not provide any particular definition of 'noise' other than that it 'includes vibration'.

Common law principles

Since noise is more likely to interfere, as a nuisance, with enjoyment, rather than cause injury or be injurious to health, interpretation as to what is a statutory noise nuisance is based on common law principles. Thus, when a court is considering whether noise constitutes a nuisance, it may consider various matters including:

(a) the duration and time of the noise;

(b) the nature of the activity;

(c) the harm suffered by the person affected; and

(d) the neighbourhood in which the noise took place.

In practice, the environmental health officer who is making the evaluation will take the above matters into consideration.

Noise in a street which is emitted from a vehicle, machinery or equipment

This particular statutory nuisance was incorporated in the EPA by the NSNA. The amendment was made largely because the EPA did not provide sufficient protection from noise in streets. (A street is defined as 'a highway and any other road, footway, square or court that is for the time being open to the public'.) A great deal of noise is generated in streets not only from vehicles but from people playing loud music, ice cream vans and car alarms. Certain exemptions are provided, however. It does not apply to noise created by traffic, the armed forces or by political demonstrations (or demonstrations supporting or opposing a campaign or cause). 'Traffic' is understood to mean vehicles in motion.

Noise Abatement Notices

Once an environmental health office is satisfied that a statutory noise nuisance exists, or is likely to occur, then a Noise Abatement Notice must be served on the person responsible for the noise nuisance. This is the person by whose act, default or at whose sufferance the noise is attributable.

Where the noise is emitted from a vehicle, the notice must be served on the owner and driver. Where the noise comes from machinery or equipment, then it should be served on the person who is, for the time being, the operator.

3

Defences

The defences available are:

(a) Reasonable excuse

It will be a defence if the defendant can prove that there was a reasonable excuse for non-compliance. A birthday party or celebration will not constitute a reasonable excuse.

(b) Best practicable means

Where a noise nuisance has been alleged in relation to noise from a vehicle, machinery or equipment, then it is a defence to show that best practicable means were used to prevent or control the noise. The defence is only available where the machinery or equipment is used for industrial, trade or business purposes.

Failure to comply with an Abatement Notice

Local authorities have three options available in the event of failure to comply with a notice. As far as noise nuisance is concerned, the environmental health officer has powers to take steps to abate the nuisance. This may mean the officer can remove stereo/audio equipment. If the noise is from a car alarm, the officer has power to open the car, if necessary by force, and immobilise the offending alarm.

Expenses incurred by a local authority in abating a nuisance can be recovered with interest. The local authority can also place a charge on the premises.

THE NOISE ACT 1996

The principal features of this Act are outlined below.

Adopting the powers

Local authorities must give three months' notice before adopting the powers under the Act.

They must advertise the introduction of the powers in a local newspaper for two consecutive weeks two months before they are introduced. They must state that a resolution has been passed, the commencement date and the general effect of the powers.

Investigation of complaints

Local authorities must take reasonable steps to investigate complaints of excessive noise emitted from another dwelling by a complainant present in a dwelling house during night hours, namely 11.00 pm to 7.00 am.

If the officer is satisfied that noise is being emitted and might exceed the permitted level, a warning notice can be served. The officer must decide whether to assess the noise inside or outside the complainant's dwelling and whether to use a measuring device.

If the source of the complaint is outside the local authority's boundaries, it can still use the powers as if the dwelling was in its area, whether or not the powers have been adopted by the other local authority.

Warning notices

Warning notices must state that the officer considers noise is being emitted from a dwelling during night hours and that it exceeds or may exceed the permitted levels measured from within the complainant's dwelling.

Notices must state that the person responsible for the noise may be guilty of an offence if noise continues for a specified period starting between ten minutes after the time the notice is served and ending the following 7.00 am.

A warning notice can be served by delivering it to any person present or near the offending dwelling who appears to be responsible for the noise or by leaving it at the offending dwelling.

Offences

A person is guilty of an offence if noise is emitted from the dwelling after a warning notice has been served during the period specified in the notice and if the noise exceeds the permitted levels measured from within a complainant's dwelling.

The Secretary of State will determine the permitted level in writing and may approve the type of device used for the measurement of noise.

Evidence

To be used in evidence a noise measurement must be made by an approved device used under the conditions to which approval was given.

The measurement should be documented. It may be appended to a statement signed by the officer, giving particulars of the measurement or the circumstances in which it was made, state the type of device used, and that the conditions set were satisfied.

An officer may also sign a document stating that a certain dwelling has been identified as the source of the noise. A copy must be served on the person charged with the offence not less than seven days before the trial.

Fixed penalty notices

A fixed penalty notice of £100 can be given if an officer believes someone has committed an offence under the Act. This offers the offender the opportunity of discharging any liability to conviction for the offence by payment of the fixed penalty.

Payment must be made within 14 days.

Sums received by a local authority must be paid to the Secretary of State.

Seizure

Seizure powers can be used when a warning notice has failed to keep noise within the permitted level. An officer can enter a dwelling and seize any equipment which is being used in the emission of noise.

A warrant can be granted by a Justice of the Peace on sworn information in writing that a warning notice has been served and breached as measured from within the complainant's dwelling and that entry to the dwelling has been refused. If premises are unoccupied, they must 'be effectively secured against trespassers' as found.

The power to abate noise causing a statutory nuisance under the Environmental Protection Act includes the power to seize and remove any equipment used in the emission of the noise in question.

THE CRIMINAL JUSTICE ACT 1991

The Criminal Justice Act (CJA) brought in provisions relating to *unit fines*. A stated aim of the CJA is to encourage the use of fines, to make their effect fairer on offenders of different means, and to decrease the incidence of fine defaulters becoming imprisoned. Sections 17–24 detail the provisions relating to unit fines and create the situation whereby the actual amount imposed by the courts as a fine will reflect two aspects, namely the seriousness of the offence and the means of the offender.

The seriousness of the offence

This is measured in units and the Act fixes a maximum limit to the number of units attributable to any offence according to the accepted concept of standard levels. These limits are indicated in Table 3.2.

Table 3.2 Units of fines for offences

Level	Maximum	Maximum number of units
1	£ 200	Up to 2 units
2	£ 500	Up to 5 units
3	£1000	Up to 10 units
4	£2500	Up to 25 units
5	£5000	Up to 50 units

When a court is determining the seriousness of an offence, relevant factors that must be considered are:

- aggravating and mitigating factors relating to the offence itself
- aggravating factors disclosed by previous convictions, such as similarities
- when considering a fine, the facts of all other associated offences.

An *associated offence* may be taken to be one for which an offender is convicted or sentenced, or which is taken into consideration at the same hearing.

Under section 29(1), an offence is not to be regarded as more serious simply because of previous convictions or any failure to respond to previous sentences. Under the CJA, therefore, previous convictions do not, *per se*, aggravate the seriousness of the current offence.

Section 29(2), however, states that where any aggravating factors of an offence are disclosed by the circumstances of other offences committed by the offender, nothing shall prevent the court from taking these factors into account for the purpose of forming an opinion as to the seriousness of the offence. On this basis, where previous offences display similarities to the case before the court, this may be taken to be an aggravating factor with regard to the seriousness of the offence under consideration.

The offender's means

The means of the offender are reflected in a court's calculations of the offender's *disposable weekly income* (DWI). This is essentially one third of what remains of an offender's income after the deduction of general living expenses. The assessment is made after the offender has submitted a *statement of means*. It is the DWI that will determine the value of the unit used for their fine. Section 18 actually stipulates a minimum and maximum value for a unit, that is, £4 and £100 respectively. On this basis, the maximum fine that could be imposed is £5000, that is, 50 units at £100. Thus, by establishing fines at a level that an offender can afford to pay, it is envisaged that few offenders will be involved in fine default courts, which reduces their risk of being sent to prison for so doing.

Section 20 grants a court the power to require an offender to submit a statement of means. Failure, without reasonable excuse, to provide this information can result in a fine not exceeding level 3 on the standard scale. On the other hand, a person who knowingly or recklessly makes a false statement or fails to disclose any material fact is punishable with up to three months' imprisonment or a level 4 fine.

Default arrangements

Section 22 deals with the maximum periods of imprisonment in the event of an offender defaulting on the payment of a fine (see Table 3.3 and 3.4).

For those cases where a fine has not been set in units, section 23 of the CJA states the maximum period of imprisonment in default for a given amount of fine (see Table 3.4)

Table 3.3 Imprisonment related to units

Units	Maximum days of imprisonment
1–2	7
3–5	14
6–10	28
11–25	45
26–50	3 months

Table 3.4 Imprisonment related to fines

Amount	Maximum days of imprisonment
Up to £200	7
Over £200 and up to £500	14
Over £500 and up to £1000	28
Over £1000 and up to £2500	45
Over £2500 and up to £5000	1–3 months

Exceptions

Unit fines do not apply in cases where the offender is a company. Certain offences (for instance under the EPA and HSWA) may not be covered by the system as the offence is committed by a corporate body. In other cases, however, identifiable individuals, such as directors, may be responsible for the offence and subject to the unit fine provisions.

Where the maximum fine under a statute or regulations is higher than the normal maximum fine stipulated under the CJA (£5000), the unit fines provisions do not apply. For instance, the maximum fine under the EPA and HSWA is £20,000. However, the unit fine system does apply under, for instance, section 23 of the EPA (obstruction), as this offence carries a maximum of a level 5 penalty.

The unit fine system does not apply to fines imposed by a Crown Court, even when cases are committed to a Crown Court for sentence. On the other hand, even in those instances where unit fines do not apply, section 19 requires that a court, when deciding a figure for a fine, must take into account the means of the offender and this may result in an increase or decrease in the fine imposed.

LEVELS OF STATUTORY DUTY

Statutory duties give rise to *criminal liability*. There are three distinct levels of statutory duty:

1 'absolute' requirements

2 'practicable' requirements

3 'reasonably practicable' requirements.

'Absolute' requirements

Where the risk of injury or disease is inevitable if safety requirements are not followed, a statutory duty may well be absolute.

The classic instance of an absolute duty is in regulation 5(1) of the Provision and Use of Work Equipment Regulations 1992 (PUWER), which states 'Every employer *shall* ensure that work equipment is so constructed or adapted as to be suitable for the purpose for which it is to be used or provided'. Absolute duties are qualified by the terms 'shall' or 'must'.

'Practicable' requirements

A statutory requirement is qualified by the phrase 'so far as is practicable' when it implies that if, *in the light of current knowledge and invention*, it is feasible to comply with this requirement, then, irrespective of the cost or sacrifice involved, such a requirement must be complied with (see *Schwalb* v. *Fass H. & Son* (1946) 175 LT 345).

'Practicable' means more than physically possible and implies a higher duty of care than a duty qualified by the phrase 'so far as is reasonably practicable'.

'Reasonably practicable' requirements

A duty qualified by the phrase 'so far as is reasonably practicable' implies a lower or lesser level of duty than one qualified by 'so far as is practicable'. 'Reasonably practicable' is a narrower term than 'physically possible' (that is, 'practicable'), and implies that a computation must be made in which the *quantum of risk* is placed on one side of the scale and the *sacrifice involved* in carrying out the measures necessary for averting that risk is placed on the other side. If it can be shown that there is a gross disproportion between these two factors, that is, that the risk is insignificant in relation to the sacrifice, then a defendant discharges the onus on themselves (see *Edwards* v. *National Coal Board* (1949) 1 AER 743). All duties under the HSWA are qualified by the term 'so far as is reasonably practicable'.

The 'reasonable man'

What *is* a reasonable person? What is it about their behaviour that makes them reasonable and how do the courts interpret this term?

The mythical 'reasonable man' was interpreted by one judge in the past as 'the man who travels to work every day on the top deck of the No. 57 Clapham omnibus'. The term is flexible and changes with time according to society and the norms prevalent at the time.

The term 'reasonable' can be found in section 7 of the HSWA, that is, in the duty on employees to take *reasonable* care for the safety of themselves and others, including members of the public, who may foreseeably be affected by their acts or omissions at work.

THE ROLES OF THE ENFORCING AUTHORITIES

The enforcing authorities under the HSWA are:

- the Health and Safety Executive (HSE), which is split into a number of specific inspectorates, for example, Factories, Agricultural, Nuclear Installations
- local authorities, principally through their environmental health departments
- for certain matters, the Fire Authority.

Actual enforcement is undertaken by inspectors appointed under the Act and authorised by a written warrant from the enforcing authority.

The powers of inspectors

Under section 20 of the HSWA, inspectors have the following powers:

- to enter premises at any reasonable time and, where obstruction is anticipated, to enlist the support of a police officer
- on entering premises,

  - to take with them any other person duly authorised by their enforcing authority
  - any equipment or materials required for any purpose for which the power of entry is being exercised

- to make such examinations and investigations as may be necessary
- to direct that premises or any part of such premises, or anything therein, shall remain undisturbed for so long as is reasonably necessary for the purpose of any examination or investigation
- to take such measurements and photographs and make such recordings as they consider necessary for the purpose of any examination or investigation
- to take samples of any articles or substances found in any premises, and of the atmosphere in or in the vicinity of such premises
- where it appears to them that an article or substance has caused or is likely to cause danger to health or safety, to cause it to be dismantled or subjected to any process or test
- to take possession of any article or substance and to detain these for as long as is necessary:

  - to examine these
  - to ensure that they are not tampered with before their examination is completed
  - to ensure that they are available for use as evidence in any proceedings for an offence under the relevant statutory provisions

- to require any person whom they have reasonable cause to believe to be able to give any information relevant to any examination or investigation to answer such questions as the inspector thinks fit to ask and to sign a declaration of the truth of their answers
- to require the production of, and inspect and take copies of or any entry in:

  - any books or documents that, by virtue of the relevant statutory provisions, are required to be kept
  - any other books or documents that it is necessary for them to see for the purposes of any examination or investigation

- to require any person to afford them such facilities and assistance with respect to any matters or things within that person's control or in relation to which that person has responsibilities that are necessary to enable the inspector to exercise any of the powers conferred on them by this section

  - any other power that is necessary for the purpose of carrying into effect the relevant statutory provisions.

After an inspector has completed an investigation or examination, they have a duty to inform safety representatives of the actual matters they have found (section 28(8)), and must give the employer similar information.

Notices

Enforcing officers may serve two types of notice:

- improvement notices
- prohibition notices.

Improvement notices

If an inspector is of the opinion that a breach has, or is likely to, occur, they may serve an improvement notice on the employer or employee. The notice must state which statutory provision the inspector believes has been contravened and the reason for this belief. It should also state a time limit within which the contravention should be remedied.

Prohibition notices

Where an inspector is of the opinion that a work activity involves or will involve a risk of serious personal injury, they may serve a prohibition notice on the owner and/or occupier of the premises or the person having control of that activity.

Such a notice will direct that the specified activities in the notice shall not be carried out by or under the control of the person on whom the notice is served unless certain specified remedial measures have been complied with.

It should be appreciated that it is not necessary for an inspector to believe that a *legal* provision is being or has been contravened, rather a prohibition notice is served where there is an immediate threat to life and in anticipation of danger. A prohibition notice may have *immediate* effect after its service by the inspector or it may be *deferred*, thereby allowing the person time to remedy the situation, carry out works and so on. The duration of a deferred prohibition notice is stated on the notice.

For a summary of the features of improvement and prohibition notices, see Table 3.5.

Prosecution

Prosecution is frequently the outcome of failure to comply with an improvement or prohibition notice. Conversely, an inspector may simply institute legal proceedings without serving a notice.

Cases are normally heard in a Magistrates' Court, but there is also provision in the HSWA regarding indictments. Much depends on the gravity of the offence.

Penalties

The Offshore Safety Act 1992 amended the HSWA and allows magistrates to impose fines of up to £20,000 for a breach of sections 2–6 of the Act, and for breach of an improvement notice or prohibition notice. The maximum fine for other offences is £5000.

This Act also widened the range of health and safety offences for which the higher courts can impose prison sentences. The two-year maximum sentence, which existed for offences concerning explosives, licensing regimes and breach of a prohibition notice was extended to cover breach of an improvement notice.

Employment Medical Advisory Service

This service operates under the *Employment Medical Advisory Service Act 1972* and by virtue of Part II of HSWA.

EMAS has the following functions:

- to advise the Secretary of State, HSC, Manpower Services Commission and others concerned with health and safety of employed persons on health related issues at work;
- to provide information and advice to those seeking or training for employment;
- to provide medical assistance, and appoint registered medical practitioners as employment medical advisers;
- to be responsible for payment of fees to medical advisers, and for remuneration of persons attending examinations, the keeping of records and accounts.

No information about a person may be disclosed to anyone other than for the efficient performance of the adviser's functions. An employee may, by consent, waive this restriction.

Comment

With the advent of the MHSWR, the role of inspectors may well change to a more management systems-orientated approach. On this basis, instead of examining premises, plant and processes to identify contraventions of current legislation, inspectors appointed under the HSWA will devote more time to reviews of management systems for health and safety. This approach may well be experienced in large multisite organisations where greater attention will be paid to documentation of safety procedures, training systems, the appointment and relative competence of competent persons, emergency procedures and so on from head office level, accompanied by spot checks to ascertain whether or not these procedures and systems are actually implemented at local unit level.

HEALTH AND SAFETY EXECUTIVE   Serial No. 1

Health and Safety at Work etc. Act 1974, Sections 21, 23 and 24

IMPROVEMENT NOTICE

Name and address (See Section 46)	To ....
	....
(a) Delete as necessary	(a) Trading as ....
(b) Inspector's full name	(b) ....
	one of (c) ....
(c) Inspector's official designation	of (d) ....
	...............................Tel No ....
(d) Official address	hereby give you notice that I am of the opinion that at..
	(e) ....
(e) Location of premises or place and activity	you, as (a) an employer/a self employed person/a person wholly or partly in control of the premises
	(f) ....
(f) Other specified capacity	(a) are contravening/have contravened in circumstances that make it likely that the contravention will continue to be repeated.
	....
	....
(g) Provisions contravened	(g) ....
	....
	The reasons for my said opinion are: ....
	....
	....
	and I hereby require you to remedy the said contraventions or, as the case may be, the matters occasioning them by
	(h) ....
(h) Date	(a) In the manner stated in the attached schedule which forms part of the notice.
	Signature..................... Date ....
	Being an Inspector appointed by an Instrument in writing made pursuant to Section 19 of the said Act and entitled to issue this notice.
	(a) An Improvement notice is also being served on
	....
	of ....
LP1	related to the matters contained in this notice.

● **FIG 3.2 (a) Example of an improvement notice**

HEALTH AND SAFETY EXECUTIVE   Serial No. P

Health and Safety at Work etc. Act 1974, Sections 22–24

PROHIBITION NOTICE

Name and address (See Section 46)
(a) Delete as necessary
(b) Inspector's full name
(c) Inspector's official designation
(d) Official address

To ....

....

(a) Trading as ....

(b) ....

one of (c) ....

of (d) ....

....tel no. ....

hereby give you notice that I am of the opinion that the following activities,

namely:– ....

....

....

which are (a) being carried on by you/about to be carried on by you/under your control

(e) Location of activity

at (e) ....

Involve, or will involve (a) a risk/an imminent risk, of serious personal injury. I am further of the opinion that the said matters involve contraventions of the following statutory provisions:– ....

....

....

....

because ....

....

....

and I hereby direct that the said activities shall not be carried on by you or under your control (a) Immediately /after

(f) Date

(f) ....

unless the said contraventions and matters included in the schedule, which forms part of this notice, have been remedied.

Signature ....Date ....

LP2

being an inspector appointed by an instrument in writing made pursuant to Section 19 of the said Act and entitled to issue this notice.

● **FIG 3.2 (b) Example of a prohibition notice**

Table 3.5 A summary of the features of notices

Type of notice	Circumstances		When notice takes effect	Effect of appeal (section 24*)	Person on whom notice is served
	Contravention of a 'relevant statutory provision'	Risk involved			
Improvement notice (section 21)	Must have been one and it is likely that it will be continued or repeated	No risk specified and covers cases where there is no risk	When specified, but not earlier than 21 days after issue	Suspends the notice until appeal is determined or the appeal is withdrawn	Person contravening provision
Immediate prohibition notice (section 22)	Not necessary	Where there is or will be immediate risk of serious personal injury†	Immediate	No suspension, unless the Employment Tribunal rules otherwise	Person under whose control the activity is carried out or by whom it is carried out
Deferred prohibition notice (section 22)	Not necessary	Risk of serious personal injury not imminent	At the end of the period specified in the notice	As in the case of an immediate notice	As in the case of an immediate notice

* Appeals against notices are dealt with in regulations made under section 24 as follows:
– The Industrial Tribunals (Improvement and Prohibition Notices Appeals) Regulations 1974 (SI 1974, No. 1925) – for notices served in England and Wales.
– The Industrial Tribunals (Improvement and Prohibition Notices Appeals) (Scotland) Regulations 1974 (SI 1974, No. 1926).
† 'Personal injury' includes any disease and any impairment of a person's physical or mental condition (section 53).

③

The principal regulations

INTRODUCTION

As we saw in the last chapter, statutes enable the Minister or Secretary of State to make regulations. All regulations, such as the Noise at Work Regulations 1989, are made under the HSWA. More recently, regulations have been made in order to implement EC Directives. For instance, the 'Council Directive on the introduction of measures to encourage improvements in the health and safety of workers at work' (the Framework Directive) was implemented in the UK as the Management of Health and Safety at Work Regulations (MHSWR) 1992. Similarly, the 'Council Directive concerning the minimum safety and health requirements for the use by workers of machines, equipment and installations' (the Machinery Safety Directive) was implemented as the Provision and Use of Work Equipment Regulations 1992.

This chapter covers the principal health and safety regulations. The depth of treatment of the various regulations varies from an in-depth treatment, in the case of the more significant regulations, to an overview in other cases, outlining the key provisions.

BUILDING REGULATIONS 2000

These regulations are made under the Building Act 1984. They revoke and replace with amendments the 1991 Regulations and consolidate all subsequent amendments to those regulations. They impose requirements on people carrying out certain building operations. The regulations are principally concerned with the various aspects of building control, such as 'material change of use' of buildings, and contain provisions directed at:

- securing the safety, health and welfare and convenience of people in and about buildings;
- furthering the conservation of fuel and power by the use of energy ratings for buildings;

- supervision of building work; and
- unauthorised building work.

Generally, building work and any material change of use must comply with the 'applicable requirements' detailed in Schedule 1. Any building work must be carried out with adequate and proper materials and in 'a workman-like manner'. Compliance with the requirements in Parts A to K of Schedule 1 does not require anything to be done 'beyond what is necessary to secure reasonable standards of health and safety'.

Schedule 1 – requirements

This Schedule incorporates specific requirements with regard to:

- Structure
- Fire safety
- Site preparation and resistance to moisture
- Toxic substances
- Resistance to the passage of sound
- Ventilation
- Hygiene
- Drainage and waste disposal
- Heat producing appliances
- Protection from falling, collision and impact
- Conservation of fuel and power
- Access and facilities for disabled people
- Glazing – safety in relation to impact, opening and cleaning.

The specific health and safety-related areas of Schedule 1 are covered below.

Part B – fire safety

Means of warning and escape

The building shall be so designed and constructed so that there are appropriate provisions for the early warning of fire, and appropriate means of escape in case of fire from the building to a place of safety outside the building capable of being safely and effectively used at all material times.

Internal fire spread (linings)

To inhibit the spread of fire within the building the internal linings shall:

(a) adequately resist the spread of flame over their surfaces; and

(b) have, if ignited, a rate of heat release which is reasonable in the circumstances.

'Internal linings' mean the materials lining any partition, wall, ceiling or other internal structure.

Internal fire spread (structure)

The building shall be so designed and constructed so that, in the event of fire, its stability will be maintained for a reasonable period.

A wall common to two or more buildings shall be designed and constructed so that it adequately resists the spread of fire between those buildings. For the purposes of this sub-paragraph a house in a terrace and a semi-detached house are each to be treated as a separate building.

To inhibit the spread of fire within the building, it shall be sub-divided with fire-resisting construction to an extent appropriate to the size and intended use of the building.

The building shall be so designed and constructed so that the unseen spread of fire and smoke within concealed spaces in its structure and fabric is inhibited.

External fire spread

The external walls of the building shall adequately resist the spread of fire over the walls and from one building to another, having regard to the height, use and position of the building.

The roof of the building shall adequately resist the spread of fire over the roof and from one building to another, having regard to the use and position of the building.

Access and facilities for the fire service

The building shall be so designed and constructed so as to provide reasonable facilities to assist fire fighters in the protection of life.

Reasonable provision shall be made within the site of the building to enable fire appliances to gain access to the building.

Part D – toxic substances

Cavity insulation

If insulating material is inserted into a cavity in a cavity wall reasonable precautions shall be taken to prevent the subsequent permeation of any toxic fumes from that material into any part of the building occupied by people.

Part E – resistance to the passage of sound

Airborne sound (walls)

A wall which:

(a) separates a dwelling from another building or from another dwelling; or

(b) separates a habitable room or kitchen within a dwelling from another part of the same building which is not used exclusively as part of the dwelling,

shall have reasonable resistance to the transmission of airborne sound.

Part F – ventilation

Means of ventilation

There shall be adequate means of ventilation provided for people in the building.

Part G – hygiene

Sanitary conveniences and washing facilities

Adequate sanitary conveniences shall be provided in rooms provided for that purpose, or in bathrooms. Any such room or bathroom shall be separated from places where food is prepared.

Adequate washbasins shall be provided in:

(a)   rooms containing water closets; or

(b)   rooms or spaces adjacent to rooms containing water closets.

Any such room or space shall be separated from places where food is prepared.

There shall be a suitable installation for the provision of hot and cold water to washbasins provided in accordance with the above paragraph.

Sanitary conveniences and washbasins to which this paragraph applies shall be designed and installed so as to allow effective cleaning.

4

Hot water storage

A hot water storage system that has a hot water storage vessel which does not incorporate a vent pipe to the atmosphere shall be installed by a person competent to do so, and there shall be precautions to:

(a)   prevent the temperature of stored water at any time exceeding 100°C; and

(b)   ensure that the hot water discharged from safety devices is safely conveyed to where it is visible but will not cause danger to persons in or about the building.

Part J – heat producing appliances

Air supply

Heat producing appliances shall be so installed that there is an adequate supply of air to them for combustion and for the efficient working of any flue-pipe or chimney.

Discharge of products of combustion

Heat producing appliances shall have adequate provision for the discharge of the products of combustion to the outside air.

Protection of building

Heat producing appliances and flue-pipes shall be so installed, and fire-places and chimneys shall be so constructed, as to reduce to a reasonable level the risk of the building catching fire in consequence of their use.

Part K – Protection from falling, collision and impact

Stairs, ladders and ramps

Stairs, ladders and ramps shall be so designed, constructed and installed as to be safe for people moving between different levels in or about the building.

Protection from falling

(a) Any stairs, ramps, floors and balconies and any roof to which people have access; and

(b) any light well, basement area or similar sunken area connected to a building,

shall be provided with barriers where it is necessary to protect people in or about the building from falling.

Vehicle barriers and loading bays

Vehicle ramps and any levels in a building to which vehicles have access shall be provided with barriers where it is necessary to protect people in or about the building.

Vehicle loading bays shall be constructed in such a way, or be provided with such features, as may be necessary to protect people in them from collision with vehicles.

Protection from collision with open windows, etc.

Provision shall be made to prevent people moving in or about the building from colliding with open windows, skylights or ventilators.

Protection against impact from and trapping by doors

Provision shall be made to prevent any door or gate:

(a) which slides or opens upwards from falling onto any person; and

(b) which is powered, from trapping any person.

Provision shall be made for powered doors and gates to be opened in the event of a power failure.

Provision shall be made to ensure a clear view of the space on either side of a swing door or gate.

Part M – access and facilities for disabled people

Interpretation

In this Part 'disabled people' means people who have:

(a) an impairment which limits their ability to walk or which requires them to use a wheelchair for mobility; or

(b) impaired hearing or sight.

Access and use

Reasonable provision shall be made for disabled people to gain access to and use the building.

Sanitary conveniences

If sanitary conveniences are provided in any building which is not a dwelling, reasonable provision shall be made for disabled people.

Part N – glazing – safety in relation to impact, opening and cleaning

Protection against impact

Glazing, with which people are likely to come into contact whilst moving in or about the building shall:

(a)   if broken on impact, break in a way which is unlikely to cause impact; or

(b)   resist impact without breaking; or

(c)   be shielded or protected from impact.

Manifestation of glazing

Transparent glazing, with which people are likely to come into contact while moving in or about the building, shall incorporate features which make it apparent.

Safe opening and closing of windows, etc

Windows, skylights and ventilators which can be opened by people in or about the building shall be so constructed or equipped that they may be opened, closed or adjusted safely.

Safe access for cleaning windows, etc.

Provision shall be made for any windows, skylights or any transparent or translucent walls, ceilings or roofs to be safely accessible for cleaning.

Approved documents

The Secretary of State is empowered by the Building Act 1984 to approve and issue documents containing practical guidance with respect to the requirements contained in these regulations.

THE CHEMICALS (HAZARD INFORMATION AND PACKAGING FOR SUPPLY) (CHIP 2) REGULATIONS 1994

These regulations cover many important aspects with regard to the classification, labelling and packaging of chemicals. The following definitions are particularly significant:

Aerosol dispenser means an article which consists of a non-reusable receptacle containing a gas compressed, liquefied or dissolved under pressure, with or without liquid, paste or powder and fitted with a release device allowing the contents to be ejected as solid or liquid particles in suspension in a gas, as a foam, paste or powder or in a liquid state.

Category of danger means in relation to a substance or preparation dangerous for supply, one of the categories of danger specified in column 1 of Part I of Schedule 1.

Classification means, in relation to a substance or preparation dangerous for supply, classification in accordance with regulation 5 (classification for supply).

Indication of danger means, in relation to a substance or preparation dangerous for supply, one or more of the indications of danger referred to in column 1 of Schedule 2 and:

- in the case of a substance dangerous for supply *listed* in Part I of the approved supply list, it is one or more indications of danger for that substance specified by a symbol-letter in column 3 of Part V of that list; or
- in the case of a substance dangerous for supply *not so listed* or a *preparation dangerous for supply*, it is one or more indications of danger determined in accordance with the classification of that substance or preparation under regulation 5 and the approved classification and labelling guide.

Package means, in relation to a substance or preparation dangerous for supply, the package in which the substance or preparation is supplied and which is liable to be individually handled during the course of the supply and includes the receptacle containing the substance or preparation and any other packaging associated with it and any pallet or other device which enables more than one receptacle containing a substance or preparation dangerous for supply to be handled as a unit, but *does not include*:

- a freight container (other than a tank container), a skip, a vehicle or other article of transport equipment; or
- in the case of supply by way of retail sale, any wrapping such as a paper or plastic bag into which the package is placed when it is presented to the purchaser.

Packaging means, in relation to a substance or preparation dangerous for supply, as the context may require, the receptacle, or any components, materials or wrappings associated with the receptacle for the purpose of enabling it to perform its containment function or both.

Poisons advisory centre means a body approved for the time being for the purposes of regulation 14 (notification of constituents of certain preparations dangerous for supply) by the Secretary of State for Health in consultation with the Secretaries of State for Scotland and Wales, the HSC and such other persons or bodies as appear to him or her to be appropriate.

Preparations means mixtures or solutions of two or more substances.

Preparation dangerous for supply means a preparation which is in one or more of the categories of danger specified in column 1 of Schedule 1.

Receptacle means, in relation to a substance or preparation dangerous for supply, a vessel, or the innermost layer of packaging, which is in contact with the substance and which is liable to be individually handled when the substance is used and includes any closure or fastener.

Risk phrase means, in relation to a substance or preparation dangerous for supply, a phrase listed in Part III of the approved supply list, and in these Regulations specific risk phrases may be designated by the letter 'R' followed by a distinguishing number or combination of numbers, but the risk phrase shall be quoted in full on any label or safety data sheet in which the risk phrase is required to be shown.

Safety phrase means, in relation to a substance or preparation dangerous for supply, a phrase listed in Part IV of the approved supply list, and in these regulations specific safety phrases may be designated by the letter 'S' followed by a distinguishing number or combination of numbers, but the safety phrase shall be quoted in full on any label or safety data sheet in which the safety phrase is required to be shown.

Substances means chemical elements and their compounds in the natural state or obtained by any production process, including any additive necessary to preserve the stability of the product and any impurity deriving from the process used, but excluding any solvent which may be separated without affecting the stability of the substance or changing its composition.

Substance dangerous for supply means:

- a substance listed in Part I of the approved supply list; or
- any other substance which is in one or more of the categories of danger specified in column 1 of Schedule 1.

Supplier means a person who supplies a substance or preparation dangerous for supply, and in the case of a substance which is imported (whether or not from a Member State), includes the importer established in Great Britain of that substance or preparation.

Supply in relation to a substance or preparation:

- means, subject to paragraph (b) or (c) below, supply of that substance or preparation, whether as principal or agent for another, in the course of or for use at work, by way of:
  - sale or offer for sale;
  - commercial sample; or
  - transfer from a factory, warehouse or other place of work and its curtilage to another place of work, whether or not in the same ownership;
- for the purposes of sub-paragraphs (a) and (b) of regulation 16(2) (HSE as enforcement agency), except in relation to regulations 7 (advertisements) and 12 (child-resistant fastenings and warning devices), in any case for which by virtue of those sub-paragraphs the enforcing authority for these Regulations is the Royal Pharmaceutical Society or the local weights and

measures authority, has the meaning assigned to it by section 46 of the Consumer Protection Act 1987 and also includes offer to supply and expose for supply; or

- in relation to regulations 7 (advertisements) and 12 (child-resistant fastenings and warning devices) shall have the meaning assigned to it by regulations 7(2) and 12(12) respectively.

Application of the regulations

These regulations apply to any substance or preparation which is dangerous for supply *except*:

- radioactive substances or preparations;
- animal feeds;
- cosmetic products;
- medicines and medicinal products;
- controlled drugs;
- substances or preparations which contain disease-producing micro-organisms;
- substances or preparations taken as samples under any enactment;
- munitions, which produce explosion or pyrotechnic effect;
- foods;
- a substance or preparation which is under customs control;
- a substance which is intended for export to a country which is not a Member State;
- pesticides;
- a substance or preparation transferred within a factory, warehouse or other place of work;
- a substance to which regulation 7 of the Notification of New Substances Regulations 1993 applies;
- substances, preparations and mixtures in the form of wastes.

The approved supply list

This is the list entitled *Information Approved for the Classification and Labelling of Substances and Preparations Dangerous for Supply* approved by the HSC comprising Parts I to VI, together with such notes and explanatory material as are requisite for the use of the list.

Classification of substances and preparations dangerous for supply (regulation 5)

A supplier shall not supply a substance or preparation dangerous for supply unless it has been classified in accordance with the following paragraphs of this regulation.

In the case of a *substance which is listed in the approved supply list*, the classification shall be that specified in the entry for that substance in column 2 of Part V of that list.

In the case of a *substance which is a new substance* within the meaning of regulation 2(1) of the Notification of New Substances Regulations 1993 and which has been notified in accordance with regulation 4 or 6(1) or (2) of those regulations, the substance shall be classified in conformity with that notification.

In the case of *any other substance dangerous for supply*, after an investigation to become aware of relevant and accessible data which may exist, the substance shall be classified by placing it in one or more of the *categories of danger* specified in column 1 of Part I of Schedule 1 corresponding to the properties specified in the entry opposite thereto in column 2 and by assigning appropriate *risk phrases* by the use of the criteria set out in the *approved classification and labelling guide*.

Subject to paragraph 6, *a preparation to which these Regulations apply* shall be classified as dangerous for supply in accordance with Schedule 3 by the use of the criteria set out in the approved classification and labelling guide.

A preparation which is intended for use as a *pesticide* (other than a pesticide which has been approved under the Food and Environment Protection Act 1985) shall be classified as dangerous for supply in accordance with Schedule 4.

Safety data sheets for substances and preparations dangerous for supply (regulation 6)

Subject to paragraphs (2) and (5), the supplier of a substance or preparation dangerous for supply shall provide the recipient of that substance or preparation with a safety data sheet containing information under the headings specified in Schedule 5 to enable the recipient of that substance or preparation to take the necessary measures relating to the protection of health and safety at work and relating to the protection of the environment, and the safety data sheet shall clearly show its date of first publication or latest revision, as the case may be.

In this regulation, *supply* shall *not* include supply by way of:

- offer for sale;
- transfer from a factory, warehouse or another place of work and its curtilage to another place of work in the same ownership; or
- returning substances or preparations to the person who supplied them, provided that the properties of that substance or preparation remain unchanged.

The supplier shall keep the safety data sheet up to date and revise it forthwith if any significant new information becomes available regarding safety or risks to human health or the protection of the environment in relation to the substance or preparation concerned and the revised safety data sheet shall be clearly marked with the word '*revision*'.

Except in the circumstances to which paragraph (5) relates, the safety data sheet shall be provided free of charge no later than the date on which the substance or preparation is first supplied to the recipient and where the safety data sheet has been revised in accordance with paragraph (3), a copy of the revised safety data sheet shall be provided free of charge to all recipients who have received the substance or preparation in the last 12 months and the changes in it shall be brought to their notice.

Safety data sheets need *not* be provided with substances or preparations dangerous for supply sold to *the general public* in circumstances to which regulation 16(2)(a) or (b) applies (relating to supply from a shop, etc.) if sufficient information is furnished to enable users to take the necessary measures as regards the protection of health and safety, except that safety data sheets shall be provided free of charge at the request of persons who intend the substance or preparation to be used at work, but in those circumstances paragraph (4) (in so far as it relates to the subsequent provision of revised data sheets) shall not apply to such requests.

The particulars required to be given in the safety data sheets shall be in English, except that where a substance or preparation is intended to be supplied to a recipient in another Member State, the safety data sheet may be in the official language of that State.

Advertisements for substances dangerous for supply (regulation 7)

A person who supplies or offers to supply a substance dangerous for supply shall ensure that the substance is not advertised unless mention is made in the advertisement of the hazard or hazards presented by the substance.

In this regulation the word '*supply*' has the same meaning as in section 46 of the Consumer Protection Act 1987.

Packaging of substances and preparations dangerous for supply (regulation 8)

The supplier of a substance or preparation which is dangerous for supply shall not supply any such substance or preparation unless it is in a package which is suitable for that purpose, and in particular, unless:

- the receptacle containing the substance or preparation and any associated packaging are *designed, constructed, maintained and closed* so as to prevent any of the contents of the receptacle from escaping when subjected to the stresses and strains of normal handling, except that this sub-paragraph shall not prevent the fitting of a suitable safety device;
- the receptacle and any associated packaging, in so far as they are likely to come into contact with the substance or preparation, are made of *materials* which are neither liable to be adversely affected by that substance nor

liable in conjunction with that substance to form any other substance which is itself a risk to the health or safety of any person; and
- where the receptacle is fitted with a *replaceable closure*, that closure is designed so that the receptacle can be repeatedly re-closed without its contents escaping.

Labelling of substances and preparations dangerous for supply (regulation 9)

Subject to regulations 9 and 10 of the Carriage of Dangerous Goods by Road and Rail (Classification, Packaging and Labelling) Regulations 1994 (which allow combined carriage and supply labelling in certain circumstances) and paragraphs (5) to (9), a supplier shall not supply a substance or preparation which is dangerous for supply unless the particulars specified in paragraph (2) relating to a substance or paragraph (3) relating to a preparation, as the case may be, are clearly shown in accordance with the requirements of regulation 11 (methods of marking or labelling of packages):

- on the *receptacle* containing the substance or preparation; and
- if that receptacle is inside one or more layers of packaging, on any such layer which is likely to be to the outermost layer of packaging during the supply or use of the substance or preparation, unless such packaging permits the particulars shown on the receptacle or other packaging to be clearly seen.

The particulars required under paragraph (1) in relation to a substance dangerous for supply shall be:

- the name, full address and telephone number of a person in a Member State who is responsible for supplying the substance, whether it be its manufacturer, importer or distributor;
- the name of the substance, being the name or one of the names for the substance listed in Part I of the approved supply list, or if it is not so listed an internationally recognised name;
- the following particulars ascertained in accordance with Part I of Schedule 6, namely:

  - the indication or indications of danger and the corresponding symbol or symbols;
  - the risk phrases (set out in full);
  - the safety phrases (set out in full);
  - the EEC number (if any) and, in the case of a substance dangerous for supply which is listed in Part I of the approved supply list, the words 'EEC label'.

The particulars required under paragraph (1) in relation to a preparation which is, or (where sub-paragraph (d) below applies) may be dangerous for supply shall be:

- the name and full address and telephone number of a person in a Member State who is responsible for supplying the preparation, whether he be its manufacturer, importer or distributor;
- the trade name or other designation of the preparation;
- the following particulars ascertained in accordance with Part I of Schedule 6, namely:

  - identification of the constituents of the preparation which result in the preparation being classified as dangerous for supply;
  - the indication or indications of danger and the corresponding symbol or symbols;
  - the risk phrases (set out in full);
  - the safety phrases (set out in full);
  - in the case of a pesticide, the modified information specified in paragraph 5 of Part I of Schedule 6; and
  - in the case of a preparation intended for sale to the general public, the nominal quantity (nominal mass or nominal volume); and

- where required by paragraph 5(5) of Part I of Schedule 3, the words specified in that paragraph.

Where the Executive receives a notification of a derogation provided for by paragraph 3(1) of Part I of Schedule 6, it shall forthwith inform the European Commission thereof.

Indications such as 'non-toxic' or 'non-harmful' or any other statement indicating that the substance or preparation is not dangerous for supply shall not appear on the package.

Except for the outermost packaging of a package in which a substance or preparation is transferred, labelling in accordance with this regulation shall not be required where a substance or preparation dangerous for supply is supplied by way of transfer from a factory, warehouse or other place of work and its curtilage to another place of work if, at that other place of work it is not subject to any form of manipulation, treatment or processing which results in the substance or preparation dangerous for supply being exposed or, for any purpose other than labelling in accordance with these regulations, results in any receptacle containing the substance or preparation being removed from its outer packaging.

Except in the case of a substance or preparation dangerous for supply for which the indication of danger is required to be *explosive, very toxic or toxic* or which is classified as *sensitising*, labelling under this regulation shall not be required for such small quantities of that substance or preparation that there is no reason to fear danger to persons handling that substance or preparation or to other persons.

Where, in the case of a substance or preparation dangerous for supply, other than a pesticide, the package in which the substance or preparation is supplied does not contain more than 125 millilitres of the substance or preparation, the risk phrases required by paragraph 2(c)(ii) or 3(c)(iii), and the safety phrases required by paragraph 2(c)(iii) or 3(c)(iv), as the case may be, need not be shown if the substance or preparation is classified only in one or more of the categories of danger, highly flammable, flammable, oxidising or irritant or in the case of substances not intended to be supplied to the public, harmful.

Where, because of the size of the label, it is not reasonably practicable to provide the safety phrases required under paragraph 2(c)(iii) or 3(c)(iv), as the case may be, on the label, that information may be given on a separate label or on a sheet accompanying the package.

Particular labelling requirements for certain preparations (regulation 10)

In the case of preparations to which Part II of Schedule 6 applies, the appropriate provisions of that Part of the Schedule shall have effect to regulate the labelling of such preparations, even if the preparations referred to in Part IIB of that Schedule would not otherwise be dangerous for supply.

In the case of preparations packaged in aerosol dispensers, the flammability criteria set out in Part II of Schedule I shall have effect for the classification and labelling of those preparations for supply in place of the categories of danger 'extremely flammable', 'highly flammable' or 'flammable' set out in Part I of that Schedule, and where a dispenser contains a substance so classified, that dispenser shall be labelled in accordance with the provisions of paragraph 2 of the said Part II.

Methods of marking or labelling packages (regulation 11)

Any package which is required to be labelled in accordance with regulations 9 and 10 may carry the particulars required to be on the label clearly and indelibly marked on a part of that package reserved for that purpose and, unless the context otherwise requires, any reference in these regulations to a label includes a reference to that part of the package so reserved.

Subject to paragraph (7), any label required to be carried on a package shall be securely fixed to the package with its entire surface in contact with it and the label shall be clearly and indelibly printed.

The colour and nature of the marking shall be such that the symbol (if any) and wording stand out from the background so as to be readily noticeable and the wording shall be of such a size and spacing as to be easily read.

The package shall be so labelled that the particulars can be read horizontally when the package is set down normally.

Subject to paragraph (7), the dimensions of the label required under regulation 9 shall be as shown in Table 4.1.

Any symbol required to be shown in accordance with regulation 9(2)(c)(i) or 9(3)(c)(ii) and specified in column 3 of Schedule 2 shall be printed in black on an orange-yellow background and its size (including the orange-yellow background) shall be at least equal to an area of one-tenth of that of a label which complies with paragraph (5) and shall not in any case be less than 100 square millimetres.

If the package is an awkward shape or so small that it is unsuitable to attach a label complying with paragraphs (2) and (5), the label shall be so attached in some other appropriate manner.

The particulars required to be shown on the label shall be in English, except that where a substance or preparation is intended to be supplied to a recipient in another Member State, the label may be in the official language of the State.

Table 4.1 Dimensions of labels for packages of substances and preparations dangerous for supply

Capacity of package	Dimensions of label
not exceeding 3 litres	if possible at least 52 x 74 mm
exceeding 3 litres but not exceeding 50 litres	at least 74 x 105 mm
exceeding 50 litres but not exceeding 500 litres	at least 105 x 148 mm
exceeding 500 litres	at least 148 x 210 mm

Child-resistant fastenings and tactile warning devices (regulation 12)

The British and International Standards referred to in this regulation are further described in Schedule 7.

This regulation shall not apply in relation to a pesticide.

Subject to paragraph (5), a person shall not supply a substance or preparation referred to in paragraph (4) in a receptacle of any size fitted with a replaceable closure unless the packaging complies with the requirements of BS EN 28317 or ISO 8317.

Paragraph 3 shall apply to:

- substances and preparations dangerous for supply which are required to be labelled with the indications of danger *'very toxic'*, *'toxic'* or *'corrosive'*;
- preparations containing methanol in a concentration equal to or more than 3 per cent by weight;
- preparations containing dichloromethane in a concentration equal to or more than 1 per cent by weight;
- liquid preparations having a kinematic viscosity measured by rotative viscometry in accordance with BS 2782 method 730B or ISO 3291 of less than 7×10^{-6} m^2 s^{-1} at 40°C and containing aliphatic or aromatic hydrocarbons or both in a total concentration equal to or more than 10 per cent by weight, except where such a preparation is supplied in an aerosol dispenser.

Paragraph (3) shall not apply if the person supplying it can show that it is obvious that the packaging in which the substance or preparation is supplied is sufficiently safe for children because they cannot obtain access to the contents without the help of a tool.

If the packaging in which the substance or preparation is supplied was approved on or before 31 May 1993 by the British Standards Institution as complying with the requirements of the British Standards Specification 6652:1989 it shall be treated in all respects as complying with the requirements of BS EN 28317.

A person shall not supply a preparation dangerous for supply if the packaging in which the preparation is supplied has:

- a shape or designation or both likely to attract or arouse the active curiosity of children or to mislead consumers;
- a presentation or designation or both used for human or animal foodstuffs, medicinal or cosmetic products.

A person shall not supply a substance or preparation referred to in paragraph (9) in a receptacle of any size, unless the packaging carries a tactile warning of danger in accordance with BS 7280 or EN Standard 272.

Paragraph (8) shall apply to substances and preparations dangerous for supply which are required to be labelled with the indication of danger 'very toxic', 'toxic', 'corrosive', 'harmful', 'extremely flammable' or 'highly flammable'.

A duly authorised officer of the enforcing authority, for the purpose of ascertaining whether there has been a concentration of paragraph (3) may require the person supplying a substance or preparation to which that paragraph applies to provide him with a certificate from a qualified test house stating that:

- the closure is such that it is not necessary to test to BS EN 28317 or ISO 8317; or
- the closure has been tested and found to conform to that standard.

For the purpose of paragraph (10) a qualified test house means a laboratory that conforms to BS7501 or EN 45000.

In this regulation, 'supply' means offer for sale, sell or otherwise make available to the general public.

Retention of classification data for substances and preparations dangerous for supply (regulation 13)

A person who classifies a substance in accordance with regulation 5(4) or a preparation dangerous for supply shall keep a record of the information used for the purposes of classifying for at least three years after the date on which the substance or preparation was supplied by him for the last time and shall make the record or a copy of it available to the appropriate enforcing authority referred to in regulation 16(2) at its request.

Notification of the constituents of certain preparations dangerous for supply to the poisons advisory centre (regulation 14)

This regulation shall apply to any preparation which is classified on the basis of one or more of its health effects referred to in column 1 of Schedule I.

Subject to regulation 17 (transitional provisions), the supplier of a preparation to which this regulation applies shall, if it was first supplied before these regulations came into force (or, if it was first supplied after that date, before first supplying it), notify the poisons advisory centre of the information required to be in the safety data sheet prepared for the purposes of regulation 6 relating to the preparation.

The supplier shall ensure that the information supplied to the poisons advisory centre in pursuance of paragraph (2) is kept up to date.

The poisons advisory centre shall only disclose any information sent to it in pursuance of paragraph (2) or (3) on request by, or by a person working under the direction of, a registered medical practitioner in connection with the medical treatment of a person who may have been affected by the preparation.

Exemption certificates (regulation 15)

Subject to paragraph (2) and to any of the provisions imposed by the Community in respect of the free movement of dangerous substances and preparations, the Executive may by a certificate in writing exempt any person or class of persons, substance or preparation to which these regulations apply, or class of such substances or preparations, from all or any of the requirements or prohibitions imposed by or under these regulations and any such exemption may be granted subject to conditions and to a limit of time and may be revoked at any time by a certificate in writing.

The Executive shall *not* grant any such exemption unless, having regard to the circumstances of the case, and in particular to:

● the conditions, if any, which it proposes to attach to the exemption; and
● any requirements imposed by or under any enactments which apply to the case,

it is satisfied that the health or safety of persons who are likely to be affected by the exemption will not be prejudiced in consequence of it.

Enforcement, civil liability and defence (regulation 16)

In so far as any provision of regulations 5–14 is made under section 2 of the European Communities Act 1972:

- subject to paragraph (2), the provisions of the Health and Safety at Work etc. Act 1974, which relate to the approval of codes of practice and their use in criminal proceedings, to enforcement and to offences shall apply to that provision as if that provision had been made under section 15 of that Act;
- a breach of duty imposed by that provision shall confer a right of action in civil proceedings, insofar as that breach of duty causes damage.

Notwithstanding regulation 3 of the Health and Safety (Enforcing Authority) Regulations 1989, the enforcing authority for these regulations shall be the Executive, except that:

- where a substance or preparation dangerous for supply is supplied in or from premises which are registered under section 75 of the Medicines Act 1968 the enforcing authority shall be the Royal Pharmaceutical Society;
- where a substance or preparation dangerous for supply is supplied otherwise than as in sub-paragraph (a) above:

  - in or from any shop, mobile vehicle, market stall or other retail outlet; or
  - otherwise to members of the public, including by way of free sample, prize or mail order,

the enforcing authority shall be the local weights and measures authority; and

- for regulations 7 and 12 the enforcing authority shall be the local weights and measures authority.

In every case where by virtue of paragraph (2) these regulations are enforced by the Royal Pharmaceutical Society or the local weights and measures authority, they shall be enforced as if they were safety regulations made under section 11 of the Consumer Protection Act 1987 and the provisions of section 12 of that Act shall apply to these regulations as if they were safety regulations and as if the maximum period of imprisonment on summary conviction specified in subsection (5) thereof were three months instead of six months.

In any proceedings for an offence under these regulations, it shall be a defence for any person to prove that he took all reasonable precautions and exercised all due diligence to avoid the commission of that offence.

Classification of hazardous substances (supply requirements)

Hazardous substances are classified according to Schedule 1 of the Chemicals (Hazard Information and Packaging for Supply) (CHIP 2) Regulations 1994 as follows. Part I is shown in Table 4.2.

Table 4.2 Classification of hazardous substances

PART I: CATEGORIES OF DANGER

Column 1 Category of danger	Column 2 Property (See Note 1)	Column 3 Symbol-letter
PHYSICO-CHEMICAL PROPERTIES		
Explosive	Solid, liquid, pasty or gelatinous substances and preparations which may react exothermically without atmospheric oxygen thereby quickly evolving gases, and which under defined test conditions detonate, quickly deflagrate or upon heating explode when partially confined	E
Oxidising	Substances and preparations which give rise to an exothermic reaction in contact with other substances, particularly flammable substances	O
Extremely flammable	Liquid substances and preparations having an extremely low flash point and a low boiling point and gaseous substances and preparations which are flammable in contact with air at ambient temperature and pressure	F+
Highly flammable	The following substances and preparations, namely: (a) substances and preparations which may become hot and finally catch fire in contact with air at ambient temperature without any application of energy; (b) solid substances and preparations which may readily catch fire after brief contact with a source of ignition and which continue to burn or to be consumed after removal of the source of ignition; (c) liquid substances and preparations having a very low flash point;	F

	(d) substances and preparations which, in contact with water or damp air, evolve highly flammable gases in dangerous quantities. (See Note 2)	
Flammable	Liquid substances and preparations having a low flash point.	None

HEALTH EFFECTS

Very toxic	Substances and preparations which in very low quantities can cause death or acute or chronic damage to health when inhaled, swallowed or absorbed via the skin.	T+
Toxic	Substances and preparations which in low quantities can cause death or acute or chronic damage to health when inhaled, swallowed or absorbed via the skin.	T
Harmful	Substances and preparations which may cause death or acute or chronic damage to health when inhaled, swallowed or absorbed via the skin.	Xn
Corrosive	Substances and preparations which may, on contact with living tissues, destroy them.	C
Irritant	Non-corrosive substances and preparations which through immediate, prolonged or repeated contact with the skin or mucous membrane, may cause inflammation.	Xi
Sensitising	Substances and preparations which, if they are inhaled or if they penetrate the skin, are capable of eliciting a reaction by hypersensitisation such that on further exposure to the substance or preparation, characteristic adverse effects are produced.	

Sensitising by inhalation		Xn
Sensitising by skin contact		Xi
Carcinogenic (See Note 3)	Substances and preparations which, if they are inhaled or ingested or if they penetrate the skin, may induce cancer or increase its incidence.	
Category 1		T
Category 2		T
Category 3		Xn
Mutagenic (See Note 3)	Substances and preparations which, if they are inhaled or ingested or if they penetrate the skin, may induce heritable genetic defects or increase their incidence.	
Category 1		T
Category 2		T
Category 3		Xn
Toxic for reproduction (See Note 3)	Substances and preparations which, if they are inhaled or ingested or if they penetrate the skin, may produce or increase the incidence of non-heritable adverse effects in the progeny and/or an impairment of male or female reproductive functions or capacity.	
Category 1		T
Category 2		T
Category 3		Xn
Dangerous for the environment (See Note 4)	Substances which, were they to enter into the environment, would present or might present an immediate or delayed danger for one or more components of the environment.	N

Notes to Table 4.2

1. As further described in the approved classification and labelling guide.
2. Preparations packed in aerosol dispensers shall be classified as flammable in accordance with the additional criteria set out in Part II of this Schedule.
3. The categories are specified in the approved classification and labelling guide.
4. ● In certain cases specified in the approved supply list and in the approved classification and labelling guide substances classified as dangerous for the environment do not require to be labelled with the symbol for this category of danger.
   ● This category of danger does not apply to preparations.

Part II: Classification of substances and preparations dangerous for supply in aerosol dispensers as flammable

A substance or preparation which is packed in aerosol dispensers shall be classified as dangerous for supply at least as 'flammable' if that dispenser contains either:

● more than 45 per cent by weight of flammable substances; or
● more than 250 grammes of flammable substances.

Where an aerosol dispenser contains a substance or preparation which is classified in accordance with paragraph 1 as *flammable* it shall show in accordance with the requirements of regulation 10 either:

● the word 'flammable'; or
● the symbol having the symbol-letter F in column 2 of Schedule 2,

or both the word 'flammable' and that symbol.

Schedule 5: headings under which particulars are to be provided in safety data sheets

Under the CHIP Regulations obligatory information under the following headings must be provided in a safety data sheet:

● Identification of the substance/preparation
● Composition/information on ingredients
● Hazards identification
● First aid measures
● Fire fighting measures
● Accidental release measures
● Handling and storage
● Exposure controls/Personal protection
● Physical and chemical properties
● Stability and reactivity
● Toxicological information
● Ecological information
● Disposal considerations
● Transport information
● Regulatory information
● Other information

The CHIP 2 package

The Chemicals (Hazard Information and Packaging for Supply) Regulations 1994 (SI 1994 No. 3247)

ACOP: *Safety data sheets for substances and preparations dangerous for supply* (HSC)

ACOP: *The approved guide to the classification and labelling of substances and preparations dangerous for supply* (HSC)

The Approved Supply List: Information approved for the classification, packaging and labelling of substances dangerous for supply (HSC)

The Popular Guide: CHIP for everyone [HS (G) 126]

The Complete Idiot's Guide to CHIP [IND (G) 181 (L)]

Why do I need a safety data sheet? [IND (G) 182 (L)]

Read the label [IND (G) 186 (L)].

THE CHEMICALS (HAZARD INFORMATION AND PACKAGING FOR SUPPLY) (AMENDMENT) REGULATIONS 2000 (CHIP 2000)

These regulations updated the information that must be given by suppliers to users when dangerous chemicals are supplied, in particular, the requirements for child-resistant fasteners. A new standard for tactile warning devices was introduced at the same time.

CHIP 2000 also updated the information on 168 dangerous substances via a sixth edition of the Approved Supply List, the document based on EC Directives, which sets out classification and labelling information for several thousand commonly supplied chemical substances.

CHEMICALS (HAZARD INFORMATION AND PACKAGING FOR SUPPLY) (CHIP 3) REGULATIONS 2002

These regulations consolidate CHIP 2 and all subsequent amendments and implement three directives:

- The Dangerous Preparations Directive (including the 1st ATP – *Adaptation to Technical Progress*), the Dangerous Substances Directive (including the 28th ATP), and
- the Safety Data Sheets Directive (including the second amendment).

Revised HSE Guidance is incorporated in 'CHIP 3 for Everyone'.

Principal changes to CHIP 2

The principal changes are as follows:

- the introduction of a new Approved Supply List (7th edition) incorporating 500 new or amended listings;
- preparations are classified on the basis of their environmental effects for

the first time; information is provided on the conventional method for cal-culating the sum effect of each environmentally hazardous component;

- Category 3 carcinogens are relabelled; the R40 risk phrase is modified as 'Limited evidence of a carcinogenic effect' and a new risk phrase R68 takes the previous R40 wording 'Possible risk of irreversible effects';
- preparations containing ≥0.1 of a sensitiser are re-labelled 'Contains ... (chemical name of sensitiser); may produce an allergic reaction';
- the introduction of a 15% component concentration limit for risk phrase R67 'Vapours may cause drowsiness and dizziness';
- the introduction of a new warning phrase for liquid preparations whose flash point would not make them classifiable as 'flammable' but which contain a flammable or highly flammable component in a >5% concentra-tion plus a halogenated hydrocarbon solvent; the new phrase is 'Can become flammable (or highly flammable) in use';
- the extension of the requirement to prepare a Safety Data sheet for some non-dangerous preparations together with a label 'Safety Data sheet avail-able to professional user on request';
- a requirement to improve the accuracy and comprehensibility of Safety Data Sheets through taking the specific needs of the user into account (including employees who will be guided by the Safety Data Sheet under the COSHH Regulations); reference is made to the competence and train-ing of the person preparing a Safety Data Sheet;
- a requirement for more information to be provided with respect to envi-ronmental effects; and
- some changes in test methods and criteria.

THE CHILDREN (PROTECTION AT WORK) REGULATIONS 1998

These regulations amend the Children and Young Persons Acts 1933 and 1963 in order to implement, with regard to children, the EC Directive on the Protection of Young People at Work (94/33/EC).

The main features of these regulations are that:

- the minimum age at which a child may be employed in any work, other than as an employee of his parent or guardian in light agricultural or hor-ticultural work on an occasional basis, is 14 years;
- anything other than light work is prohibited; 'light work' is work which does not jeopardise a child's safety, health, development, attendance at school or participation in work experience;
- the employment of children over the age of 13 years in categories of light work specified in local authority byelaws is permitted;
- the hours which a child over the age of 14 years may work, and the rest periods which are required, are specified; a child must have at least one two-week period in his school holidays free from any employment;

- the 1933 Act is amended to extend the prohibition against a child going abroad for the purposes of performing for profit without a local authority licence, to further cover a child going abroad for the purpose of taking part in sport or working as a model in circumstances where payment is made; and
- where children take part in public performances, existing requirements for a local authority licence are extended to require such a licence to be obtained before a child may take part in sport or work as a model in circumstances where payment is made either to the child or to someone else.

THE CONFINED SPACES REGULATIONS 1997

Work in confined spaces has always been a high risk activity and a major source of deaths in construction activities, chemical processing operations, agriculture and the public utilities, such as water undertakings.

These regulations require employers to:

- avoid entry to confined spaces, for example, by doing the work from outside;
- follow a safe system of work, e.g. a Permit-to-Work system, if entry to a confined space is unavoidable; and
- put in place adequate emergency arrangements before work starts, which will also safeguard rescuers.

The regulations are accompanied by an ACOP and HSE Guidance.

THE CONSTRUCTION (DESIGN AND MANAGEMENT) REGULATIONS 1994

These regulations impose requirements and prohibitions with respect to the design and management aspects of *construction work* as defined. They give effect to Council Directive 92/57/EEC on the implementation of minimum safety and health requirements at temporary or mobile construction sites.

Regulation 2: Interpretation

The more significant definitions in the regulations are given below:

Agent in relation to any client means any person who acts as agent for a client in connection with the carrying on by the person of a trade, business or other undertaking (whether for profit or not).

Cleaning work means the cleaning of any window or any transparent or translucent wall, ceiling or roof in or on a structure where such cleaning involves a risk of a person falling more than 2 metres.

Client means any person for whom a project is carried out, whether it is carried out by another person or carried out in-house.

Construction phase means the period of time starting when construction work in any project starts and ending when construction work in that project is completed.

Construction work means the carrying out of any building, civil engineering or engineering construction work and includes any of the following:

- the construction, alteration, conversion, fitting out, commissioning, renovation, repair, upkeep, redecoration or other maintenance (including cleaning which involves the use of water or an abrasive at high pressure or the use of substances classified as corrosive or toxic for the purposes of regulation 7 of the Chemicals (Hazard Information and Packaging for Supply) Regulations 1994, de-commissioning, demolition or dismantling of a structure;
- the preparation for an intended structure, including site clearance, exploration, investigation (but not site survey) and excavation, and laying or installing the foundations of the structure;
- the assembly of prefabricated elements to form a structure or the disassembly of prefabricated elements which, immediately before such disassembly, formed a structure;
- the removal of a structure or part of a structure or of any product or waste resulting from demolition or dismantling of a structure or from disassembly of prefabricated elements which, immediately before such disassembly, formed a structure;
- the installation, commissioning, maintenance, repair or removal of mechanical, electrical, gas, compressed air, hydraulic, telecommunications, computer or similar services which are normally fixed within or to a structure,

but does *not* include the exploration for or extraction of mineral resources or activities preparatory thereto carried out at a place where such exploration or extraction is carried out.

Contractor means any person who carried on a trade or business or other undertaking (whether for profit or not) in connection with which he:

- undertakes to or does carry out or manage construction work;
- arranges for any person at work under his control (including where he is an employer, an employee of his) to carry out or manage construction work.

Design in relation to any structure includes drawing, design details, specification and bill of quantities (including specification of articles or substances) in relation to the structure.

Designer means any person who carries on a trade, business or other undertaking in connection with which he:

- prepares a design; or
- arranges for any person under his control (including, where he is an employer, any employee of his) to prepare a design,

relating to a structure or part of a structure.

Domestic client means a client for whom a project is carried out, not being a project carried out in connection with the carrying on by the client of a trade, business or other undertaking (whether for profit or not).

Project means a project which includes or is intended to include construction work.

Structure means:

- any building, steel or reinforced concrete structure (not being a building), railway line or siding, tramway line, dock, harbour, inland navigation, tunnel, shaft, bridge, viaduct, waterworks, reservoir, pipe or pipe-line (whatever, in either case, it contains or is intended to contain), cable, aqueduct, sewer, sewage works, gasholder, road, airfield, sea defence works, river works, drainage works, earthworks, lagoon, dam, wall, caisson, mast, tower, pylon, underground tank, earth retaining structure, or structure designed to preserve or alter any natural feature, and any other structure similar to the foregoing; or
- any formwork, falsework, scaffold or other structure designed or used to provide support or means of access during construction work; or
- any fixed plant in respect of work which is installation, commissioning, de-commissioning or dismantling and where any such work involves a risk of a person falling more than 2 metres.

In determining whether any person arranges for a person (the *relevant person*) to prepare a design or to carry out or manage construction work, regard shall be had to the following, namely:

- a person does arrange for the relevant person to do a thing where:

  - he specifies in, or in connection with, any arrangement with a third person that the relevant person shall do that thing (whether by nominating the relevant person as a subcontractor to the third person or otherwise); or
  - being an employer, it is done by any of his employees in-house.

- a person does not arrange for the relevant person to do a thing where:

  - being a self-employed person, he does it himself or, being in partnership it is done by any of his partners;
  - being a firm carrying on its business anywhere in Great Britain whose principal place of business is in Scotland, it is done by any partner in the firm; or
  - having arranged for a third person to do the thing, he does not object to the third person for it to be done by the relevant person,

and the expressions *arrange* and *arranges* shall be construed accordingly.

For the purpose of these regulations:

- *a project is carried out in-house* where any employer arranges for the project to be carried out by an employee of his who acts, or by a group of employees who act, in either case, in relation to such a project as a separate part of the undertaking of the employer distinct from the part for which the project is carried out;

- *construction work is carried out or managed in-house* where an employer arranges for the construction work to be carried out or managed by an employee of his who acts, or by a group of employees who act, in either case, in relation to such construction work as a separate part of the undertaking of the employer distinct from the part for which the construction work is carried out or managed; or
- *a design is prepared in-house* where any employer arranges for the design to be prepared by an employee of his who acts, or by a group of employees who act, in either case, in relation to such design as a separate part of the undertaking of the employer distinct from the part for which the design is prepared.

Regulation 16(1) (co-operation between all contractors) shall not apply to projects in which no more than one contractor is involved.

Where construction work is carried out or managed in-house or a design is prepared in-house, then, for the purposes of paragraphs 5 and 6, each part of the undertaking shall be treated as a person and shall be counted as a designer, or as the case may be, contractor, accordingly.

Except where regulation 5 (requirements on developers) applies, regulations 4, 6, 8 to 12 and 14 to 19 shall not apply to or in relation to construction work included or intended to be included in a project carried out for a domestic client.

Clients and agents of clients (regulation 4)

A client may appoint an agent or another client to act as the only client in respect of a project and where such an appointment is made the provisions of paragraph 2 shall apply.

No client shall appoint any person as his agent unless the client is reasonably satisfied that the person he intends to appoint has the competence to perform the duties imposed on a client by these regulations.

Where the person appointed under paragraph 1 makes a declaration in accordance with paragraph 4, then, from the date of receipt of the declaration by the Executive (HSE), such requirements and prohibitions as are imposed upon a client shall apply to the person so appointed (so long as he remains as such) as if he were the only client in respect of that project.

A *declaration* in accordance with this paragraph:

- is a declaration in writing, signed by or on behalf of the person referred to in paragraph 3, to the effect that the client or agent who makes it will act as client for the purposes of these regulations; and
- shall include the name of the person by or on behalf of whom it is made, the address where documents may be served on that person and the address of the construction site; and
- shall be sent to the HSE.

Where the HSE receives a declaration in accordance with paragraph 4, it shall give notice to the person by or on behalf of whom the declaration is made and the notice shall include the date the declaration was received by the HSE.

Where the person referred to in paragraph 3 does not make a declaration in accordance with paragraph 4, any requirement or prohibition imposed by these regulations on a client shall also be imposed on him but only to the extent it relates to any matter within his authority.

Requirements on developer (regulation 5)

This regulation applies where the project is carried out for a domestic client and the client enters into an arrangement with the person (the developer) who carries on a trade, business or other undertaking (whether for profit or not) in connection with which:

- land or an interest in land is granted or transferred to the client; and
- the developer undertakes that construction work will be carried out on the land; and
- following the construction work, the land will include premises which, as intended by the client, will be occupied as a residence.

Where this regulation applies, with effect from the time the client enters into the arrangement referred to in paragraph 1, the requirements of regulations 6 and 8 to 12 shall apply to the developer as if he were the client.

Appointments of planning supervisor and principal contractor (regulation 6)

Subject to paragraph 6(b), every client shall appoint:

- a planning supervisor; and
- a principal contractor,

in respect of each project.

The client shall *not* appoint as principal contractor any person who is not a contractor.

The planning supervisor shall be appointed as soon as is practicable after the client has such information about the project and the construction work involved in it as will enable him to comply with the requirements imposed on him by regulations 8(1) and 9(1).

The principal contractor shall be appointed as soon as is practicable after the client has such information about the project and the construction work involved in it as will enable him to comply with the requirements imposed on him by regulations 8(3) and 9(3) when making an arrangement with a contractor where such arrangement consists of the appointment of the principal contractor.

The appointments mentioned in paragraph 1 shall be terminated, changed or renewed as necessary to ensure that those appointments remain filled at all times until the end of the construction phase.

Paragraph 1 does not prevent:

- the appointment of the same person as planning supervisor or as principal contractor provided that person is competent to carry out the functions under these regulations of both appointments; or
- the appointment of the client as planning supervisor or as principal contractor or as both, provided the client is competent to perform the relevant functions under these regulations.

Notification of project (regulation 7)

The planning supervisor shall ensure that notice of the project in respect of which he is appointed is given to the HSE in accordance with paragraphs 2 and 4 unless the planning supervisor has reasonable grounds for believing that the project is not notifiable.

Any notice required by paragraph 1 shall be in writing or in such other manner as the HSE may from time to time approve in writing and shall contain the particulars specified in paragraph 3 or, where applicable, paragraph 4 and shall be given at the times specified in those paragraphs.

Notice containing such of the particulars specified in Schedule 1 as are known or can reasonably be ascertained shall be given as soon as is practicable after the appointment of the planning supervisor.

Where any particulars specified in Schedule 1 have not been notified under paragraph 3, notice of such particulars shall be given as soon as is practicable after the appointment of the principal contractor and, in any event, before the start of construction work.

Where a project is carried out for a domestic client then, except where regulation 5 applies, every contractor shall ensure that notice of the project is given to the HSE in accordance with paragraph 6 unless the contractor has reasonable grounds for believing that the project is not notifiable.

Competence of planning supervisor, designers and contractors (regulation 8)

No client shall appoint any person as planning supervisor in respect of a project unless the client is reasonably satisfied that the person he intends to appoint has the competence to perform the functions of planning supervisor under the regulations in respect of that project.

No person shall arrange for a designer to prepare a design unless he is reasonably satisfied that the designer has the competence to prepare that design.

No person shall arrange for a contractor to carry out or manage construction work unless he is reasonably satisfied that the contractor has the competence to carry out or, as the case may be, manage, that construction work.

Any reference in this regulation to a person having competence shall extend only to his competence:

- to perform any requirement; and
- to conduct his undertaking without contravening any prohibition, imposed on him by or under any of the relevant statutory provisions.

Provision for health and safety (regulation 9)

No person shall appoint any person as planning supervisor in respect of a project unless the client is reasonably satisfied that the person he intends to appoint has allocated or, as appropriate, will allocate adequate resources to enable him to perform the functions of planning supervisor under these regulations in respect of that project.

● **FIG 4.1 How to decide if your project has to be notified to the Health and Safety Executive**

No person shall arrange for a designer to prepare a design unless he is reasonably satisfied that the designer has the competence to prepare that design.

No person shall arrange for a contractor to carry out or manage construction work unless he is reasonably satisfied that the contractor has allocated or, as appropriate, will allocate, adequate resources to enable the contractor to comply with the requirements and prohibitions imposed upon him by or under the relevant statutory provisions.

Notification of project

Note

1. This form can be used to notify any project covered by the Construction (Design and Management) Regulations 1994 which will last longer than 30 days or 500 person days. It can also be used to provide additional details that were not available at the time of initial notification of such projects. (Any day on which construction work is carried out (including holidays and weekends) should be counted, even if the work on that day is of short duration. A person day is one individual, including supervisors and specialists, carrying out construction work for one normal working shift.)

2. The form should be completed and sent to the HSE area office covering the site where construction work is to take place. You should send it as soon as possible after the planning supervisor is appointed to the project.

3. The form can be used by contractors working for domestic clients. In this case only parts 4–8 and 11 need to be filled in.

HSE – For official use only

Client	V	PV	NV	Planning supervisor	V	PV	NV
Focus serial number				Principal contractor	V	PV	NV

---

1 Is this the initial notification of this project or are you providing additional information that was not previously available

Initial notification ☐   Additional notification ☐

2 **Client:** name, full address, postcode and telephone number (*if more than one client, please attach details on separate sheet*)

Name: Telephone number:
Address:
Postcode:

3 **Planning Supervisor:** name, full address, postcode and telephone number

Name: Telephone number:
Address:
Postcode:

4 **Principal Contractor** (*or contractor when project for a domestic client*): name, full address, postcode and telephone number

Name: Telephone number:
Address:
Postcode:

5 **Address of site:** where construction work is to be carried out

Address:
Postcode

F10 (rev 03.95)

● **FIG 4.2 Example of a Notification of Project Form**

6 **Local Authority:** name of the local government district council or island council within whose district the operations are to be carried out

> [ ]

7 **Please give your estimates on the following:** Please indicate if these estimates are

original ☐   revised ☐   *(tick relevant box)*

   a. The planned date for the commencement of the construction work [ ]

   b. How long the construction work is expected to take *(in weeks)* [ ]

   c. The maximum number of people carrying out construction work on site at any one time [ ]

   d. The number of contractors expected to work on site [ ]

8 **Construction work:** give brief details of the type of construction work that will be carried out

> [ ]

9 **Contractors:** name, full address and postcode of those who have been chosen to work on the project *(if required continue on a separate sheet). (Note this information is only required when it is known at the time notification is first made to HSE. An update is not required)*

> [ ]

Declaration of planning supervisor

10 I hereby declare that...*(name of organisation)* has been appointed as planning supervisor for the project

   Signed by or on behalf of the organisation...............................*(print name)* ....

   Date...

Declaration of principal contractor

11 I hereby declare that ....*(name of principal contractor)* has been appointed as principal contractor for the project. *(or contractor undertaking project for domestic client)*

   Signed by or on behalf of the organisation...............................*(print name)*.....................................

   Date...

● **FIG 4.2 cont'd**

Start of construction phase (regulation 10)

Every client shall ensure, so far as is reasonably practicable, that the construction phase of any project does not start unless a health and safety plan complying with regulation 15(4) has been prepared in respect of that project.

Client to ensure information is available (regulation 11)

Every client shall ensure that the planning supervisor for any project carried out for the client is provided (as soon as is reasonably practicable but in any event before the commencement of the work to which the information relates) with all information mentioned in paragraph 2 about the state or condition of any premises at or on which construction work included or intended to be included in the project is or is intended to be carried out.

The information required to be provided by paragraph 1 is information which is relevant to the functions of the planning supervisor under these regulations and which the client has or could ascertain by making enquiries which it is reasonable for a person in his position to make.

Client to ensure health and safety file is available for inspection (regulation 12)

Every client shall take such steps as it is reasonable for a person in his position to take to ensure that the information in any health and safety file which has been delivered to him is kept available for inspection by any person who may need information in the file for the purpose of complying with the requirements and prohibitions imposed upon him by or under the relevant statutory provisions.

It shall be sufficient compliance with paragraph 1 by a client who disposes of his entire interest in the property of the structure if he delivers the health and safety file for the structure to the person who acquires his interest in the property of the structure and ensures such person is aware of the nature and purpose of the health and safety file.

Requirements on designer (regulation 13)

Except where a design is prepared in-house, no employer shall cause or permit any employee of his to prepare, and no self-employed person shall prepare, a design in respect of any project unless he has taken reasonable steps to ensure that the client for that project is aware of the duties to which the client is subject by virtue of these regulations and of any practical guidance issued from time to time by the Commission (HSC) with respect to the requirements of these regulations.

Every designer shall:

- ensure that any design he prepares and which he is aware will be used for the purposes of construction work includes among the design considerations adequate regard to the need:

- to avoid foreseeable risks to the health and safety of any person at work carrying out construction work or cleaning work in or on the structure at any time, or of any person who may be affected by the work of such a person at work;
- to combat at source risks to the health and safety of any person at work carrying out construction work or cleaning work in or on the structure at any time, or of any person who may be affected by the work of such a person at work; and
- to give priority to measures which will protect all persons at work who may carry out construction work or cleaning work at any time and all persons who may be affected by the work of such persons over measures which only protect each person carrying out such work;

● ensure that the design includes adequate information about any aspect of the project or structure or materials (including articles or substances) which might affect the health or safety of any person at work carrying out construction work or cleaning work in or on the structure at any time or of any person who may be affected by the work of such a person at work; and
● co-operate with the planning supervisor and with any other designer who is preparing any design in connection with the same project or structure so far as is necessary to enable each of them to comply with the requirements and prohibitions placed on him in relation to the project by or under the relevant statutory provisions.

Sub-paragraphs (1) and (b) of paragraph 2 shall require the design to include only the matters referred to therein to the extent that it is reasonable to expect the designer to address them at the time the design is prepared and to the extent that it is otherwise reasonably practicable to do so.

Requirements on planning supervisor (regulation 14)

The planning supervisor appointed for any project shall:

● ensure, so far as is reasonably practicable, that the design of any structure comprised in the project:

- includes among the design considerations adequate regards to the needs specified in heads (i) to (iii) of regulations 13(2) (a); and
- includes information as specified in regulation 13(2) (b);

● take such steps as it is reasonable for a person in his position to take to ensure co-operation between designers so far as is necessary to enable each designer to comply with the requirements placed on him by regulation 13;
● in a position to give adequate advice to:

- any client and any contractor with a view to enabling each of them to comply with regulations 8(2) and 9(2); and
- any client with a view to enabling him to comply with regulations 8(3), 9(3) and 10;

- ensure that a health and safety file is prepared in respect of each structure comprised in the project containing:

  - information included with the design by virtue of regulation 13(2) (b);
  - any other information relating to the project which it is reasonably foreseeable will be necessary to ensure the health and safety of any person at work who is carrying out or will carry out construction work or cleaning work in or on the structure of or any persons who may be affected by the work of such a person at work;

- review, amend or add to the health and safety file prepared by virtue of sub-paragraph (d) of this regulation as necessary to ensure that it contains the information mentioned in that sub-paragraph when it is delivered to the client in accordance with sub-paragraph (f) of this regulation; and
- ensure that, on the completion of construction work on each structure comprised in the project, the health and safety file in respect of that structure is delivered to the client.

Requirements relating to the health and safety plan (regulation 15)

4

The planning supervisor appointed for any project shall ensure that a health and safety plan in respect of the project has been prepared no later than the time specified in paragraph 2 and contains the information specified in paragraph 3.

The information required by paragraph 1 to be contained in the health and safety plan is:

- a general description of the construction work comprised in the project;
- details of the time within which it is intended that the project, and any intermediate stages, will be completed;
- details of the risks to health and safety of any person carrying out the construction work so far as such risks are known to the planning supervisor or are reasonably foreseeable;
- any other information which the planning supervisor knows or could ascertain by making reasonable enquiries and which it would be necessary for any contractor to have if he wished to show:

  - that he has the competence on which any person is required to be reasonably satisfied by regulation 8; or
  - that he has allocated or, as appropriate, will allocate, adequate resources on which any person is required to be reasonably satisfied by regulation 9;

- such information as the planning supervisor knows or could ascertain by making reasonable enquiries and which it is reasonable for the planning supervisor to expect the principal contractor to need in order for him to comply with the requirement imposed on him by paragraph 4; and

- such information as the planning supervisor knows or could ascertain by making reasonable enquiries and which it would be reasonable for any contractor to know in order to understand how he can comply with any requirements placed upon him in respect of welfare by or under the relevant statutory provisions.

The principal contractor shall take such measures as it is reasonable for a person in his position to take to ensure that the health and safety plan contains until the end of the construction phase the following features:

- arrangements for the project (including where necessary, for management of construction work and monitoring of compliance with the relevant statutory provisions) which will ensure, so far as is reasonably practicable, the health and safety of all persons at work carrying out the construction work and of all persons who may be affected by the work of such persons at work, taking account of:

  - risks involved in the construction work;
  - any activity specified in paragraph 5; and

- sufficient information about arrangements for the welfare of persons at work by virtue of the project to enable any contractor to understand how he can comply with any requirements placed upon him in respect of welfare by or under the relevant statutory provisions.

Any activity is an activity referred to in paragraph 4(a) (ii) if:

- it is an activity of persons at work;
- it is carried out in or on premises where construction work is or will be carried out; and
- either:

  - the activity may affect the health or safety of persons at work carrying out the construction work or persons who may be affected by the work of such persons at work; or
  - the health or safety of the persons at work carrying out the activity may be affected by the work of persons at work carrying out the construction work.

Requirements on and powers of principal contractor (regulation 16)

The principal contractor appointed for any project shall:

- take reasonable steps to ensure co-operation between all contractors (whether they are sharing the construction site for the purpose of regulation 9 of the Management of Health and Safety at Work Regulations 1992 or otherwise) so far as is necessary to enable each of those contractors to comply with the requirements and prohibitions imposed on him by or under the relevant statutory provisions relating to construction work;

- ensure, so far as is reasonably practicable, that every contractor, and every employee at work in connection with the project complies with any rules contained in the health and safety plan;
- takes reasonable steps to ensure that only authorised persons are allowed into any premises or part of premises where construction work is being carried out;
- ensure that any particulars required to be in any notice given under regulation 7 are displayed in readable condition in a position where they can be read by any person at work on construction work in connection with the project; and
- promptly provide the planning supervisor with any information which:
  - is in the possession of the principal contractor or which he could ascertain by making reasonable enquiries of a contractor; and
  - it is reasonable to believe the planning supervisor would include in the health and safety file in order to comply with the requirements imposed upon him in respect thereof in regulation 14; and
  - is not in the possession of the planning supervisor.

The principal contractor may:

- give reasonable directions to any contractor so far as is necessary to enable the principal contractor to comply with his duties under these regulations;
- include in the health and safety plan rules for the management of the construction work which are reasonably required for the purposes of health and safety.

Any rules contained in the health and safety plan shall be in writing and shall be brought to the attention of persons who may be affected by them.

Information and training (regulation 17)

The principal contractor appointed for any project shall ensure, so far as is reasonably practicable, that every contractor is provided with comprehensible information on the risks to the health or safety of that contractor or of any employees or other persons under the control of that contractor arising out of or in connection with the construction work.

The principal contractor shall ensure, so far as is reasonably practicable, that every contractor who is an employer provides any of his employees at work carrying out the construction work with:

- any information which the employer is required to provide to those employees in respect of that work by virtue of regulation 8 of the Management of Health and Safety at Work Regulations 1992; and
- any health and safety training which the employer is required to provide to those employees in respect of that work by virtue of regulation 11(2) (b) of the Management of Health and Safety at Work Regulations 1992.

Advice from, and views of, persons at work (regulation 18)

The principal contractor shall:

- ensure that employees and self-employed persons at work on the construction work are able to discuss, and offer advice to him on, matters connected with the project which it can reasonably be foreseen will affect their health or safety; and
- ensure that there are arrangements for the co-ordination of the views of employees at work on construction work, or of their representatives, where necessary for reasons of health and safety having regard to the nature of the construction work and the size of the premises where the construction work is carried out.

Requirements and prohibitions on contractors (regulation 19)

Every contractor shall, in relation to the project:

- co-operate with the principal contractor so far as is necessary to enable each of them to comply with his duties under the relevant statutory provisions;
- so far as is reasonably practicable, promptly provide the principal contractor with any information (including any relevant part of any risk assessment made by virtue of the Management of Health and Safety at Work Regulations 1992) which might affect the health or safety of any person at work carrying out the construction work or of any person who may be affected or which might justify a review of the health and safety plan;
- comply with any directions of the principal contractor given to him under regulation 16(2);
- comply with any rules applicable to him in the health and safety plan;
- promptly provide the principal contractor with the information in relation to any injury, death, condition or dangerous occurrence which the contractor is required to notify under the Reporting of Injuries, Diseases and Dangerous Occurrences Regulations 1995; and
- promptly provide the principal contractor with any information which:

  - is in the possession of the contractor or which he could ascertain by making reasonable enquiries of persons under his control; and
  - it is reasonable to believe the principal contractor would provide to the planning supervisor in order to comply with the requirements imposed on the principal contractor in respect thereof by regulation 16(1) (e); and
  - which is not in the possession of the principal contractor.

No employer shall cause or permit any employee of his to work on construction work unless the employer has been provided with the information mentioned in paragraph 4.

No self-employed person shall work on construction work unless he has been provided with the information mentioned in paragraph 4.

The information referred to in paragraphs 2 and 3 is:

- the name of the planning supervisor for the project;
- the name of the principal contractor for the project; and
- the contents of the health and safety plan or such part of it as is relevant to the construction work which any such employer or, as the case may be, which the self-employed person, is to carry out.

It shall be a defence in any proceedings for contravention of paragraph 2 or 3 for the employer or self-employed person to show that he made all reasonable enquiries and reasonably believed:

- that he had been provided with the information mentioned in paragraph 4; or
- that, by virtue of any provision in regulation 3, this regulation did not apply to the construction work.

Exclusion of civil liability (regulation 21)

Breach of a duty imposed by these regulations, other than those imposed by regulation 10 and regulation 16(1) (c), shall not confer a right of action in any civil proceedings.

Enforcement (regulation 22)

The enforcing authority for these regulations shall be the Health and Safety Executive.

Schedule 1: Particulars to be notified to the HSE

- Date of forwarding
- Exact address of the construction site
- Name and address of the client or clients (see note)
- Type of project
- Name and address of the planning supervisor
- A declaration signed by or on behalf of the planning supervisor that he has been appointed as such
- Name and address of the principal contractor
- A declaration signed by or on behalf of the principal contractor that he has been appointed as such
- Date planned for the start of the construction phase
- Planned duration of the construction phase
- Estimated maximum number of people at work on the construction site
- Planned number of contractors on the construction site
- Name and address of any contractor or contractors already chosen.

Note: Where a declaration has been made in accordance with regulation 4(4), item 3 above refers to the client or clients on the basis that the declaration has not yet taken effect.

Summary of the main duties

Client

The client must be satisfied that each of the four categories below are competent and ensure the allocation of sufficient resources, including time, to the project. He must further ensure that work does not begin until a satisfactory health and safety plan has been prepared.

Designer

The designer must ensure that structures are designed to avoid or minimise risks during construction and maintenance. He must provide adequate information where risks cannot be avoided and alert clients to their duties.

Planning supervisor

The planning supervisor has overall responsibility for co-ordinating the health and safety aspects of the design and planning stage. He is responsible for the early stages of the health and safety plan, and must ensure that the health and safety file is prepared and delivered to the client at the end of the project.

Principal contractor

The principal contractor must develop and implement the health and safety plan. He must take account of health and safety issues when preparing and presenting tenders, and co-ordinate the activities of contractors to ensure compliance with health and safety legislation. He must check on the provision of information and training for employees, and on consultation arrangements with employees and the self-employed. He must further ensure only authorised personnel are allowed on site.

Contractors

All contractors must co-operate with the principal contractor and provide relevant information on the risks to health and safety arising from their work and on the means of control. They must provide information to the principal contractor and to employees. The self-employed have similar duties to contractors.

Documents

Health and safety plan

The plan covers two phases. At pre-tender stage it must include a general description of the work, timings, details of risks to workers, and information for the principal contractors and on welfare arrangements. At the construction stage it must include arrangements for the health and safety of all those affected by the construction work, arrangements for management of the work and for monitoring of legal compliance, and information about welfare arrangements.

Health and safety file

This document contains information for the client/user of the building on the risks present during maintenance, repair or renovation.

THE CONSTRUCTION (HEAD PROTECTION) REGULATIONS 1989

These regulations impose requirements for the provision of suitable head protection for, and the wearing of suitable head protection by, people at work on building operations or works of engineering construction within the meaning of the FA.

Regulation 3 requires employers to provide suitable head protection (see the definition of this below) for their employees, and for the self-employed to provide it for themselves. It also imposes requirements for the maintenance and replacement of that head protection. Employees are required to report to their employers losses of or defects in their head protection.

Employers, and others having control over workers, must ensure that head protection is worn. The regulations allow rules to be made or directions given as to the wearing of head protection in specified circumstances. There is a specific requirement on people at work to wear head protection.

'Suitable head protection', under the regulations, means head protection that:

- is designed to provide protection, so far as is reasonably practicable, against foreseeable risks of injury to the head to which the wearer may be exposed
- after any necessary adjustment, fits the wearer
- is suitable with regard to the work or activity in which the wearer may be engaged.

THE CONSTRUCTION (HEALTH, SAFETY AND WELFARE) REGULATIONS 1996

These regulations impose requirements with respect to the health, safety and welfare of persons at work carrying out 'construction work' as defined and of others who may be affected by that work. Specified regulations apply in respect of 'construction work carried out on a construction site' as defined. Where a workplace on a construction site is set aside for purposes other than construction, the regulations do not apply.

Subject to specific exceptions, the regulations impose requirements on duty holders, i.e. employers, the self-employed and others who control the way in which construction work is carried out. Employees have duties in respect of their own actions. Every person at work has duties as regards co-operation with others and the reporting of danger.

The principal duties of employers, the self-employed and controllers

Safe places of work

A general duty to ensure a safe place of work and safe means of access to and from that place of work.

Specific provisions include the following:

Precautions against falls

(a) prevention of falls from heights by physical precautions or, where this is not possible, provision equipment that will arrest falls;
(b) provision and maintenance of physical precautions to prevent falls through fragile materials;
(c) erection of scaffolding, access equipment, harnesses and nets under the supervision of a competent person; and
(d) specific criteria for using ladders.

Falling objects

(a) where necessary to protect people at work and others, taking steps to prevent materials or objects from falling;
(b) where it is not reasonably practicable to prevent falling materials, taking precautions to prevent people from being struck, e.g. covered walkways;
(c) prohibition of throwing any materials or objects down from a height if they could strike someone; and
(d) storing materials and equipment safely.

Work on structures

(a) prevention of accidental collapse of new or existing structures or those under construction;
(b) ensuring any dismantling or demolition of any structure is planned and carried out in a safe manner under the supervision of a competent person; and
(c) only firing explosive charges after steps have been taken to ensure that no one is exposed to risk or injury from the explosion.

Excavations, cofferdams and caissons

(a) prevention of the collapse of ground both in and above excavations;
(b) identification and prevention of risk from underground cables and other services; and ensuring cofferdams and caissons are properly designed, constructed and maintained.

Prevention or avoidance of drowning

(a) taking steps to prevent people falling into water or other liquid so far as is reasonably practicable;
(b) ensuring that personal protective and rescue equipment is immediately available for use and maintained, in the event of a fall; and
(c) ensuring sure transport by water is under the control of a competent person.

Traffic routes, vehicles, doors and gates

(a) ensuring construction sites are organised so that pedestrians and vehicles can both move safely and without risks to health;
(b) ensuring routes are suitable and sufficient for the people or vehicles using them;
(c) prevention or control of the unintended movement of any vehicle;
(d) ensuring arrangements for giving a warning of any possible dangerous movement, e.g. reversing vehicles;
(e) ensuring safe operation of vehicles including prohibition of riding or remaining in unsafe positions; and
(f) ensuring doors and gates which could prevent danger, e.g. trapping risk of powered doors and gates, are provided with suitable safeguards.

Prevention and control of emergencies

(a) prevention of risk from fire, explosion, flooding and asphyxiation;
(b) provision of emergency routes and exits;
(c) provision of arrangements for dealing with emergencies, including procedures for evacuating the site; and
(d) where necessary, provision of fire-fighting equipment, fire detectors and alarm systems.

Welfare facilities

(a) provision of sanitary and washing facilities and an adequate supply of drinking water;
(b) provision of rest facilities, facilities to change and store clothing.

Site-wide issues

(a) ensuring sufficient fresh or purified air is available at every workplace, and that associated plant is capable of giving visible or audible warning of failure;
(b) ensuring a reasonable working temperature is maintained at indoor workplaces during working hours;
(c) provision of facilities for protection against adverse weather conditions;
(d) ensuring suitable and sufficient emergency lighting is available;
(e) ensuring suitable and sufficient lighting is available, including the

provision of secondary lighting where there would be a risk to health or safety if the primary or artificial lighting failed;

(f) maintaining construction sites in good order and in a reasonable state of cleanliness;

(g) ensuring the perimeter of a construction site to which people, other than those working on the site could gain access, is marked by suitable signs so that its extent can be easily identified; and

(h) ensuring all plant and equipment used for construction work is safe, of sound construction and used and maintained so that it remains safe and without risks to health.

Training, inspection and reports

(a) ensuring construction activities where training, technical knowledge or experience is necessary to reduce risks of injury are only carried out by people who meet these requirements or, if not, are supervised by those with appropriate training, knowledge or experience;

(b) before work at height, on excavations, cofferdams or caissons begins, ensuring the place of work is inspected (and at subsequent specified periods), by a competent person, who must be satisfied that the work can be done safely; and

(c) following inspection, ensuring written reports are made by the competent person.

CONSTRUCTION (HEALTH, SAFETY AND WELFARE) REGULATIONS 1996

SCHEDULE 7

PLACES OF WORK REQUIRING INSPECTION

Column 1 Place of work		Column 2 Time of inspection
1. Any working platform or part thereof or any personal suspension equipment	(i) (ii) (iii) (iv)	Before being taken into use for the first time; and after any substantial addition, dismantling or other alteration; and after any event likely to have affected its strength and stability; and at regular intervals not exceeding 7 days since the last inspection.
2. Any excavation which is supported	(i) (ii) (iii)	Before any person carries out work at the start of every shift; and after any event likely to have affected the strength or stability of the excavation or any part thereof; and after any accidental fall of rock or earth or other material.
3. Cofferdams and caissons	(i) (ii)	Before any person carries out work at the start of every shift; and after any event likely to have affected the strength or stability of the cofferdam or caisson or any part thereof.

4

CONTROL OF ASBESTOS AT WORK REGULATIONS 2002

These regulations apply to all workplaces where asbestos products are manufactured, used and handled and cover all persons at risk from work with asbestos.

They are accompanied by an ACOP, *The control of asbestos*, and the following HSE Guidance Notes provide further information:

EH 10   Asbestos – Control Limits: Measurement of airborne dust concentrations and assessment of control measures

EH 35   Probable asbestos dust concentrations in construction processes

EH 36   Work with asbestos insulating board

EH 41   Respiratory protective equipment for use against asbestos

EH 47   The provision, use and maintenance of hygiene facilities for work with asbestos insulation and coatings.

Interpretation (regulation 2)

Action level means one of the following cumulative exposures to asbestos over a continuous 12-week period when measured or calculated by a method approved by the HSC, namely:

(a)   where the exposure is solely to chrysotile, 72 fibre-hours/ml of air;

(b)   where exposure is to any other form of asbestos either alone or in mixtures including mixtures of chrysotile with any other form of asbestos, 48 fibre-hours/ml of air; or

(c)   where both types of exposure occur separately during the 12-week period concerned, a proportionate number of fibre-hours/ml of air.

Asbestos means the following minerals, that is to say, crocidolite, amosite, chrysotile, fibrous actinolite, fibrous anthophyllite or fibrous tremolite or any mixture containing any of those minerals.

Control limit means one of the following concentrations of asbestos in the atmosphere when measured or calculated by the reference method described in Annex 1 to Council Directive 83/477/EEC or by a method giving equivalent results approved by the HSC, namely:

(a)   for chrysotile:

     (i)   0.3 fibres/ml of air averaged over a continuous period of 4 hours;

     (ii)   0.9 fibres/ml of air averaged over a continuous period of 10 minutes;

(b)   for any other form of asbestos either alone or in mixtures including mixtures of chrysotile with any other form of asbestos:

     (i)   0.2 fibres/ml of air averaged over a continuous period of 4 hours;

     (ii)   0.6 fibres/ml of air averaged over a continuous period of 10 minutes.

Control measure means a measure taken to prevent or reduce exposure to asbestos (including the provision of systems of work and supervision, the cleaning of workplaces, premises, plant and equipment, the provision and use of engineering controls and personal protective equipment.

Medical examination includes any laboratory tests and X-rays that a relevant doctor may require).

Personal protective equipment means all equipment (including clothing) which is intended to be worn or held by a person at work and which protects that person against one or more risks to his health, and any addition or accessory designed to meet that objective.

For the purpose of these regulations, except as specified in regulation 10(4), in determining whether an employee is exposed to asbestos or whether the extent of such exposure exceeds the action level or relevant control limit, no account shall be taken of respiratory protective equipment which, for the time being, is being worn by that employee.

Duty to manage asbestos in non-domestic premises (regulation 4)

In this regulation, *the dutyholder* means:

4

- every person who has, by virtue of a contract or tenancy, an obligation to any extent in relation to the maintenance or repair of non-domestic premises or any means of access thereto or egress therefrom; or
- in relation to any part of non-domestic premises where there is no such contract or tenancy, every person who has, to any extent, control of that part of those non-domestic premises or any means of access thereto or egress therefrom, and where there is more than one dutyholder, the relative contribution to be made by each such person in complying with the requirements of this regulation will be determined by the nature and extent of the maintenance and repair obligation owed by that person.

Every person shall co-operate with the dutyholder so far as is necessary to enable the dutyholder to comply with his duties under this regulation.

In order to enable him to manage the risk from asbestos in non-domestic premises, the dutyholder shall ensure that a suitable and sufficient assessment is carried out as to whether asbestos is or is liable to be present in the premises.

In making the assessment:

- such steps as are reasonable in the circumstances shall be taken; and
- the condition of any asbestos which is, or has been assumed to be, present in the premises shall be considered.

Without prejudice to the generality of the above paragraph, the dutyholder shall ensure that:

- account is taken of building plans or other relevant information and of the age of the premises; and

- an inspection is made of those parts of the premises which are reasonably accessible.

The dutyholder shall ensure that the assessment is reviewed forthwith if:

- there is reason to suspect that the assessment is no longer valid; or
- there has been a significant change in the premises to which the assessment relates.

The dutyholder shall ensure that the conclusions of the assessment and every review are recorded.

Where the assessment shows that asbestos is or is liable to be present in any part of the premises the dutyholder shall ensure that:

- a determination of the risk from that asbestos is made;
- a written plan identifying those parts of the premises concerned is prepared; and
- the measures which are to be taken for managing the risk are specified in the written plan.

The measures to be specified in the plan for managing the risk shall include adequate measures for:

- monitoring the condition of any asbestos or any substance containing or suspected of containing asbestos;
- ensuring any asbestos or any such substance is properly maintained or where necessary safely removed; and
- ensuring that information about the location and condition of any asbestos or any such substance is:

  - provided to every person liable to disturb it; and
  - made available to the emergency services.

The dutyholder shall ensure that:

- the plan is reviewed and revised at regular intervals, and forthwith if:

  - there is reason to suspect the plan is no longer valid; or
  - there has been a significant change in the premises to which the plan relates;

- the measures specified in the plan are implemented; and
- the measures taken to implement the plan are recorded.

Identification of the type of asbestos (regulation 5)

An employer shall not carry out work which is liable to expose his employees to asbestos unless either:

- before commencing that work, he has identified, by analysis or otherwise, the type of asbestos involved in the work; or
- he has assumed that the asbestos is not chrysotile alone and for the purposes of these regulations has treated it accordingly.

Assessment of work which exposes employees to asbestos (regulation 6)

An employer shall not carry out work which is liable to expose his employees to asbestos unless he has:

- made a suitable and sufficient assessment of the risk created by that exposure to the health of those employees and of the steps that need to be taken to meet the requirements of these regulations;
- recorded the significant findings of that risk assessment as soon as is practicable after the risk assessment is made; and
- implemented the steps referred to in sub-paragraph (a) above.

Without prejudice to the generality of paragraph 1 above, the risk assessment shall:

- subject to regulation 5, identify the type of asbestos to which employees are liable to be exposed;
- determine the nature and degree of exposure which may occur in the course of the work;
- consider the effect of control measures which have been or will be taken in accordance with regulation 10;
- consider the results of monitoring of exposure in accordance with regulation 18;
- set out the steps to be taken to prevent that exposure or reduce it to the lowest level reasonably practicable;
- consider the results of relevant medical surveillance; and
- include such additional information as the employer may need in order to complete the risk assessment.

The risk assessment shall be reviewed regularly, and forthwith if:

- there is reason to suspect that the existing risk assessment is no longer valid;
- there is a significant change in the work to which the risk assessment relates; or
- the results of any monitoring carried out pursuant to regulation 18 show it to be necessary,

and where, as a result of the review, changes to the risk assessment are required, those changes shall be made.

Where the risk assessment has determined that the exposure of his employees to asbestos may exceed the action level, the employer shall keep a copy of the significant findings of that risk assessment of those premises at which, and for such time as, the work to which that risk assessment relates is being carried out.

Plans of work (regulation 7)

An employer shall not undertake any work with asbestos unless he has prepared a suitable written plan of work detailing how that work is to be carried out.

The employer shall keep a copy of the plan of work:

- at those premises at which; and
- for such time as,

the work to which the plan relates is being carried out.

The plan of work shall include in particular details of:

- the nature and probable duration of the work;
- the location of the place where the work is to be carried out;
- the methods to be applied where the work involves the handling of asbestos or materials containing asbestos;
- the characteristics of the equipment to be used for:

  - protection and decontamination of those carrying out the work; and
  - protection of other persons on or near the worksite; and

- the measures which the employer intends to take in order to comply with the requirements of regulation 10.

The employer shall ensure, so far as is reasonably practicable, that the work to which the plan of work relates is carried out in accordance with that plan.

Notification of work with asbestos (regulation 8)

This regulation shall apply to work at the premises of the employer in which an employee is or is liable to be exposed to asbestos unless:

- the extent of that exposure neither exceeds nor is liable to exceed the action level; or
- the work is work to which the Asbestos (Licensing) Regulations 1983 apply.

An employer shall not carry out work to which this regulation applies for the first time unless he has notified the enforcing authority in writing of the particulars specified in Schedule 1 at least 14 days before commencing that work or before such shorter time as the enforcing authority may agree.

Where an employer has notified work in accordance with the above paragraph and there is a material change in that work which might affect the particulars so notified (including the cessation of the work), the employer shall forthwith notify the enforcing authority of that change.

Information, instruction and training (regulation 9)

Every employer shall ensure that adequate information, instruction and training is given to those of his employees:

- who are liable to be exposed to asbestos, or who supervise such employees, so that they are aware of:

  - the significant findings of the risk assessment;
  - the risks to health from asbestos;
  - the precautions which should be observed; and
  - the relevant control limit and action level,

  in order to safeguard themselves and other employees; and

- who carry out work in connection with the employer's duties under these regulations, so that they can carry out that work effectively.

  The information, instruction and training shall be:

- given at regular intervals;
- adapted to take account of significant changes in the type of work carried out or methods of work used by the employer; and
- provided in a manner appropriate to the nature and degree of exposure identified by the risk assessment.

Prevention or reduction of exposure to asbestos (regulation 10)

4

Every employer shall:

- prevent the exposure of his employees to asbestos so far as is reasonably practicable;
- where it is not reasonably practicable to prevent such exposure:

  - reduce the exposure of his employees to asbestos to the lowest level reasonably practicable by measures other than the use of respiratory protective equipment (RPE); and
  - ensure that the number of his employees who are exposed to asbestos is as low as is reasonably practicable.

Where employees may be exposed to asbestos in a manufacturing process or in the installation of a product, prevention of such exposure to asbestos shall be achieved, where it is practicable, by substituting for asbestos a substance which, under the conditions of its use, does not create a risk to the health of his employees or creates a lesser risk than that created by asbestos.

Where it is not practicable to avoid the use of asbestos by substitution in accordance with the above paragraph, the measures referred to in paragraph 1(b)(i) shall include, in order of priority:

- the design and use of appropriate work processes, systems and engineering controls and the provision and use of suitable work equipment and materials in order to avoid or minimise the release of asbestos; and
- the control of exposure at source, including adequate ventilation systems and appropriate organisational measures.

Where it is not reasonably practicable to reduce the exposure of an employee to below the control limits which apply to that exposure, then, in addition to taking the measures required by that paragraph, the employer shall provide that employee with suitable RPE which will reduce the concentration of asbestos in the air inhaled by the employee (after taking into account the effect of that RPE) to a concentration which is as low as is reasonably practicable and in any event below those control limits.

Personal protective equipment (PPE) provided by an employer shall be suitable for the purpose and shall:

- comply with any provision in the Personal Protective Equipment at Work Regulations 2002 which is applicable to that item of PPE; or
- in the case of RPE, where no provision referred to in the above sub-paragraph applies, be of a type approved or shall conform to a standard approved, in either case, by the HSE.

Every employer shall take immediate steps to remedy the situation where the concentration of asbestos in air inhaled by any employee exceeds the relevant control limit.

Where asbestos is used in a work process, or is produced by a work process, the employer shall ensure that the quantity of asbestos and materials containing asbestos at the premises where that work is carried out is reduced to as low a level as is reasonably practicable.

Use of control measures (regulation 11)

Every employer who provides any control measure, other thing or facility pursuant to these regulations shall ensure so far as is reasonably practicable that it is properly used or applied as the case may be.

Every employee shall make full and proper use of any control measure, other thing or facility provided pursuant to these regulations and, where relevant, shall:

- take all reasonable steps to ensure it is returned after use to any accommodation provided for it; and
- if he discovers a defect therein report it forthwith to his employer.

Maintenance of control measures etc. (regulation 12)

Every employer who provides any control measure shall ensure that, where relevant, it is maintained in an efficient state, in efficient working order, in good repair and in a clean condition.

Where exhaust ventilation equipment or RPE (except disposable RPE) is provided, the employer shall ensure that thorough examinations and tests of that equipment are carried out at suitable intervals by a competent person.

Every employer shall keep a suitable record of the examinations and tests carried out and of repairs carried out as a result of those examinations and tests, and that record or a suitable summary thereof shall be kept available for at least 5 years from the date on which it was made.

Provision and cleaning of protective clothing (regulation 13)

Every employer shall provide adequate and suitable protective clothing for such of his employees as are exposed or are liable to be exposed to asbestos, unless no significant quantity of asbestos is liable to be deposited on the clothes of the employee while he is at work.

The employer shall ensure that protective clothing provided is either disposed of as asbestos waste or adequately cleaned at suitable intervals.

The cleaning shall be carried out either on the premises where the exposure to asbestos has occurred, where those premises are suitably equipped for such cleaning, or in a suitably equipped laundry.

The employer shall ensure that protective clothing which has been used and is to be removed from the premises (whether for cleaning, further use or disposal) is packed, before being removed, in a suitable container which shall be labelled in accordance with the provisions of Schedule 2 as if it were a product containing asbestos or, in the case of protective clothing intended for disposal as waste, in accordance with regulation 23(3).

Where, as a result of the failure or improper use of the protective clothing, a significant quantity of asbestos is deposited on the personal clothing of an employee, that personal clothing shall be treated as if it were protective clothing.

Arrangements to deal with accidents, incidents and emergencies (regulation 14)

In order to protect the health of his employees from accident, incident or emergency related to the use of asbestos in a work process or to the removal or repair of asbestos-containing materials at the workplace, the employer shall ensure that:

- procedures, including the provision of relevant safety drills (which shall be tested at regular intervals), have been prepared which can be put into effect when such an event occurs;
- information on emergency arrangements including:

  - details of relevant work hazards and hazard identification arrangements; and
  - specific hazards likely to arise at the time of an accident, incident or emergency,

is available; and

- suitable warning and other communication systems are established to enable an appropriate response, including remedial actions and rescue operations, to be made immediately when such an event occurs.

The employer shall ensure that information on the procedure and systems is:

- made available to relevant accident and emergency services to enable those services, whether internal or external to the workplace, to prepare their own response procedures and precautionary measures; and
- displayed at the workplace, if this is appropriate.

Paragraph 1 shall not apply where:

- the results of the risk assessment show that, because of the quantity of the asbestos present at the workplace, there is only a slight risk to the health of the employees; and
- the measures taken by the employer to comply with the duty under regulation 10(1) are sufficient to control that risk.

In the event of an accident, incident or emergency related to the unplanned release of asbestos at the workplace, the employer shall ensure that:

- immediate steps are taken to:
  - mitigate the effects of the event;
  - restore the situation to normal; and
  - inform any person who may be affected; and
- only those persons who are responsible for the carrying out of repairs and other necessary work are permitted in the affected area and they are provided with:
  - appropriate RPE and protective clothing; and
  - any necessary specialised safety equipment and plant,

which shall be used until the situation is restored to normal.

Duty to prevent or reduce the spread of asbestos (regulation 15)

Every employer shall prevent or, where this is not reasonably practicable, reduce to the lowest level reasonably practicable, the spread of asbestos from any place where work under his control is carried out.

Cleanliness of premises and plant (regulation 16)

Every employer who undertakes work which exposes or is liable to expose his employees to asbestos shall ensure that:

- the premises, or those parts of the premises where that work is carried out, and the plant used in connection with that work are kept in a clean state; and
- where such work has been completed, the premises, or those parts of the premises where the work was carried out, are thoroughly cleaned.

Where a manufacturing process which gives rise to asbestos dust is carried out in a building, the employer shall ensure that any part of the building in which the process is carried out is:

- so designed and constructed as to facilitate cleaning; and
- equipped with an adequate and suitable vacuum cleaning system which shall, where reasonably practicable, be a fixed system.

Designated areas (regulation 17)

Every employer shall ensure that any area in which work under his control is carried out is designated as:

- an asbestos area, where the exposure to asbestos of an employee who worked in that area for the whole of his working time would exceed or would be liable to exceed the action level; and
- a respirator zone, where the concentration of asbestos in that area would exceed or would be liable to exceed a relevant control limit.

Asbestos areas and respirator zones shall be clearly and separately demarcated and identified by notices indicating:

- that the area is an asbestos area or a respirator zone or both, as the case may be; and
- in the case of a respirator zone, that the exposure of an employee who enters it is liable to exceed a relevant control limit and, if it does, that RPE must be worn.

The employer shall not permit any employee, other than an employee who by reason of his work is required to be in an area designated in accordance with paragraph 1, to enter or remain in any such area and only employees who are so permitted shall enter or remain in any such area.

Every employer shall ensure that:

- his employees do not eat, drink or smoke in an area designated as an asbestos area or respirator zone; and
- arrangements are made for such employees to eat or drink in some other place.

Air monitoring (regulation 18)

Every employer shall monitor the exposure of his employees to asbestos by measurement of asbestos fibres present in the air:

- at regular intervals; and
- when a change occurs which may affect the exposure.

The above paragraph shall not apply where:

- the exposure of an employee is not liable to exceed the action level; or
- the employer is able to demonstrate by another method of evaluation that the requirements of regulation 10(1) have been complied with.

The employer shall keep a suitable record of:

- monitoring carried out in accordance with paragraph 1; or
- where he decides that monitoring is not required because paragraph 2(b) applies, the reason for that decision.

The record required by paragraph 3, or a suitable summary thereof, shall be kept:

- in a case where exposure is such that a health record is required to be kept under regulation 21, for at least 40 years; or
- in any other case, for at least 5 years,

from the date of the last entry made in it.

In relation to the record required by paragraph 3, the employer shall:

- on reasonable notice being given, allow an employee access to his personal monitoring record;
- provide the HSE with copies of such monitoring records as the HSE may require; and
- if he ceases to trade, notify the HSE forthwith in writing and make available to the HSE all monitoring records kept by him.

Standards for air testing (regulation 19)

Any employer who carries out any measurement of the concentration of asbestos fibres present in the air shall ensure that he meets criteria equivalent to those set out in the paragraphs of ISO 17025 which cover organisation, quality systems, control of records, personnel, accommodation and environmental conditions, test and calibration methods, method validation, equipment, handling of test and calibration items, and reporting results.

Every employer who requests a person to carry out any measurement of the concentration of asbestos fibres present in the air shall ensure that that person is accredited by an appropriate body as complying with ISO 17025.

Standards for analysis (regulation 20)

Every employer who analyses a sample of any material to determine whether it contains asbestos shall ensure that he meets criteria equivalent to those set out in the paragraphs of ISO 17025 which cover organisation etc. (as per regulation 19).

Every employer who requests a person to analyse a sample of any material to determine whether it contains asbestos shall ensure that that person is accredited by the appropriate body as complying with ISO 17025.

Health records and medical surveillance (regulation 21)

Every employer shall ensure that:

- a health record, containing particulars approved by the HSE, relating to each of his employees who is exposed to asbestos is maintained unless the exposure of that employee does not exceed the action level; and
- that record or a copy thereof is kept available in a suitable form for at least 40 years from the date of the last entry made in it.

Every employer shall ensure that each of his employees who is exposed to asbestos is under adequate medical surveillance by a relevant doctor unless the exposure of that employee does not exceed the action level.

The medical surveillance required shall include:

- a medical examination not more than 2 years before the beginning of such exposure; and
- periodic medical examinations at intervals of not more than 2 years or such shorter time as the relevant doctor may require while such exposure continues,

and each such medical examination shall include a specific examination of the chest.

Where an employee has been examined in accordance with the above, the relevant doctor shall issue a certificate to the employer and employee stating:

- that the employee has been so examined; and
- the date of the examination,

and the employer shall keep that certificate or a copy thereof for at least 40 years from the date on which it was issued.

An employee to whom this regulation applies shall, when required by his employer and at the cost of the employer, present himself during his working hours for such examination and tests as may be required and shall furnish the relevant doctor with such information concerning his health as the relevant doctor may reasonably require.

Where, for the purpose of carrying out his functions under these regulations, a relevant doctor requires to inspect any record kept for the purpose of these regulations, the employer shall permit him to do so.

Where medical surveillance is carried out on the premises of the employer, the employer shall ensure that suitable facilities are made available for the purpose.

The employer shall:

- on reasonable notice being given, allow an employee access to his personal health record;
- provide the HSE with copies of such personal health records as the HSE may require; and
- if he ceases to trade, notify the HSE forthwith in writing and make available to the HSE all personal health records kept by him.

Where, as a result of medical surveillance, an employee is found to have an identifiable disease or adverse health effect which is considered by a relevant doctor to be the result of exposure to asbestos at work the employer of that employee shall:

- ensure that a suitable person informs the employee accordingly and provides the employee with information and advice regarding further medical surveillance;
- review the risk assessment;
- review any measure taken to comply with regulation 10, taking into account any advice by a relevant doctor or by the HSE;
- consider assigning the employee to alternative work where there is no risk of further exposure to asbestos, taking into account any advice given by a relevant doctor; and
- provide a review of the health of every other employee who has been similarly exposed, including a medical examination (which shall include a specific examination of the chest) where such an examination is recommended by a relevant doctor or by the HSE.

Washing and changing facilities (regulation 22)

Every employer shall ensure that, for any of his employees who is exposed or liable to be exposed to asbestos, there are provided:

- adequate washing and changing facilities;
- where he is required to provide protective clothing, adequate facilities for the storage of:
  - personal protective clothing; and
  - personal clothing not worn during working hours;
- where he is required to provide RPE, adequate facilities for the storage of that equipment.

The facilities for the storage of personal protective clothing, personal clothing not worn during working hours and RPE shall be separate from each other.

Storage, distribution and labelling of raw asbestos and asbestos waste (regulation 23)

Every employer who undertakes work with asbestos shall ensure that raw asbestos or waste which contains asbestos is not:

- stored;
- received into or despatched from any place of work; or
- distributed within any place of work except in a totally enclosed distribution system, unless it is in a sealed container clearly marked in accordance with paragraphs 2 and 3 showing that it contains asbestos.

Raw asbestos shall be marked in accordance with the provision of Schedule 2.

Waste containing asbestos shall be labelled:

- where the Carriage of Dangerous Goods (Classification, Packaging and Labelling) and Use of Transportable Pressure Receptacles Regulations 1996 apply, in accordance with those regulations;
- where the waste is conveyed by road in a road tanker or tank container in circumstances where the Carriage of Dangerous Goods by Road Regulations 1996 apply, in accordance with those regulations; and
- in any other case, in accordance with the provisions of Schedule 2.

Supply of products containing asbestos for use at work (regulation 24)

Subject to paragraph 2, a person shall not supply a product which contains asbestos, being an article or substance for use at work, unless that product is labelled in accordance with the provisions of Schedule 2.

Where a component of a product contains asbestos, it shall be sufficient compliance with this regulation if that component is labelled in accordance with the provisions of Schedule 2 except that where the size of the component makes it impossible for a label to be fixed to it neither that component nor the product need be labelled.

Defence (regulation 28)

Subject to regulation 21 of the Management of Health and Safety at Work Regulations 1999, in any proceedings for an offence consisting of a contravention of these regulations it shall be a defence for any person to prove that he took all reasonable precautions and exercised all due diligence to avoid the commission of that offence.

Schedule 1

Particulars to be included in a notification

The following particulars are to be included in a notification made in accordance with regulation 8(2), namely:

- the name of the notifier and the address and telephone number of his usual place of business;
- a brief description of:
  - the type(s) of asbestos to used and handled (crocidolite, amosite, chrysotile or other);
  - the maximum quantity of asbestos to be held at any one time on the premises at which the work is to take place;
- the activities or processes involved;
- the products to be manufactured, where applicable, and
- the date of commencement of the work activity.

Schedule 2

The labelling of raw asbestos, asbestos waste and products containing asbestos

Subject to sub-paragraphs 2 and 3 of this paragraph, the label to be used on:

- raw asbestos (together with the labelling required under the Chemicals (Hazard Information and Packaging for Supply) Regulations 2002 and the Carriage of Dangerous Goods (Classification, Packaging and Labelling and Use of Transportable Pressure Receptacles Regulations 1996),
- asbestos waste, when required to be so labelled by regulation 23(3), and
- products containing asbestos, including used protective clothing to which regulation 13(2) applies,

shall be in the form and in the colours of the following diagram and shall comply with the specifications set out in paragraphs 2 and 3.

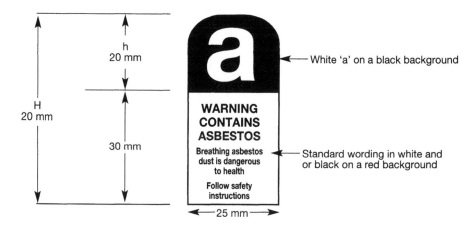

In the case of a product containing crocidolite, the words 'contains asbestos' shown in the diagram shall be replaced by the words 'contains crocidolite/blue asbestos'.

Where the label is printed directly onto a product, a single colour contrasting with the background colour may be used.

The dimensions in millimetres of the label referred to in paragraph 1(1) shall be those shown on the diagram in that paragraph, except that larger measurements may be used, but in that case the dimension indicated as h on the diagram shall be 40% of the dimension indicated as H.

The label shall be clearly and indelibly printed so that the words in the lower half of the label can be easily read, and those words shall be printed in black or white.

Where a product containing asbestos may undergo processing or finishing it shall bear a label containing safety instructions appropriate to the particular product and in particular the following instructions:

'operate if possible out of doors or in a well ventilated place';

'preferably use hand tools or low speed tools equipped, if necessary, with an appropriate dust extraction facility. If high speed tools are used, they should always be so equipped.';

'if possible, dampen before cutting or drilling'; and

'dampen dust, place it in a properly closed receptacle and dispose of it safely'.

Additional safety information given on a label shall not detract from or contradict the safety information given in accordance with sub-paragraph 1.

Labelling of packaged and unpackaged products containing asbestos in accordance with the foregoing paragraphs shall be effected by means of:

- an adhesive label firmly affixed to the product or its packaging, as the case may be;
- a tie-on label firmly attached to the product or its packaging, as the case may be; or
- direct printing onto the product or its packaging, as the case may be.

Where, in the case of an unpackaged product containing asbestos, it is not reasonably practicable to comply with the provisions of sub-paragraph 1 the label shall be printed on a suitable sheet accompanying the product.

Labelling of raw asbestos and asbestos waste shall be effected in accordance with sub-paragraph 1(a) or (c).

For the purposes of this Schedule but subject to sub-paragraph 5, a product supplied in loose plastic or other similar wrapping (including plastic and paper bags) but no other packaging, shall be treated as being supplied in a package whether the product is placed in such wrapping at the time of its supply or was already so wrapped previously.

No wrapping in which a product is placed at the time of its supply shall be regarded as packaging if any product contained in it is labelled in accordance with the requirements of this Schedule or any other packaging in which that product is contained is so labelled.

THE CONTROL OF MAJOR ACCIDENT HAZARDS (COMAH) REGULATIONS 1999

The COMAH Regulations replaced the Control of Industrial Major Accident Hazard (CIMAH) Regulations 1984. As such, COMAH emphasises the control of risks both to people and the environment.

Under the regulations the HSE and the Environment Agency (EA) in England, together with the Scottish Environment Protection Agency (SEPA), are required to form a joint 'competent authority' for enforcing the regulations.

The regulations apply to defined 'establishments' where dangerous substances are present in quantities equal to or exceeding those specified in column 2 of Parts 2 or 3 or Schedule 1, with certain exceptions.

Definition of 'establishment'

An establishment means the whole area under the control of the same person where dangerous substances are present in one or more installations, and for this purpose two or more areas under the control of the same person and separated only by a road, railway or inland waterway shall be treated as one whole area.

Major accident

This is defined as:

> an occurrence (including, in particular, a major emission, fire or explosion) resulting from uncontrolled developments in the course of the operation of any establishment and leading to serious danger to human health or the environment, immediate or delayed, inside or outside the establishment, and involving one or more dangerous substances.

Application of COMAH to individual workplaces is based on the presence of specified quantities of dangerous substances. Under the COMAH Regulations, substances are both listed and classified by type.

Principal requirements of the regulations

The regulations:

- impose a duty on the operator of an establishment to take all measures necessary to prevent major accidents and limit their consequences for persons and the environment
- impose a duty on the operator to prepare a major accident prevention policy document containing specified information and to revise it in specified circumstances
- require the operator to notify the competent authority of specified matters at specific times
- require the operator to send at specified times a safety report to the competent authority containing specified information, to revise that report in specified circumstances, and not to start the construction of operation of the establishment until he has received from the competent authority the conclusions of its examination of the report
- require the operator to review and revise the safety report in specified circumstances
- require the operator to prepare an on-site emergency plan for specified purpose and containing specified information
- require, subject to any exemption that may be granted by the competent authority, the local authority to prepare an off-site emergency plan for specified purposes and containing specified information and require the operator to supply the local authority the information necessary to enable the plan to be prepared
- require on-site and off-site emergency plans to be reviewed, tested and implemented, and empower the local authority to charge the operator a

fee for preparing, reviewing and testing the off-site emergency plan
- require the operator to provide, after consulting the local authority, specified information to specified persons at specified times and to make that information publicly available
- require the operator to demonstrate to the competent authority that he has taken all measures to comply with regulations
- require the operator to provide, when requested to do so by the competent authority, information to the authority for specified purposes
- require the operator to notify major accidents to the competent authority
- require the operator to provide information to, and co-operate with, other establishments in a group designated by the competent authority
- impose functions on the competent authority with respect to:

  - its consideration of the safety report sent by the operator;
  - prohibiting the operation of the establishment;
  - inspections and investigations;
  - enforcement; and
  - the provision of information

- provide for fees to be payable by the operator to the HSE in relation to the performance of specified functions of the Executive or competent authority
- amend the Petroleum Consolidation Act 1928, the Petroleum-Spirit (Motor Vehicles etc) Regulations 1929 and the Petroleum-Spirit (Plastic Containers) Regulations 1982 so as to disapply those regulations to establishments subject to these regulations and to sites subject to the Notification of Installations Handling Hazardous Substances Regulations 1982.

Site classification and reporting

All high hazard sites are covered by the regulations, being classified into 'top tier' and 'lower tier' according to their inventories of hazardous substances held.

Operators of *low tier* sites must notify the competent authorities and prepare a Major Accident Prevention Policy (MAPP). *Top tier* sites may include their MAPP document in the safety report covering, for instance, preparation of on-site emergency plans and the provision of certain information to the public.

Charging arrangements

The HSE is empowered to charge organisations for inspections carried out under the regulations.

Enforcement committees

Because COMAH covers both health and safety and environmental issues, an enforcement committee must be formed for each site which includes experts from both the HSE and the EA.

Appointment of lead person

One person from this committee (the lead person) must take the lead in liaising with the organisation in order to prevent confusion. The identity of the lead person is determined by the dominant hazards on site. Where the hazards are mainly to human life, the HSE takes the lead, and where the hazards are mainly environmental, the EA takes the lead.

THE CONTROL OF LEAD AT WORK REGULATIONS 2002

These regulations define 'lead' as meaning lead (including lead alkyls, lead alloys, any compounds of lead and lead as a constituent of any substance or material) which is liable to be inhaled, ingested or otherwise absorbed by persons except where it is given off from the exhaust system of a vehicle on a road within the meaning of section 192 of the Road Traffic Act 1988.

The regulations:

- specify occupational exposure limits for lead and lead alkyls (i.e. tetraethyl lead and tetramethyl lead);
- specify suspension levels with respect to:

  - blood-lead concentrations (in respect of women of reproductive capacity, young persons and other employees); and
  - urinary lead concentrations (in respect of women of reproductive capacity and other employees);prohibit the use of glazes in the manufacture of pottery, other than a leadless glaze or a low solubility glaze;

- prohibit the employment of young persons and women of reproductive capacity in any activity specified in Schedule 1 (involving specified lead smelting and refining process and lead-acid battery manufacturing processes);
- require an employer to make a sufficient assessment of the risk created by the work to the health of employees and of the steps that need to be taken to meet the requirements of the regulations; a risk assessment shall include consideration of:

  - the hazardous properties of the lead;
  - information on health effects provided by the supplier, including information contained in any relevant safety data sheet;
  - the level, type and duration of exposure;
  - the circumstances of the work, including the amount of lead involved;
  - activities, such as maintenance, where there is the potential for high level of exposure;
  - any relevant occupational exposure limit, action level and suspension level;
  - the effect of prevention and control measures which have been or will be taken in accordance with regulation 6;
  - the results of relevant medical surveillance;

- the results of monitoring of exposure in accordance with regulation 9;
- in circumstances where the work will involve exposure to lead and another substance hazardous to health, the risk presented by exposure to those substances in combination;
- whether the exposure of any employee to lead is liable to be significant; and
- such additional information as the employer may need in order to complete the risk assessment;

- require an employer to ensure that the exposure of his employees to lead is either prevented or, where this is not reasonably practicable, adequately controlled; where the substitution of lead with a safer substance cannot be undertaken, the employer must apply protective measures appropriate to the activity and consistent with the risk assessment;
- require an employer to take measures to prevent eating, drinking or smoking in any place which is or is liable to be contaminated by lead;
- require any physical control measure to be maintained in an efficient state, in efficient working order, in good repair and in a clean condition; local exhaust ventilation systems to be examined and tested at least once every 14 months;
- where there is risk of 'significant' exposure, the employer must operate a suitable procedure for air monitoring, maintaining a suitable record of such monitoring;
- require an employer to ensure that each of his employees who is or is liable to be exposed to lead is under suitable medical surveillance by a relevant doctor;
- require an employer to provide suitable and sufficient information, instruction and training to employees who are liable to be exposed to lead;
- require employers to have appropriate procedures for dealing with accidents, incidents and emergencies.

Schedule 1 specifies activities in which the employment of young persons and women of reproductive capacity is prohibited, e.g. in lead smelting and refining processes.

THE CONTROL OF PESTICIDES REGULATIONS 1986

The regulations apply to substances used for:

- protecting plants from harmful organisms
- regulating plant growth
- giving protection, for example, against harmful insects, and rendering them harmless
- controlling organisms with harmful effects, such as in buildings
- protecting animals, against parasites, for example.

Prohibitions (regulation 4)

There is a general prohibition on advertising, selling, supplying, storing or using a pesticide without the approval of the Minister of Agriculture, Fisheries and Food and the Secretary of State, as per regulation 5, below.

Approvals (regulation 5)

This regulation authorises the Minister and Secretary of State to give approval to a pesticide of any description.

Consents (regulation 6)

This regulation authorises the Minister and Secretary of State to consent to the advertising, sale, supply, storage and use of pesticides under certain conditions.

The seizure or disposal of pesticides (regulation 7)

Under this regulation, the Minister and Secretary of State have the power to seize or dispose of pesticides used in breach of regulation 4.

The release of information to the public (regulation 8)

Evaluations of pesticides may be revealed to any member of the public who requests them, for a fee.

Schedules 1–3

These Schedules detail the conditions operating with regard to giving consent to advertise, sell, store or use pesticides.

Schedule 4

This Schedule details the conditions that apply to regulate crop spraying by air.

THE CONTROL OF SUBSTANCES HAZARDOUS TO HEALTH (COSSH) REGULATIONS 2002

The COSSH Regulations apply to every form of workplace and every type of work activity involving the use of substances which may be hazardous to health to people at work. The regulations are supported by a number of HSC Approved Codes of Practice, including:

'Control of substances hazardous to health'
'Control of carcinogenic substances' and
'Control of biological agents'.

Regulation 2 – Definitions

The following definitions are important:

Biological agent means any micro-organism, cell, culture or human endoparasite, including any which have been genetically modified, which may cause any infection, allergy, toxicity or otherwise create a hazard to human health.

Carcinogen means:

(a)   a substance or preparation which is classified in accordance with the classification provided for by regulation 4 of the CHIP Regulations would be in the category of danger, carcinogenic (category 1) or carcinogenic (category 2) whether or not the substance or preparation would be required to be classified under those Regulations; or

(b)   a substance or preparation:
   (i)   listed in Schedule 1; or
   (ii)   arising from a process specified in Schedule 1 which is a substance hazardous to health.

Control measure means a measure taken to reduce exposure to a substance hazardous to health (including the provision of systems of work and supervision, the cleaning of workplaces, premises, plant and equipment, the provision and use of engineering controls and personal protective equipment).

Hazard, in relation to a substance, means the intrinsic property of that substance which has the potential to cause harm to the health of a person, and 'hazardous' shall be construed accordingly.

Health surveillance means the assessment of the state of health of an employee, as related to exposure to substances hazardous to health, and includes biological monitoring.

Inhalable dust means airborne material which is capable of entering the nose and mouth during breathing, as defined by BS EN 481 1993.

Maximum exposure limit for a substance hazardous to health means the maximum exposure limit approved by the HSC for that substance in relation to a specified reference period when calculated by a method approved by the HSC.

Micro-organism means a microbiological entity, cellular or non-cellular, which is capable of replication or of transferring genetic material.

Occupational exposure standard for a substance hazardous to health means the standard approved by the HSC for that substance in relation to the specified reference period when calculated by a method approved by the HSC.

Preparation means a mixture or solution of two or more substances.

Respirable dust means airborne material which is capable of penetrating the gas exchange region of the lung, as defined by BS EN 481 1993.

Risk, in relation to the exposure of an employee to a substance hazardous to health, means the likelihood that the potential for harm to the health of a person will be attained under the conditions of use and exposure and also the extent of that harm.

Substance means any natural or artificial substance whether in solid or liquid form or in the form of a gas or vapour (including micro-organism).

Substance hazardous to health means a substance (including any preparation):

(a) which is listed for supply in Part 1 of the Approved Supply List as dangerous for supply within the meaning of the CHIP Regulations and for which an indication of danger specified for the substance is very toxic, toxic, harmful, corrosive or irritant;

(b) for which the HSC has approved a maximum exposure limit or an occupational exposure standard;

(c) which is a biological agent;

(d) which is dust of any kind, except dust which is a substance within paragraph (a) or (b) above, when present at a concentration in air equal to or greater than:

   (i)   10 mg/m³, as a time-weighted average over an 8-hour period, of inhalable dust; or

   (ii) 4 mg/m³, as a time-weighted average over an 8-hour period, of respirable dust;

(e) which, not being a substance falling within sub-paragraphs (a) to (d), because of its chemical or toxicological properties and the way it is used or is present at the workplace creates a risk to health.

Regulation 3 – Duties under these regulations

Where a duty is placed by these regulations on an employer in respect of his employees, he shall, so far as is reasonably practicable, be under a like duty in respect of any other person, whether at work or not, who may be affected by the work carried out by the employer except that the duties of the employer:

- under regulations 11 (health surveillance) shall not extend to persons who are not his employees; and
- under regulations 10, 12(1) and (2) and 13 (which relate respectively to monitoring, information and training and dealing with accidents) shall not extend to persons who are not his employees, unless those persons are on the premises where the work is being carried out.

Regulation 4 – Prohibitions relating to certain substances

Those substances described in Column 1 of Schedule 2 are prohibited to the extent set out in the corresponding entry in Column 2 of that Schedule.

Regulation 5 – Application of regulations 6 to 13

Regulations 6 to 13 shall have effect with a view to protecting persons against a risk to their health, whether immediate or delayed, arising from exposure to substances hazardous to health except:

- where and to the extent that the following Regulations apply, namely:
  - the Coal Mines (Respirable Dust) Regulations 1975;
  - the Control of Lead at Work Regulations 2002;
  - the Control of Asbestos at Work Regulations 2002;
- where the substance is hazardous to health solely by virtue of its radio-active, explosive or flammable properties, or solely because it is at a high or low temperature or a high pressure;
- where the risk to health is a risk to the health of a person to whom the substance is administered in the course of his medical treatment.

Regulation 6 – Assessment of the risk to health created by work involving substances hazardous to health

An employer shall not carry out work which is liable to expose any employees to any substance hazardous to health unless he has:

- made a suitable and sufficient assessment of the risk created by that work to the health of those employees and of the steps that need to be taken to meet the requirements of these regulations; and
- implemented the steps referred to in sub-paragraph (a) above.

The risk assessment shall include consideration of:

- the hazardous properties of the substance;
- information on health effects provided by the supplier, including information contained in any relevant safety data sheet;
- the level, type and duration of exposure;
- the circumstances of the work, including the amount of the substance involved;
- activities, such as maintenance, where there is a potential for a high level of exposure;
- any relevant occupational exposure standard, maximum exposure limit or similar occupational exposure limit;
- the effect of preventive and control measures which have been or will be taken in accordance with regulation 7;
- the results of relevant health surveillance;
- the results of monitoring of exposure in accordance with regulation 10;
- in circumstances where the work will involve exposure to more than one substance hazardous to health, the risk presented by exposure to such substances in combination;
- the approved classification of any biological agent; and
- such additional information as the employer may need in order to complete the risk assessment.

The risk assessment shall be reviewed regularly and forthwith if:

- there is reason to suspect that the risk assessment is no longer valid;
- there has been a significant change in the work to which the assessment relates; or
- the results of any monitoring carried out in accordance with regulation 10 show it to be necessary,

and where, as a result of the review, changes to the risk assessment are required, those changes shall be made.

Where the employer employs five or more employees, he shall record:

- the significant findings of the risk assessment as soon as is practicable after the risk assessment is made; and
- the steps he has taken to meet the requirements of regulation 7.

Regulation 7 – Prevention or control of exposure to substances hazardous to health

Every employer shall ensure that the exposure of his employees to substances hazardous to health is either prevented or, where this is not reasonably practicable, adequately controlled.

In complying with his duty under the above paragraph, substitution shall by preference be undertaken, whereby the employer shall avoid, so far as is reasonably practicable, the use of a substance hazardous to health at the workplace by replacing it with a substance or process which, under the conditions of its use, either eliminates or reduces the risk to the health of his employees.

Where it is not reasonably practicable to prevent exposure to a substance hazardous to health, the employer shall comply with his duty of control under paragraph 1 by applying protection measures appropriate to the activity and consistent with the risk assessment, including in order of priority:

- the design and use of appropriate work processes, systems and engineering controls and the provision and use of suitable work equipment and materials;
- the control of exposure at source, including adequate ventilation systems and appropriate organisational measures; and
- where adequate control of exposure cannot be achieved by other means, the provision of suitable personal protective equipment in addition to the measures required by the above sub-paragraphs.

The measures referred to in the above paragraph shall include:

- arrangements for the safe handling, storage and transport of substances hazardous to health, and waste containing such substances, at the workplace;
- the adoption of suitable maintenance procedures;
- reducing, to the minimum required for the work concerned:

- the number of employees subject to exposure;
- the level and duration of exposure; and
- the quantity of substances hazardous to health present at the workplace;

- the control of the working environment, including appropriate general ventilation; and
- appropriate hygiene measures including adequate washing facilities.

Without prejudice to the generality of paragraph 1, where it is not reasonably practicable to prevent exposure to a carcinogen, the employer shall apply the following measures in addition to those required by paragraph 3 above:

- totally enclosing the process and handling systems, unless this is not reasonably practicable;
- the prohibition of eating, drinking and smoking in areas that may be contaminated by carcinogens;
- cleaning floors, walls and other surfaces at regular intervals and whenever necessary;
- designating those areas and installations which may be contaminated by carcinogens and using suitable and sufficient warning signs; and
- storing, handling and disposing of carcinogens safely, including closed and clearly labelled containers.

Without prejudice to the generality of paragraph 1, where it is not reasonably practicable to prevent exposure to a biological agent, the employer shall apply the following measures in addition to those required by paragraph 3 above:

- displaying suitable and sufficient warning signs, including the biohazard sign shown in Part IV of Schedule 3;
- specifying appropriate decontamination and disinfection procedures;
- instituting means for the safe collection, storage and disposal of contaminated waste, including the use of secure and identifiable containers, after suitable treatment where appropriate;
- testing, where it is necessary and technically possible, for the presence, outside the primary physical confinement, of biological agents used at work;
- specifying procedures for working with, and transporting at the workplace, a biological agent or material that may contain such an agent;
- where appropriate, making available effective vaccines for those employees who are not already immune to the biological agent to which they are exposed or are liable to be exposed;
- instituting hygiene measures compatible with the aim of preventing or reducing the accidental transfer or release of a biological agent from the workplace, including:

  - the provision of appropriate and adequate washing and toilet facilities; and
  - where appropriate, the prohibition of eating, drinking, smoking and the

application of cosmetics in working areas where there is a risk of contamination by biological agents;

and

• where there are human patients or animals which are, or are suspected of being, infected with a Group 3 or 4 biological agent, the employer shall select the most suitable control and containment measures from those listed in Part II of Schedule 3 with a view to controlling adequately the risk of infection.

Without prejudice to the generality of paragraph 1, where there is exposure to a substance for which a maximum exposure limit has been approved, control of exposure shall, so far as the inhalation of that substance is concerned, only be treated as being adequate if the level of exposure is reduced so far as is reasonably practicable and in any case below the maximum exposure limit.

Without prejudice to the generality of paragraph 1, where there is exposure to a substance for which an occupational exposure standard has been approved, control of exposure shall, so far as the inhalation of that substance is concerned, only be treated as being adequate if:

• that occupational exposure standard is not exceeded; or
• where that occupational exposure standard is exceeded, the employer identifies the reasons for the standard being exceeded and takes appropriate action to remedy the situation as soon as is reasonably practicable.

Personal protective equipment provided by an employer in accordance with this regulation shall be suitable for the purpose and shall:

• comply with any provision in the Personal Protective Equipment Regulations 2002 which is applicable to that item of personal protective equipment; or
• in the case of respiratory protective equipment, where no provision referred to in the above sub-paragraph applies, be of a type approved or shall conform to a standard approved, in either case, by the HSE.

Without prejudice to the provisions of this regulation, Schedule 3 shall have effect in relation to work with biological agents.

In this regulation *adequate* means adequate having regard only to the nature of the substance and the nature and degree of exposure to substances hazardous to health.

Regulation 8 – Use of control measures, etc.

Every employer who provides any control measure, other thing or facility in accordance with these Regulations shall take all reasonable steps to ensure that it is properly used or applied as the case may be.

Every employee shall make full and proper use of any control measure, other thing or facility provided in accordance with these Regulations and, where relevant, shall:

- take all reasonable steps to ensure it is returned after use to any accommodation provided for it; and
- if he discovers a defect therein, report it forthwith to his employer.

Regulation 9 – Maintenance, examination and testing of control measures

Every employer who provides any control measure to meet the requirements of regulation 7 shall ensure that, where relevant, it is maintained in efficient state, in efficient working order and in good repair and in a clean condition.

Where engineering controls are provided to meet the requirements of regulation 7, the employer shall ensure that thorough examination and testing of those controls is carried out:

- in the case of local exhaust ventilation plant, at least once every 14 months, or for local exhaust ventilation plant used in conjunction with a process specified in Column 1 of Schedule 4, at not more than the interval specified in the corresponding entry in Column 2 of that Schedule; or
- in any other case, at suitable intervals.

Where respiratory protective equipment (other than disposable respiratory protective equipment) is provided to meet the requirements of regulation 7, the employer shall ensure that thorough examination and, where appropriate, testing of that equipment is carried out at suitable intervals.

Every employer shall keep a suitable record of the examination and tests carried out in accordance with paragraphs 2 and 3 above and of repairs carried out as a result of those examinations and tests, and that record or a suitable summary thereof shall be kept for at least 5 years from the date on which it was made.

Every employer shall ensure that personal protective equipment, including protective clothing, is:

- properly stored in a well-defined place;
- checked at suitable intervals; and
- when discovered to be defective, repaired or replaced before further use.

Personal protective equipment which may be contaminated by a substance hazardous to health shall be removed on leaving the working area and kept apart from uncontaminated clothing and equipment.

The employer shall ensure that the equipment referred to in paragraph 6 is subsequently decontaminated and cleaned or, if necessary, destroyed.

Regulation 10 – Monitoring exposure at the workplace

Where the risk assessment indicates:

- it is a requisite for ensuring the maintenance of adequate control of the exposure of employees to substances hazardous to health; or

- it is otherwise requisite for protecting the health of employees,

the employer shall ensure the exposure of employees to substances hazardous to health is monitored in accordance with a suitable procedure.

Paragraph 1 shall not apply where the employer is able to demonstrate by another method of evaluation that the requirements of regulation 7(1) have been complied with.

The monitoring referred to in paragraph 1 shall take place:

- at regular intervals; and
- when any change occurs which may affect that exposure.

Where a substance or process is specified in Column 1 of Schedule 5, monitoring shall be carried out at least at the frequency specified in the corresponding entry in Column 2 of that Schedule.

The employer shall ensure that a suitable record of monitoring carried out for the purpose of this regulation is made and maintained and that that record or a suitable summary thereof is kept available:

- where the record is representative of the personal exposures of identifiable employees, for at least 40 years; or
- in any other case, for at least 5 years,

from the date of the last entry made in it
The employer shall:

- on reasonable notice being given, allow an employee access to his personal monitoring record;
- provide the HSE with copies of such monitoring records as the HSE may require; and
- if he ceases to trade, notify the HSE forthwith in writing and make available to the HSE all monitoring records kept by him.

Regulation 11 – Health surveillance

Where it is appropriate for the protection of the health of his employees who are, or are liable to be exposed to a substance hazardous to health, the employer shall ensure that such employees are under suitable health surveillance.

Health surveillance shall be treated as being appropriate where:

- the employee is exposed to one of the substances specified in Column 1 of Schedule 6 and is engaged in a process specified in Column 2 of that Schedule, and there is a reasonable likelihood that an identifiable disease or adverse health effect will result from that exposure; or
- the exposure of the employee to a substance hazardous to health is such that:

  - an identifiable disease or adverse health effect may be related to the exposure;
  - there is a reasonable likelihood that the disease or effect may occur

under the particular conditions of his work; and

- there are valid techniques for detecting indications of the disease or effect,

and the technique of investigation is of low risk to the employee.

The employer shall ensure that a health record, containing particulars approved by the HSE, in respect of each of his employees to whom paragraph 1 applies, is made and maintained and that that record or a copy thereof is kept available in a suitable form for at least 40 years from the date of the last entry made in it.

The employer shall:

- on reasonable notice being given, allow an employee access to his personal health record;
- provide the HSE with copies of such health records as the HSE may require; and
- if he ceases to trade, notify the HSE forthwith in writing and make available to the HSE all health records kept by him.

If an employee is exposed to a substance specified in Schedule 6 and is engaged in a process specified therein, the health surveillance required under paragraph 1 shall include medical surveillance under the supervision of a relevant doctor at intervals of not more than 12 months or at such shorter intervals as the relevant doctor may require.

Where an employee is subject to medical surveillance and a relevant doctor has certified by an entry in the health record of that employee that in his professional opinion that employee should not be engaged in work which exposes him to that substance or that he should only be so engaged under conditions specified in the record, the employer shall not permit the employee to be engaged in such work except in accordance with the conditions, if any, specified in the health record, unless that entry has been cancelled by a relevant doctor.

An employee to whom this regulation applies shall, when required by his employer and at the cost of the employer, present himself during his working hours for such health surveillance procedures as may be required for the purposes of paragraph 1 and, in the case of an employee who is subject to medical surveillance in accordance with paragraph 5, shall furnish the relevant doctor with such information concerning his health as the relevant doctor may reasonably require.

Where, as a result of health surveillance, an employee is found to have an identifiable disease or adverse health effect which is considered by a relevant doctor or other occupational health professional to be the result of exposure to a substance hazardous to health the employer of that employee shall:

- ensure that a suitably qualified person informs the employee accordingly and provides the employee with information and advice regarding further health surveillance;
- review the risk assessment;

- review any measures taken to comply with regulation 7, taking into account any advice given by a relevant doctor, occupational health professional or by the HSE;
- consider assigning the employee to alternative work where there is no risk of further exposure to that substance, taking into account any advice given by a relevant doctor or occupational health professional; and
- provide for a review of the health of any employee who has been similarly exposed, including a medical examination where such an examination is recommended by a relevant doctor, occupational health professional or by the HSE.

Where, for the purpose of carrying out his functions under these regulations, a relevant doctor requires to inspect any workplace or any record kept for the purposes of these regulations, the employer shall permit him to do so.

Where an employee or an employer is aggrieved by a decision recorded in the health record by a relevant doctor to suspend an employee from work which exposes him to a substance hazardous to health (or to impose conditions on such work), he may, by an application in writing to the HSE within 28 days of the date on which he was notified of the decision, apply for that decision to be reviewed in accordance with a procedure approved for the purposes of this paragraph by the HSC, and the result of that review shall be notified to the employee and employer and entered in the health record in accordance with the approved procedure.

Regulation 12 – Information, instruction and training for persons who may be exposed to substances hazardous to health

Every employer who undertakes work which is liable to expose an employee to a substance hazardous to health shall provide that employee with suitable and sufficient information, instruction and training.

Without prejudice to the generality of paragraph 1, the information, instruction and training provided under that paragraph shall include:

- details of the substances hazardous to health to which the employee is liable to be exposed including:
  - the names of those substances and the risk which they present to health;
  - any relevant occupational exposure standard, maximum exposure limit or similar occupational exposure limit;
  - access to any relevant safety data sheet; and
  - other legislative provisions which concern the hazardous properties of those substances;

- the significant findings of the risk assessment;
- the appropriate precautions and action to be taken by the employee in order to safeguard himself and other employees at the workplace;

- the results of any monitoring of exposure in accordance with regulation 10 and, in particular, in the case of a substance hazardous to health for which a maximum exposure limit has been approved, the employee or his representatives shall be informed forthwith, if the results of such monitoring show that the maximum exposure limit has been exceeded;
- the collective results of any health surveillance undertaken in accordance with regulation 11 in a form calculated to prevent those results from being identified as relating to a particular person; and
- where employees are working with a Group 4 biological agent or material that may contain such an agent, the provision of written instructions and, if appropriate, the display of notices which outline the procedures for handling such an agent or material.

The information, instruction and training required by paragraph 1 shall be:

- adapted to take account of significant changes in the type of work carried out or methods of work used by the employer; and
- provided in a manner appropriate to the level, type and duration of exposure identified by the risk assessment.

Every employer shall ensure that any person (whether or not his employee) who carries out work in connection with the employer's duties under these regulations has suitable and sufficient information, instruction and training.

Where containers and pipes for substances hazardous to health used at work are not marked in accordance with any relevant legislation listed in Schedule 7, the employer shall, without prejudice to any derogations provided for in that legislation, ensure that the contents of those containers and pipes, together with the nature of those contents and any associated hazards, are clearly identifiable.

Regulation 13 – Arrangements to deal with accidents, incidents and emergencies

Subject to paragraph 4 and without prejudice to the relevant provisions of the MHSWR, in order to protect the health of his employees from an accident, incident or emergency related to the presence of a substance hazardous to health at the workplace, the employer shall ensure that:

- procedures, including the provision of appropriate first aid facilities and relevant safety drills (which shall be tested at regular intervals), have been prepared which can be put into effect when such an event occurs;
- information on emergency arrangements, including:

  - details of relevant work hazards and hazard identification arrangements; and
  - specific hazards likely to arise at the time of an accident, incident or emergency,

is available;

- suitable warning and other communication systems are established to enable an appropriate response, including remedial actions and rescue operations, to be made immediately when such an event occurs.

The employer shall ensure that information on the procedures and systems and on the information required is:

- made available to relevant accident and emergency services to enable those services, whether internal or external to the workplace, to prepare their own response procedures and precautionary measures; and
- displayed at the workplace, if this is appropriate.

Subject to paragraph 4 below, in the event of an accident, incident or emergency related to the presence of a substance hazardous to health at the workplace, the employer shall ensure that:

- immediate steps are taken to:

  - mitigate the effects of the event;
  - restore the situation to normal; and
  - inform those of his employees who may be affected.

- only those persons who are essential for the carrying out of repairs and other necessary work are permitted in the affected area and they are provided with:

  - appropriate personal protective equipment; and
  - any necessary specialised safety equipment and plant;

which shall be used until the situation is restored to normal; and

- in the case of an incident or accident which has or may have resulted in the release of a biological agent which could cause severe human disease, as soon as practicable thereafter his employees or their representatives are informed of:
- the causes of the incident or accident; and
- the measures taken or to be taken to rectify the situation.

Paragraph 1 and, provided the substance hazardous to health is not a carcinogen or biological agent, paragraph 3 above shall not apply where:

- the results of the risk assessment show that, because of the quantity of each substance hazardous to health present at the workplace, there is only a slight risk to the health of employees; and
- the measures taken by the employer to comply with the duty under regulation 7(1) are sufficient to control that risk.

An employee shall report forthwith to his employer or to any other employee of that employer with specific responsibility for the health and safety of his fellow employees, any accident or incident which has or may have resulted in the release of a biological agent which could cause severe human disease.

Provisions relating to certain fumigations (regulation 14)

This regulation lays down specific requirements with respect to fumigations in which the fumigant used or intended to be used is hydrogen cyanide, phosphine or methyl bromide. It should be read in conjunction with Schedule 8 (Fumigations excepted from regulation 14) and Schedule 9 (Notification of certain fumigations).

Extension of the meaning of 'work' (regulation 19)

For the purposes of Part I of the HSWA, the meaning of 'work' shall be extended to include any activity involving the consignment, storage or use of a Group 2, 3 or 4 biological agent and the meaning of 'at work' shall be extended accordingly, and in that connection the references to employer in paragraphs 5 and 6 of Schedule 3 include references to any persons carrying out such an activity.

Defence (regulation 21)

Subject to regulation 21 of the MHSWR 1999, in any proceedings for an offence consisting of a contravention of these Regulations, it shall be a defence for any person to prove that he took all reasonable precautions and exercised all due diligence to avoid the commission of that offence.

4

Schedules

1. Other substances and processes to which the definition of 'carcinogen' exists
2. Prohibition of certain substances hazardous to health for certain purposes
3. Additional provisions relating to work with biological agents
4. Frequency of thorough examination and test of local exhaust ventilation plant used in certain processes
5. Specific substances and processes for which monitoring is required
6. Medical surveillance
7. Legislation concerned with the labelling of containers and pipes
8. Fumigations excepted from regulation 14
9. Notification of certain fumigations.

TRANSPORT OF DANGEROUS GOODS

Carriage of Dangerous Goods and Use of Transportable Pressure Equipment Regulations 2004

These regulations ('the Carriage Regulations') implement:

- the European Agreement concerning the International Carriage of Dangerous Goods by Road 2003 (ADR 2003) with a number of exceptions; and

- the Council Directive 96/49/EEC on the approximation of the laws of the Member States with regard to the transport of dangerous goods by rail (as amended); this Directive applies the Regulations concerning the International Carriage of Dangerous Goods by Rail (RID) which form Annex 1 to Appendix B to the Convention concerning International Carriage by Rail (RID 2003).

ADR 2003 sets out the conditions under which dangerous goods can be carried by road, with detailed requirements incorporated in two Annexes to the Agreement.

The Carriage Regulations also implement a further Directive on transportable pressure equipment and replace 12 previous statutory instruments that regulated the carriage of dangerous goods by road and rail in the United Kingdom. Detailed guidance on the requirements of ADR and the Carriage Regulations is incorporated in the HSE publication *Working with ADR: an introduction to the carriage of dangerous goods by road*.

The scope of the Carriage Regulations is extensive. Part II of the regulations details the requirements for RID and ADR, including training requirements, the safety obligations of those involved in the carriage of dangerous goods, the functions of safety advisers, reporting procedures for accidents and incidents, emergency plans for marshalling yards, classification of goods and the use, construction and testing of packagings and packages.

The regulations take account of national derogations agreed with the European Commission in respect of certain transport operations involving local journeys or the movement of dangerous goods in small quantities. They retain the mandatory Emergency Action Code system for placarding certain GB-registered vehicles on journeys in Great Britain. The regulations will be supported by a number of letters of authorisation allowing time-limited relaxations to some requirements.

The main changes include:

- revised limited quantity and load thresholds for transporting dangerous goods;
- packaging requirements which are based on RID or ADR;
- the transport of diesel, gas oil and heating oil come within the scope of the regulations;
- revised reporting requirements; and
- the phasing out of the competent person periodic inspection regime for transportable pressure equipment by 1 July 2006.

The functions of the 'competent authority' are specified in Part III, together with requirements for transportable pressure equipment in Part IV. Additional requirements to RID and ADR are dealt with in Part V.

There are 14 Schedules to the Carriage Regulations, which include detailed requirements for placards, marks and plate marking for carriage within Great Britain.

Classification of dangerous goods

There are nine classifications of dangerous goods, thus:

1. *Explosives*
These are sub-divided into six groups, Classes 1.1 to 1.3 being the most dangerous. Different substances are also assigned a 'compatibility group' which help a carrier determine which substances may be carried together.

2. *Gases*
This classification covers all compressed gases including those classified as 'flammable' and 'toxic', thus:

2.1 Flammable gases
2.2 Non-toxic, non-flammable compressed gases
2.3 Toxic gases.

3. *Flammable liquids*
These are liquids which give off a vapour and which burn in air if ignited.

4. *Other flammables*
These are sub-divided into:

4.1 Flammable solids which ignite and burn when heated
4.2 Spontaneously combustible (pyrophoric) which ignite when they come into contact with air
4.3 Dangerous when wet, which ignite or give off flammable gas when wet.

4

5. *Oxidisers*
These are sub-divided into:

5.1 Oxidising agents which produce oxygen helping to sustain any fire that may start
5.2 Organic peroxides – unstable compounds which give off oxygen and may heat up; can explode.

6. *Toxics*
These are sub-divided into:

6.1 Poisonous/toxic substances which may be inhaled, swallowed or absorbed through eyes
6.2 Infectious substances which may represent a biological hazard.

7. *Radioactives*
These are not sub-divided but there are three levels of danger. The most dangerous, 'Radioactive', should be accompanied by a sign with three red bars upon it.

8. *Corrosives*
These substances may cause serious burns.

9. *Miscellaneous*
This classification covers any other substance regarded as dangerous by the UN which does not readily fit into the classes above.

Warning diamonds

Each class or sub-class is assigned a warning diamond depicting the nature of the associated hazard. Warning diamonds are colour coded depending upon the nature of the hazard and incorporate a pictorial representation of the hazard within the diamond and its class.

THE DANGEROUS SUBSTANCES (NOTIFICATION AND MARKING OF SITES) REGULATIONS 1990

These regulations are principally directed at the safety of fire authority personnel attending incidents at premises containing large quantities of dangerous substances, and apply where the aggregate quantity of dangerous substances on a site is 25 tonnes or more.

The person in control of a site must notify both the HSE or the local authority and the fire authority prior to storage in excess of the aggregate quantity on site and mark the site accordingly.

All sites to which the regulations apply (with the exception of petrol filling stations) must display the appropriate signs, namely access signs and location signs, which must be kept clean and free from obstruction.

THE DANGEROUS SUBSTANCES AND EXPLOSIVE ATMOSPHERES REGULATIONS 2002

These regulations impose requirements for the purpose of eliminating or reducing risks to safety from fire, explosion or other events arising from the hazardous properties of a 'dangerous substance'! in connection with work.

The regulations implement:

- Council Directive 98/24/EC *on the protection of the health and safety of workers from the risks related to chemical agents at work,* so far as that Directive relates to safety; and
- Council Directive 99/92/EC *on minimum requirements for improving the safety and health protection of workers potentially at risk from explosive atmospheres.*

Important definitions

Dangerous substance means:

(a) a substance or preparation which meets the criteria in the approved classification and labelling guide for classification as a substance or preparation which is explosive, oxidising, extremely flammable, highly flammable or flammable, whether or not that substance or preparation is classified under the CHIP Regulations;

(b) a substance or preparation which because of its physico-chemical or chemical properties and the way it is used or is present at the workplace creates a risk, not being a substance or preparation falling within sub-paragraph (a) above; or

(c) any dust, whether in the form of solid particles or fibrous materials or otherwise, which can form an explosive mixture with air or an explosive atmosphere, not being a substance or preparation falling within sub-paragraphs (a) or (b) above.

Explosive atmosphere means a mixture, under atmospheric conditions, of air and one or more dangerous substances in the form of gases, vapours, mists or dusts in which, after ignition has occurred, combustion spreads to the entire unburned mixture.

Hazard means the physico-chemical or chemical property of a dangerous substance which has the potential to give rise to fire, explosion, or other events which can result in harmful physical effects of a kind similar to those which can be caused by fire or explosion, affecting the safety of a person.

Preparation means a mixture or solution of two or more substances.

Risk means the likelihood of a person's safety being affected by harmful physical effects being caused to him from fire, explosion or other events arising from the hazardous properties of a dangerous substance in connection with work and also the extent of that harm.

Safety data sheet means a safety data sheet within the meaning of regulation 5 of the CHIP Regulations.

Substance means any natural or artificial substance whether in solid or liquid form or in the form of a gas or vapour.

Work processes means all technical aspects of work involving dangerous substances and includes:

(a) appropriate technical means of supervision;

(b) connecting devices;

(c) control and protection systems;

(d) engineering controls and solutions;

(e) equipment;

(f) materials;

(g) machinery;

(h) plant;

(i) protective systems; and

(j) warning and other communication systems.

Regulation 4 – Duties under these regulations

1. Where a duty is placed by these regulations on an employer in respect of his employees, he shall, so far as is reasonably practicable, be under a like duty in respect of any other person, whether at work or not, who may be affected by the work carried on by the employer, except that:

(a) the duties of the employer under regulations 6(5)(f) and 7(5) (which relate, respectively, to the provision of suitable personal protective equipment and the provision of appropriate work clothing) shall not extend to the persons who are not his employees; and

(b) the duties of the employer under regulations 8 and 9 (which relate, respectively, to dealing with accidents and to provision of information, instruction and training) shall not extend to persons who are not his employees, unless those persons are at the workplace where the work is being carried on and subject to the following, namely, that, in relation to the application of regulation 9 to such persons, regulation 9 shall apply to the extent that is required by the nature and degree of that risk.

2. These regulations shall apply to a self-employed person as they apply to an employer and an employee and as if that self-employed person were both an employer and employee.

Regulation 5 – Risk assessment

1. Where a dangerous substance is or is liable to be present at the workplace, the employer shall make a suitable and sufficient assessment of the risks to his employees which arise from that substance.
2. The risk assessment shall include consideration of:

(a) the hazardous properties of the substance;

(b) information on safety provided by the supplier, including information contained in any relevant safety data sheet;

(c) the circumstances of the work including:
  (i) the work processes and substances used and their possible interactions;
  (ii) the amount of the substance involved;
  (iii) where work will involve more than one dangerous substance, the risk presented by such substances in combination; and
  (iv) the arrangements for the safe handling, storage and transport of dangerous substances and of waste containing dangerous substances;

(d) activities, such as maintenance, where there is the potential for a high level of risk;

(e) the effect of measures which have been or will be taken pursuant to these regulations;

(f) the likelihood that an explosive atmosphere will occur and its persistence;

(g) the likelihood that ignition sources, including electrostatic discharges, will be present and become active and effective;

(h) the scale of the anticipated effects of a fire or an explosion;

(i) any places which are or can be connected via openings to places in which explosive atmospheres may occur; and

(j) such additional information as the employer may need in order to complete the risk assessment.

3. The risk assessment shall be reviewed by the employer regularly so as to keep it up to date and particularly if:

(a) there is reason to suspect that the risk assessment is no longer valid; or

(b) there has been a significant change in the matters to which the risk assessment relates including when the workplace, work processes or organisation of the work undergoes significant changes, extensions or conversions;

and where, as a result of the review, changes to the risk assessment are required, those changes shall be made.

4. Where the employer employs five or more employees, the employer shall record the significant findings of the risk assessment as soon as is practicable after that assessment is made, including in particular:

(a) the measures which have been or will be taken by him pursuant to these regulations;

(b) sufficient information to show that the workplace and work processes are designed, operated and maintained with due regard for safety and that, in accordance with the Provision and Use of Work Equipment Regulations 1998, adequate arrangements have been made for the safe use of work equipment;

(c) where an explosive atmosphere may occur at the workplace, sufficient information to show:

(i) those places which have been classified into zones pursuant to regulation 7(1);

(ii) equipment which is required for, or helps to ensure, the safe operation of equipment located in places classified as hazardous pursuant to regulation 7(1);

(iii) that any verification of overall explosive safety required by regulation 7(4) has been carried out; and

(iv) the aim of any co-ordination required by regulation 11 and the measures and procedures for implementing it.

5. No new work activity involving a dangerous substance shall commence unless:

(a) an assessment has been made; and

(b) the measures required by these regulations have been implemented.

Regulation 6 – Elimination or reduction of risks from dangerous substances

1. Every employer shall ensure that risk is either eliminated or reduced so far as is reasonably practicable.
2. In complying with his duty under paragraph 1, substitution shall by preference be undertaken, whereby the employer shall avoid, so far as is reasonably practicable, the presence or use of a dangerous substance at the workplace by replacing it with a substance or process which either eliminates or reduces the risk.
3. Where it is not reasonably practicable to eliminate risk, the employer shall, so far as is reasonably practicable, apply measures, consistent with the risk assessment and appropriate to the nature of the activity or operation:

    (a) to control risks, including the measures specified in paragraph 4; and

    (b) to mitigate the detrimental effects of a fire or explosion or the other harmful physical effects arising from dangerous substances, including the measures specified in paragraph 5.

4. The following measures are, in order of priority, those specified for the purposes of paragraph 3(a):

    (a) the reduction of the quantity of dangerous substances to a minimum;

    (b) the avoidance or minimising of the release of a dangerous substance;

    (c) the control of the release of a dangerous substance at source;

    (d) the prevention of the formation of an explosive atmosphere, including the application of appropriate ventilation;

    (e) ensuring that any release of a dangerous substance which may give rise to risk is suitably collected, safely contained, removed to a safe place, or otherwise rendered safe, as appropriate;

    (f) the avoidance of:

        (i) ignition sources including electrostatic discharges; and

        (ii) adverse conditions which could cause dangerous substances to give rise to harmful physical effects; and

    (g) the segregation of incompatible dangerous substances.

5. The following measures are those specified for the purposes of paragraph 3(b):

    (a) the reduction to a minimum of the number of employees exposed;

    (b) the avoidance of the propagation of fires or explosions;

    (c) the provision of explosion pressure relief arrangements;

    (d) the provision of explosion suppression equipment;

    (e) the provision of plant which is constructed so as to withstand the pressure likely to be produced by an explosion; and

    (f) the provision of suitable personal protective equipment.

6. The employer shall arrange for the safe handling, storage and transport of dangerous substances and waste containing dangerous substances.
7. The employer shall ensure that any conditions necessary pursuant to these regulations for ensuring the elimination or reduction of risk are maintained.
8. The employer shall, so far as is reasonably practicable, take the general safety measures specified in Schedule 1, subject to those measures being consistent with the risk assessment and appropriate to the nature of the activity or operation.

Regulation 7 – Places where explosive atmospheres may occur

1. Every employer shall classify places at the workplace where an explosive atmosphere may occur into hazardous or non-hazardous places in accordance with paragraph 1 of Schedule 2 and shall classify those places so classified as hazardous into zones in accordance with paragraph 2 of that Schedule; and that Schedule shall have effect subject to the notes at the end of that Schedule.
2. The employer shall ensure that the requirements specified in Schedule 3 are applied to equipment and protective systems in the places classified as hazardous.
3. Where necessary, places classified as hazardous shall be marked by the employer with signs at their points of entry in accordance with Schedule 4.
4. Before a workplace containing places classified as hazardous is used for the first time, the employer shall ensure that its overall explosion safety is verified by a person who is competent in the field of explosion protection as a result of his experience and or professional training or both.
5. The employer shall ensure that appropriate work clothing which does not give rise to electrostatic discharges is provided for use in places classified as hazardous.

Regulation 8 – Arrangements to deal with accidents, incidents and emergencies

1. Subject to paragraph 4, in order to protect the safety of his employees from an accident, incident or emergency related to the presence of a dangerous substance at the workplace, the employer shall ensure that:

   (a) procedures, including the provision of appropriate first aid facilities and relevant safety drills (which shall be tested at regular intervals) have been prepared which can be put into effect when such an event occurs;

   (b) information on emergency arrangements, including:

      (i) details of relevant work hazards and hazard identification arrangements; and

    (ii) specific hazards likely to arise at the time of an accident, incident or emergency

is available;

(c) suitable warning and other communication systems are established to enable an appropriate response, including remedial actions and rescue operations, to be made immediately when such an event occurs;

(d) where necessary, before any explosion conditions are reached, visual, or audible, warnings are given and employees withdrawn; and

(e) where the risk assessment indicates it is necessary, escape facilities are provided and maintained to ensure that, in the event of danger, employees can leave endangered places promptly and safely.

2. Subject to paragraph 4, the employer shall ensure that information on the matters referred to in paragraph 1(a), (c) to (e) and the information required by paragraph 1(b) is:

(a) made available to relevant accident and emergency services to enable those services, whether internal or external to the workplace, to prepare their own response procedures and precautionary measures; and

(b) displayed at the workplace, unless the results of the risk assessment make this unnecessary.

3. Subject to paragraph 4, in the event of an accident, incident or emergency related to the presence of a dangerous substance at the workplace, the employer shall ensure that:

(a) immediate steps are taken to:
    (i) mitigate the effects of the event;
    (ii) restore the situation to normal; and
    (iii) inform those of his employees who may be affected; and

(b) only those persons who are essential for the carrying out of repairs and other necessary work are permitted in the affected area and that they are provided with:
    (i) appropriate personal protective equipment and protective clothing; and
    (ii) any necessary specialised safety equipment and plant,
    which shall be used until the situation is restored to normal.

4. Paragraphs 1 to 3 shall not apply where:

(a) the results of the risk assessment show that, because of the quantity of each dangerous substance at the workplace, there is only a slight risk to employees; and

(b) the measures taken by the employer to comply with his duty under regulation 6(1) are sufficient to control that risk.

Regulation 9 – Information, instruction and training

1. Where a dangerous substance is present at the workplace, the employer shall provide his employees with:

   (a) suitable and sufficient information, instruction and training on the appropriate precautions and actions to be taken by the employee in order to safeguard himself and other employees at the workplace:

   (b) the details of any such substance including:

      (i) the name of the substance and the risk which it presents;

      (ii) access to any relevant safety data sheet; and

      (iii) legislative provisions which concern the hazardous properties of the substance; and

   (c) the significant findings of the risk assessment.

2. The information, instruction and training required shall be:

   (a) adapted to take account of significant changes in the type of work carried out or the methods of work used by the employer; and

   (b) provided in a manner appropriate to the risk assessment.

Regulation 10 – Identification of hazardous contents of containers and pipes

Where containers and pipes used at work for dangerous substances are not marked in accordance with relevant requirements of the legislation listed in Schedule 5, the employer shall, subject to any derogations provided for in that legislation, ensure that the contents of those containers and pipes, together with the nature of those contents and any associated hazards, are clearly identifiable.

Regulation 11 – Duty of co-ordination

Where two or more employers share the same workplace (whether on a temporary or permanent basis) where an explosive atmosphere may occur, the employer responsible for the workplace shall co-ordinate the implementation of all the measures required by these regulations to be taken to protect employees from any risk from the explosive atmosphere.

Schedule 1 (regulation 6(8))

General safety measures

1. The following measures are those specified for the purposes of regulation 6(8).

Workplace and work processes

2. Ensuring that the workplace is designed, constructed and maintained so as to reduce risk.

3. Designing, constructing, assembling, installing, providing and using suitable work processes so as to reduce risk.

4. Maintaining work processes in an efficient state, in efficient working order and in good repair.

5. Ensuring that equipment and protective systems meet the following requirements:

   (a) where power failure can give rise to the spread of additional risk, equipment and protective systems must be able to be maintained in a safe state of operation independently of the rest of the plant in the event of power failure;

   (b) means for manual override must be possible, operated by employees competent to do so, for shutting down equipment and protective systems incorporated within automatic processes which deviate from the intended operating conditions, provided that the provision or use of such means does not compromise safety;

   (c) on operation of emergency shutdown, accumulated energy must be dissipated as quickly and as safely as possible or isolated so that it no longer constitutes a hazard; and

   (d) necessary measures must be taken to prevent confusion between connecting devices.

Organisational measures

6. The application of appropriate systems of work including:

   (a) the issuing of written instructions for the carrying out of the work; and

   (b) a system of permits to work with such permits being issued by a person with responsibility for this function prior to the commencement of the work concerned,

   where the work is carried out in hazardous places or involves hazardous activities.

Schedule 2 (regulation 7(1))

Classification of places where explosive atmospheres may occur (which substantially reproduces the provisions of Annex 1 of Council Directive 99/92/EC)

1. Places where explosive atmospheres may occur

A place in which an explosive atmosphere may occur in such quantities as to require special precautions to protect the health and safety of the workers concerned is deemed to be hazardous within the meaning of these Regulations.

A place in which an explosive atmosphere is not expected to occur in such quantities as to require special precautions is deemed to be non-hazardous within the meaning of these Regulations.

2. Classification of hazardous places

Hazardous places are classified in terms of zones on the basis of the frequency and duration of the occurrence of an explosive atmosphere.

Zone 0   A place in which an explosive atmosphere consisting of a mixture with air of dangerous substances in the form of gas, vapour or mist is present continuously or for long periods or frequently.

Zone 1   A place in which an explosive atmosphere consisting of a mixture with air of dangerous substances in the form of gas, vapour or mist is likely to occur in normal operation occasionally.

Zone 2   A place in which an explosive atmosphere consisting of a mixture with air of dangerous substances in the form of gas, vapour or mist is not likely to occur in normal operation but, if it does occur, will persist for a short period only.

Zone 20   A place in which an explosive atmosphere in the form of a cloud of combustible dust in air is present continuously, or for long periods or frequently.

Zone 21   A place in which an explosive atmosphere in the form of a cloud of combustible dust in air is likely to occur in normal operation occasionally.

Zone 22   A place in which an explosive atmosphere in the form of a cloud of combustible dust in air is not likely to occur in normal operation but, if it does occur, will persist for a short period only.

Notes

1. Layers, deposits and heaps of combustible dust must be considered as any other source which can form an explosive atmosphere.
2. 'Normal operation' means the situation when installations are used within their design parameters.

SCHEDULE 3 (regulation 7(2))

Criteria for the selection of equipment and protective systems

1. Equipment and protective systems for all places in which explosive atmospheres may occur must be selected on the basis of the requirements set out in the Equipment and Protective Systems Intended for Use in Potentially Explosive Atmospheres Regulations 1996 unless the risk assessment finds otherwise.
2. In particular, the following categories of equipment must be used in the zones indicated, provided they are suitable for gases, vapours, mists, dusts or mists and dusts, as appropriate:

   – in zone 0 or zone 20, category 1 equipment
   – in zone 1 or zone 21, category 1 or 2 equipment
   – in zone 2 or zone 22, category 1, 2 or 3 equipment.

3. For the purposes of this Schedule and regulations 7(2) and 17(1) –

   (a) *equipment* means machines, apparatus, fixed or mobile devices, control components and instrumentation thereof and detection or prevention systems which, separately or jointly, are intended for the generation, transfer, storage, measurement, control and conversion of energy and the processing of material, as the case may be, and which are capable of causing an explosion through their own potential sources of ignition;

(b) *protective systems* means devices other than components of equipment which are intended to halt incipient explosions immediately or limit the effective range of an explosion or both, as the case may be, and which systems are separately placed on the market for use as autonomous systems;

(c) *devices* means safety devices, controlling devices and regulating devices intended for use outside potentially explosive atmospheres but required for or contributing to the safe functioning of equipment and protective systems with respect to the risks of explosion;

(d) *component* means any item essential to the safe functioning of equipment and protective systems but with no autonomous function; and

(e) *potentially explosive atmosphere* means an atmosphere which could become explosive due to local and operational conditions.

SCHEDULE 4 (regulation 7(3))

(This schedule substantially reproduces the provisions of Annex III of Council Directive 99/92/EC)

Warning sign for places where explosive atmospheres may occur

Distinctive features:

(a) triangular shape;

(b) black letters on a yellow background with black edging (the yellow part to take up at least 50% of the area of the sign).

THE ELECTRICITY AT WORK REGULATIONS 1989

These regulations are made under the HSWA and apply to all work associated with electricity. The purpose of the regulations is to require that precautions be taken against the risk of death or personal injury from electricity when it is involved in work activities.

The regulations impose duties on duty holders (see regulation 3) in respect of systems, electrical equipment and conductors, and work activities on or near electrical equipment. They replace the 1908 and 1944 regulations and extend to all premises.

The regulations also establish general principles for electrical safety rather than state detailed requirements. Further detailed advice is given in the supporting Memorandum of Guidance and other authoritative documents, such as the IEE Wiring Regulations, British and European standards and HSE publications.

As the regulations state principles of electrical safety in a form that may be applied to any work activity having a bearing on electrical safety, they apply to *all* electrical equipment and systems, wherever manufactured, purchased, installed or taken into use, even if this predates them. If, however, electrical equipment *does* predate the regulations, it does not, of itself, mean that the continued use of such equipment is prohibited. The equipment may continue to be used, provided the requirements of the regulations can continue to be satisfied. In other words, the equipment need only be replaced if it becomes unsafe or there is a risk of injury when it is used.

The duties in some of the regulations are subject to the qualifying term 'reasonably practicable'. Where qualifying terms are absent, the requirement is said to be *absolute*, which means that it must be met, regardless of cost or any other consideration.

Regulation 29 provides a defence for a duty holder who can establish that they took *all reasonable steps and exercised all due diligence* to avoid committing an offence under certain regulations.

Part I: Introduction

Interpretation (regulation 2)

This regulation incorporates a number of important definitions, including:

- *circuit conductor*, meaning any conductor in a system that is intended to carry electric current in normal conditions, or to be energised in normal conditions, and includes a combined neutral and earth conductor, but does not include a conductor provided solely to perform a protective function by connection to earth or other reference point
- *conductor* means a conductor of electrical energy
- *danger* means risk of injury
- *electrical equipment* includes anything used, intended to be used or installed for use, to generate, provide, transmit, transform, rectify, convert, conduct, distribute, control, store, measure or use electrical energy

- *injury* means death or personal injury from electric shock, electric burns, electrical explosion or arcing, or from fire or explosion initiated by electrical energy, where any such death or injury is associated with the generation, provision, transmission, rectification, conversion, conduction, distribution, control, storage, measurement or use of electrical energy
- *system* means any electrical system in which all the electrical equipment is, or may be, electrically connected to a common source of electrical energy, and includes such source and such equipment.

Those on whom duties are imposed (regulation 3)

It shall be the duty of every:

- employer and self-employed person to comply with the provisions of these regulations, in so far as they relate to matters that are within their control
- manager of a mine or quarry to ensure that all requirements or prohibitions imposed by or under these regulations are complied with in so far as they related to the mine or quarry or part of a quarry of which they are the manager and to matters that are within their control

It shall be the duty of every employee while at work to:

- co-operate with their employer so far as is necessary to enable any duty placed on that employer by the provisions of these regulations to be complied with
- comply with the provisions of these regulations in so far as they relate to matters that are within their control.

Part II: General

Systems, work activities and protective equipment (regulation 4)

All systems shall, at all times, be of such construction as to prevent danger, so far as is reasonably practicable.

As may be necessary to prevent danger, all systems shall be maintained so as to prevent, so far as is reasonably practicable such danger.

Every work activity, including operation, use and maintenance of a system and work near a system, shall be carried out in such a manner as not to give rise, so far as is reasonably practicable, to danger.

Any equipment provided under these regulations for the purpose of protecting those at work on or near electrical equipment shall be suitable for this purpose, be maintained in a condition suitable for this use and be properly used.

The strength and capability of electrical equipment (regulation 5)

No electrical equipment shall be put into use where its strength and capability may be exceeded in such a way as to give rise to danger.

Adverse or hazardous environments (regulation 6)

Electrical equipment that may, reasonably foreseeably, be exposed to:

- mechanical danger
- the effects of the weather, natural hazards, temperature or pressure
- the effects of wet, dirty, dusty or corrosive conditions
- any flammable or explosive substance including dusts, vapours or gases,
  shall be of such construction or, as necessary, protected so as to prevent,
  so far as is reasonably practicable, danger arising from such exposure.

Insulation, protection and placing of conductors (regulation 7)

All conductors in a system that may give rise to danger shall either:

- be suitably covered with insulating material and, as necessary, protected
  so as to prevent danger, so far as is reasonably practicable
- have such precautions taken in respect of them (including, where appropriate, their being suitably placed) as will prevent danger, so far as is reasonably practicable.

Earthing or other suitable precautions (regulation 8)

Precautions shall be taken, either by earthing or other suitable means, to prevent danger arising when any conductor (other than a circuit conductor) that may, reasonably foreseeably, become charged as a result of either the use of the system or a fault in a system becomes so charged. Also, for the purposes of complying with this regulation, a conductor shall be regarded as earthed when it is connected to the general mass of earth by conductors of sufficient strength and current-carrying capability to discharge electrical energy to earth.

Integrity of referenced conductors (regulation 9)

If a circuit conductor is connected to earth or to any other reference point, nothing that might reasonably be expected to give rise to danger by breaking the electrical continuity or introducing high impedance shall be placed in that conductor unless suitable precautions are taken to prevent danger.

Connections (regulation 10)

Where necessary to prevent danger, every joint and connection in a system shall be mechanically and electrically suitable for use.

The means for protecting from excess current (regulation 11)

Efficient means, suitably located, shall be provided for protecting every part of a system from excess current as may be necessary to prevent danger.

The means for cutting off the supply and for isolation (regulation 12)

Where necessary, to prevent danger, suitable means (including, where appropriate, methods of identifying circuits) shall be available for:

- cutting off the supply of electrical energy to any electrical equipment
- the isolation of any electrical equipment.

Isolation here means the disconnection and separation of the electrical equipment from every source of electrical energy in such a way that this disconnection and separation is secure.

These means shall not apply to electrical equipment that is itself a source of electrical energy, but, in such a case, as is necessary, precautions shall be taken to prevent danger, so far as is reasonably practicable.

Precautions for work on equipment made dead (regulation 13)

Adequate precautions shall be taken to prevent electrical equipment that has been made dead in order to prevent danger while work is carried out on or near that equipment from becoming electrically charged during that work if danger may arise as a result.

Work on or near live conductors (regulation 14)

No one shall be engaged in any work activity on or so near any live conductor (other than one suitably covered with insulating material so as to prevent danger) such that danger may arise unless:

- it is unreasonable in all circumstances for it to be dead
- it is reasonable in all circumstances for the person concerned to be at work on or near it while it is live
- suitable precautions (including, where necessary, the provision of suitable protective equipment) are taken to prevent injury.

Working space, access and lighting (regulation 15)

For the purposes of enabling injury to be prevented, adequate working space, adequate means of access and adequate lighting shall be provided at all electrical equipment on which, or near which, work is being done in circumstances that may give rise to danger.

People to be competent to prevent danger and injury (regulation 16)

No one shall be engaged in any work activity where technical knowledge or experience is necessary to prevent danger or, where appropriate, injury, unless they possess such knowledge or experience, or are under such degree of supervision as may be appropriate, having regard to the nature of the work.

Part III: Regulations applying only to mines

Part III of the regulations makes specific provision for electrical safety in mines.

Part IV: Miscellaneous and general

Regulation 29 defence

In any proceedings for an offence consisting of a contravention of regulations 4(4), 5, 8, 9, 10, 11, 12, 13, 14, 15, 16 or 25, it shall be a defence for any person to prove that they *took all reasonable precautions and exercised all due diligence* to avoid the commission of that offence.

IEE Regulations (the wiring regulations)

These regulations are produced by the Institution of Electrical Engineers as a code of best practice. They establish safety standards for people who design and work with electrical installations, e.g. designers, installers, erectors and testers of both permanent and temporary installations.

Designed to protect employees and people generally from the hazards associated with electricity, the regulations apply only to installations operating at up to 1000 volts AC. They emphasise the needs for sound standards of workmanship and the competence of operators, and the use of suitable and correct materials to the relevant standard/specification.

The regulations define:

- extra-low voltage – not exceeding 50 volts AC or 120 volts DC;
- low voltage – exceeding extra low but not over 1000 volts AC or 1500 volts DC between conductors, or 600 volts AC or 900 volts DC between conductors and earth.

They also provide guidance on factors which installation designers must consider including the purpose of the installation, any external influences, maintainability, maximum demand, the number of conductors, earthing arrangements, type of supply and possible future extensions to the installation.

Extensive guidance is provided on methods of protection against the hazards of shock, fire, burns and excess current.

The selection and erection of equipment to ensure fitness for the purpose is an important feature of the regulations.

Procedures for inspection and testing to ensure an installation is safe, including the necessity for undertaking tests in a certain order to identify faults before an installation is energised by the supply are further incorporated in the regulations.

THE ELECTRICAL EQUIPMENT (SAFETY) REGULATIONS 1994

These regulations consolidate with amendments the Low Voltage Electrical Equipment (Safety) Regulations 1989 and implement the requirements of Council Directive No. 73/23/EEC on the harmonisation of the laws of Member States relating to electrical equipment designed for use within certain voltage limits, as amended by Council Directive No. 93/68/EEC (the 'CE marking' Directive).

The regulations, which are made under the Consumer Protection Act 1987, require:

- the affixing to all electrical equipment (or its packaging, instruction sheet or guarantee certificate) of the CE marking by way of confirmation that the equipment satisfies all the requirements of the regulations which relate to it;
- that a written declaration of conformity comprising certain information relating to the electrical equipment be drawn up and kept available for inspection by an enforcement authority for a period of ten years after manufacture of electrical equipment of that model has ceased;
- that certain technical documentation relating to electrical equipment be compiled and kept available for inspection by an enforcement authority for a period of ten years after manufacture of electrical equipment of that model has ceased;

and provide:

- for the issuing of a compliance notice in respect of electrical equipment to which the CE marking has been unduly affixed, save where the electrical equipment in question is likely to damage the health or safety of any person;
- that secondhand electrical equipment or equipment which is hired out (save where its first hiring out is its first supply to an end user) must be safe but need not comply with the requirements of the regulations relating to CE marking, the EC declaration of conformity and internal production control;
- that electrical equipment which satisfies the safety provisions of harmonised standards (or where appropriate international or national safety provisions) shall be taken to comply with the safety requirements of the regulations unless there are reasonable grounds for suspecting that it does not so comply;
- that the HSE may make arrangements for the enforcement of the regulations in relation to equipment for use in the workplace; and
- that a person who supplies electrical equipment which does not bear the CE marking shall, if required, provide certain information to an enforcement authority.

THE ENVIRONMENTAL PROTECTION (DUTY OF CARE) REGULATIONS 1991

The EPA brought in a statutory duty of care in relation to waste with effect from 1 April 1992. This duty of care is set out in section 34 of the Act and is supplemented by the Environmental Protection (Duty of Care) Regulations 1991 and a Code of Practice.

Fundamentally, any person who fails to comply with the duty under section 34 of the EPA or with the regulations commits a criminal offence. It is not necessary for any environmental damage to have been caused. All that is required is that there has been a breach of the duty of care. On summary conviction, a breach of this duty of care can lead to a fine of £5000 or, on indictment, an unlimited fine.

The duty of care

Section 34(1) provides that the duty of care applies to any person who imports, produces, carries, keeps, treats or disposes of controlled waste or, as a broker, has control of such waste, i.e. a 'waste holder'.) A broker is a person who may exercise control over waste, but may not necessarily hold it. Any person bound by the duty must take all such measures applicable to him in that capacity as are reasonable in the circumstances:

- to prevent any other person committing the offences in section 33, e.g. unlawful deposit of waste on land
- to prevent the escape of the waste from his control or that of any other person
- to ensure that if the waste is transferred, it is transferred only to an authorised person or to a person for authorised transport purposes
- when waste is transferred, to make sure that it is accompanied by a written description of the waste which will enable other persons to avoid a contravention of section 33 of the Act and to comply with a duty under section 34(1)(b) to prevent the escape of waste.

The regulations

These regulations (made under section 34(5) of the EPA) established a system of transfer notes and record keeping of waste transfers to help waste holders comply with their elements of the duty of care.

The aim of this system is to enable the *Waste Regulation Authority* (WRA) to keep track of the movement of all wastes within their area and a written description is needed to be provided when waste is transferred. The regulations place responsibilities on the transferor and transferee of waste to keep records of all waste transfers. On completion of the transfer of the controlled waste (household, commercial or industrial waste), both the giver and receiver must complete and sign a transfer note. The transfer note must incorporate the following details:

- identification of the waste
- quantity
- whether it is loose or in a container at the time of transfer
- place and time of transfer
- name and address of both transferor and transferee
- whether the transferor is the producer or importer
- if the transferee is authorised for transport purposes.

All parties involved in the transfer must keep a copy of the transfer note and the written description for at least two years. The WRA may serve a notice demanding copies of transfer notes which must be supplied within seven days.

The Code of Practice *Waste Management, The Duty of Care* recommends a series of steps which would normally be sufficient to satisfy the requirements of this duty of care. This Code of Practice is admissible in evidence and, if any provision of the Code appears to the court to be relevant to any question arising in the proceedings, it shall be taken into account in determining that question.

Penalties

Any person who fails to comply with the duty of care or the documentation requirements laid down in the regulations commits a criminal offence and liable on summary conviction to a fine not exceeding £5000. On indictment, the Crown Court can impose an unlimited fine. Directors and senior managers may also be personally liable for a breach of section 34 of the EPA.

Breach of these regulations does not give rise to civil liability, although damages may be available through common law actions such as nuisance.

FIRE AND RESCUE AUTHORITIES

The Fire and Rescue Services Act 2004, which revoked the Fire Services Act 1947, created Fire and Rescue Authorities (FRAs). Under the Act, a FRA has the power 'to do anything which is calculated to facilitate, or is conducive or incidental to, the discharge of any of its functions'.

Functions of FRAs

FRAs have certain 'core' functions with respect to making provision for fire safety and fire-fighting and with dealing with road traffic accidents and emergencies. Other functions include providing directions relating to particular fires and emergencies and responding to other eventualities. They are enabled to run training courses and to charge for their services in certain cases. The FRA can further act as a negotiating body and make provision for pensions.

The FRA has a duty to secure water supply and to liaise with water undertakers to ensure supplies of water, particularly in an emergency.

Powers of authorised employees of a FRA

Enforcement powers of FRAs and their officers are incorporated in Part VI of the Act.

In an emergency, an employee authorised in writing by the FRA may do 'anything he reasonably believes to be necessary':

- for the purpose of extinguishing or preventing fire or protecting life or property;
- for the purpose of rescuing people or protecting them from serious harm in the event of a road traffic accident;
- for dealing with any other type of emergency;
- for the purpose of preventing or limiting damage to property resulting from the above.

In particular, an authorised employee may:

- enter a premises or place, if necessary by force, without the consent of the owner or occupier;
- move or break into a vehicle without the consent of the owner;
- close a highway;
- stop and regulate traffic; and
- restrict the access of persons to a premises or place.

A person commits an offence if without reasonable excuse he obstructs or interferes with an employee of a FRA taking authorised action.

Obtaining information and investigating fires (section 45)

Authorised employees have extensive powers with respect to obtaining information and investigating fires. They may at any reasonable time enter premises:

(a) for the purpose of obtaining of information; and

(b) if there has been a fire in the premises, for the purpose of investigating the causes of the fire and why it progressed as it did.

An authorised officer, that is, an employee of a FRA who is authorised in writing by the authority, may apply to a JP for a warrant to enter premises for the purpose of the above if unable to do so or considers that he is likely to be unable to do so, otherwise than by force. (Section 45)

Supplementary powers of authorised officers (section 46)

If an authorised officer exercises a power of entry under section 45(1)(a) above, he may:

(a) take with him any other persons, and any equipment, that he considers necessary;

(b) require any person present on the premises to provide him with any facilities, information, documents or records, or other assistance, that he may reasonably request.

If an authorised officer exercises a power of entry under section 45(1)(b) he may:

(a) take with him any other persons, and any equipment, that he considers necessary;

(b) inspect and copy any documents or records on the premises or remove them from the premises;

(c) carry out any inspections, measurements and tests in relation to the premises, or to any article or substance found on the premises, that he considers necessary;

(d) take samples of any article or substance found on the premises, but not so as to destroy it or damage it unless it is necessary to do so for the purpose of the investigation;

(e) dismantle any article found on the premises, but not so as to destroy it or damage it unless it is necessary to do so for the purpose of the investigation;

(f) take possession of any article or substance found on the premises and to detain it for as long as is necessary for any of these purposes:

  (i) to examine it and to do anything he has power to do under paragraph (c) or (e);

  (ii) to ensure that it is not tampered with before his examination of it is completed;

  (iii) to ensure that it is available for use as evidence in proceedings for an offence relevant to the investigation;

(g) require a person present on the premises to provide him with any facilities, information, documents or records, or other assistance, that he may reasonably request.

If an authorised officer exercises the powers in subsection 2(d) above he must:

(a) leave a notice at the premises (either with a responsible person or if that is impracticable fixed in a prominent position) giving particulars of the article or substance and stating that he has taken a sample of it; and

(b) if it is practicable to do so, give a responsible person at the premises a portion of the sample marked in a manner sufficient to identify it.

If an authorised officer exercises the power in subsection 2(f) above he must leave a notice at the premises (either with a responsible person or if that is impracticable fixed in a prominent position) giving particulars of the article or substance and stating that he has taken possession of it.

False alarms (section 49)

A person commits an offence if he knowingly gives or causes to be given a false alarm of fire to a person acting on behalf of a FRA.

THE FIRE CERTIFICATES (SPECIAL PREMISES) REGULATIONS 1976

Under normal circumstances a fire authority is empowered to issue a premises with a fire certificate under the Fire Precautions Act 1971. However, in certain high-risk situations, this certificate is issued by the HSE. These regulations make special provisions for certain premises or 'special premises' storing quantities of hazardous substances and the procedures for fire certification by the HSE.

Fourteen 'special premises' are listed in Schedule 1 to the regulations including those which:

- use over 50 tonnes of highly flammable liquid in the manufacturing process;
- manufacture 50 tonnes or more of expanded cellular plastics per week;
- can store 100 tonnes or more of LPG unless required as a fuel or for the heat treatment of metals;
- can store 100 tonnes or more of liquefied natural gas unless kept solely as a fuel;
- can store 100 tonnes of liquefied flammable gas consisting predominantly of methyl acetylene unless kept solely as a fuel, etc.

Temporary buildings on construction sites are also covered unless the following conditions are satisfied, namely:

- not more than 20 persons are employed at any time in the building;
- not more than 10 persons are employed at any time elsewhere than on the ground floor;
- no explosive or highly flammable materials are stored or used in or under the building;
- means of escape and apparatus for fire fighting are provided.

The person having control of the premises, i.e. the 'responsible person' must make formal application for a fire certificate, taking into account the particulars listed in Schedule 2, i.e. nature of processes, nature and quantity of explosive or highly flammable substances, and maximum number of people likely to be in any building.

In certain cases, an applicant may be required to furnish plans of the premises and those of adjacent premises.

On receipt of an application for a fire certificate, the HSE must arrange for an inspection of the site, to ensure satisfactory means of escape and means for fighting fire, prior to the issue of a certificate.

An occupier must keep a copy of the certificate available for inspection and display a notice at a suitable place on the premises stating that a certificate has been issued, where it may be inspected and the date of posting of the notice.

Any material changes to a building or process likely to affect matters specified in the fire certificate must be notified to the HSE prior to the implementation of such changes.

It is an offence to undertake work on such premises without a fire certificate and to fail to comply with the conditions attached to same.

THE FIRE PRECAUTIONS (WORKPLACE) REGULATIONS 1997

These regulations apply to all workplaces (see definition of 'workplace' under the WHSWR) other than 'excepted workplaces'.

Moreover, these regulations do not stand on their own, extending duties under existing health and safety and fire safety legislation. As such, they must be read in conjunction with a range of duties on employers and controllers of workplaces under the HSWA, FPA, FSSPSA, WHSWR and, in particular, the MHSWR. The majority of the duties are of an absolute nature.

Part II (Regulations 3 to 6) deals with the general application of the regulations with regard to fire precautions in the workplace.

Application of Part II (regulation 3)

Regulation 3 places a general duty on employers to comply with this Part of the regulations in respect of any workplace, other than an excepted workplace, which is to any extent under his control, so far as the requirements relate to matters within his control.

Every person who has, to any extent, control of a workplace, other than an excepted workplace, shall ensure that, so far as the requirements relate to matters within his control, the workplace complies with any applicable requirement of this Part.

Similar provisions apply in respect of persons who, by virtue of any contract or tenancy, have an obligation in respect of the maintenance or repair, or safety, of a workplace.

Any reference to a person having control of any workplace is a reference to a person having control in connection with the carrying on by him of a trade, business or undertaking (whether for profit or not).

Excepted workplaces

Regulation 5 defines an excepted workplace thus:

- any workplace which comprises premises for which a fire certificate is in force or for which an application is pending under the FPA 1971;
- any workplace which comprises premises in respect of which there is a safety certificate under the Safety of Sports Grounds Act 1975 or under Part III of the FSSPSA 1987, and which are in use for the activities specified;
- any workplace which comprises premises to which the Fire Precautions (Sub-surface Railway Stations) Regulations 1989 apply;

- any workplace which is or is on a construction site to which the Construction (Health, Safety and Welfare) Regulations 1996 apply;
- any workplace which is in or on a ship within the meaning of the Docks Regulations 1988, including any such ship which is in the course of construction or repair;
- any workplace which comprises premises to which the Fire Certificates (Special Premises) Regulations 1976 apply;
- any workplace which is deemed to form part of a mine for the purposes of the Mines and Quarries Act 1954;
- any workplace which is or is in an offshore installation within the meaning of the Offshore Installations and Pipelines Works (Management and Administration) Regulations 1995;
- any workplace which is or is in or on an aircraft, locomotive or rolling stock, trailer or semi-trailer used as a means of transport or a vehicle for which a licence is in force under the Vehicle Excise and Registration Act 1994 or a vehicle exempted from duty under that Act;
- any workplace which is in fields, woods or other land forming part of an agricultural or forestry undertaking but which is not inside a building and is situated away from the undertaking's main buildings.

4

Fire-fighting and fire detection (regulation 4)

Where necessary (due to features of a workplace, activities undertaken, hazards present or other relevant circumstances) in order to safeguard the safety of employees in case of fire, workplaces must be equipped with appropriate fire-fighting equipment and with fire detectors and alarms. Any non-automatic fire-fighting equipment must be easily accessible, simple to use and indicated by signs. What is appropriate must be determined by the dimensions and use of the building housing the workplace, the equipment it contains, the physical and chemical properties of substances likely to be present and the maximum number of people present at any one time.

An employer must take measures for fire-fighting which are adapted to the nature of the activities undertaken and the size of his undertaking and of the workplace concerned, taking into account persons other than employees who may be present.

He must nominate employees to implement these measures and ensure that the number of such employees, their training and the equipment available to them are adequate, taking into account the size of, and specific hazards involved in, the workplace concerned, and arrange any necessary contacts with external emergency services.

Emergency routes and exits (regulation 5)

Emergency routes and exits must be kept clear at all times. Where necessary, the following requirements must be complied with:

- emergency routes and exits shall lead as directly as possible to a place of safety;
- in the event of danger, it must be possible for employees to evacuate the workplace quickly and as safely as possible;
- the number, distribution and dimensions of emergency routes and exits shall be adequate having regard to the use, equipment and dimensions of the workplace and the maximum number of persons that may be present there at any one time;
- emergency doors shall open in the direction of escape;
- sliding and revolving doors shall not be used for exits specifically intended as emergency exits;
- emergency doors shall not be so locked or fastened that they cannot be easily and immediately opened by any person who may require to use them in an emergency;
- emergency routes and exits must be indicated by signs; and
- emergency routes and exits requiring illumination shall be provided with emergency lighting of adequate intensity in the case of failure of the normal lighting.

Maintenance (regulation 6)

The workplace and any equipment and devices provided under regulations 4 and 5 shall be subject to a suitable system of maintenance and be maintained in an efficient state, in efficient working order and in good repair.

Amendment of the MHSWR

Part III covers amendments to the general and specific provisions of the MHSWR.

Regulation 7 states that the general provisions of the MHSWR must be taken into account when interpreting these regulations. Regulation 8 requires that specific provisions of Part II be inserted into the following provisions of the MHSWR, namely:

- definition of 'the preventive and protective measures';
- risk assessment;
- health and safety assistance;
- co-operation and co-ordination (shared workplaces); and
- persons working in host employers' undertakings.

Reference to these regulations must be made in regulation 8 (information to employees) of the MHSWR.

The requirements of Part II of these regulations must be considered in the interpretation of regulation 9 (co-operation and co-ordination) of the MHSWR.

The workplace fire precautions legislation (regulation 9)

Specific provisions with regard to enforcement, disclosure of information, provisions as to offences, service of notices and civil liability under the HSWA do not apply, and these regulations do not form part of the relevant statutory provisions.

The *workplace fire precautions legislation* means:

- Part II of these regulations;
- regulations 1 to 4, 6 to 10 and 11(2) and (3) of the MHSWR as amended by Part III of these regulations in so far as those regulations:

  - impose requirements concerning general fire precautions to be taken or observed by an employer; and
  - have effect in relation to a workplace in GB other than an excepted workplace,

  and for this purpose *general fire precautions* means measures which are to be taken or observed in relation to the risk to the safety of employees in the case of fire in a workplace, other than any special precautions in connection with the carrying on of any manufacturing process.

Part IV – Enforcement and offences

Part IV provides fire authorities and their inspectors with a range of enforcement procedures according to the severity of the situation. This may be by way of:

- a written opinion from the authority, stating the breach of the fire precautions legislation and the action that could be taken to remedy it (regulation 10);
- prosecution, where:
  - (i) a person fails to comply with the fire precautions legislation;
  - (ii) that failure places employees at serious risk; and
  - (iii) that failure is intentional or is due to the person being reckless as to whether he complies or not (regulation 11);
- service of a prohibition notice (regulation 12);
- (except where an enforcement notice cannot be delayed) following a written notice of intent, service of an enforcement notice where there is a failure to comply with any provision of the workplace fire precautions legislation and that failure places employees at serious risk in the case of fire (regulation 13); and
- application to a court for an enforcement order where a person has failed to comply with any requirement imposed upon him by the workplace fire precautions legislation (regulation 14).

A person guilty of an offence under regulation 11 shall be liable:

- on summary conviction to a fine not exceeding the statutory maximum; or

- on conviction on indictment, to a fine, or to imprisonment for a term not exceeding two years, or both.

Enforcement notices: rights of appeal (regulation 14)

A person on whom an enforcement notice is served may, within 21 days from the date of the notice, appeal to the court. Such an appeal may either cancel or affirm the enforcement notice, with or without modifications.

Enforcement notices: offences (regulation 15)

It is an offence for any person to contravene any requirement imposed by an enforcement notice. A person guilty of such an offence shall be liable:

- on summary conviction to a fine not exceeding the statutory maximum; or
- on conviction on indictment, to a fine, or to imprisonment for a term not exceeding two years, or both.

In any proceedings for an offence under this regulation it shall be a defence for the person charged to prove that he took all reasonable precautions and exercised all due diligence to avoid the commission of the offence.

THE GAS APPLIANCES (SAFETY) REGULATIONS 1992

Under these regulations gas appliances and fittings must comply with Schedule 3 which specifies that:

- appliances and fittings must be safe when properly used and present no danger to persons;
- there must be comprehensive technical instructions for the installer when sold;
- all necessary information and instructions must be provided for the user;
- all materials must be capable of withstanding stresses imposed during foreseeable use;
- the manufacturer or supplier must guarantee that correct materials are used;
- appliances must be constructed to ensure safe use; and
- appliances must have a valid EC-type examination or certificate.

It is a specific offence to put at risk the health and safety of domestic animals or property by breach of these regulations.

THE GAS SAFETY (INSTALLATION AND USE) REGULATIONS 1998

These regulations apply to all natural gas, LPG, methane and other gas within the definition of gas in the Gas Act 1986. Certain non-refillable gas containers, such as disposable gas cylinders used in camping, are exempted.

The regulations are directed principally at domestic premises. Mines and quarries, factories, agricultural land and buildings, and building and engineering sites are excluded. The regulations are accompanied by an Approved Code of Practice *Safety in the Installation and Use of Gas Systems and Appliances.*

The regulations are concerned with ensuring safety in the installation of gas fires and gas water heaters in bedrooms, and there are certain provisions covering gas regulators. A primary purpose of the regulations is to reduce the incidence of gas-related carbon monoxide poisoning which, on average, claims around 30 lives a year, many in the private sector.

Breaches of the regulations are a criminal offence. A defence that the accused took 'all reasonable steps to prevent the contravention' is available.

The regulations provide, among other matters, that gas installers and maintenance work in domestic and commercial premises (includng shops, offices and hotels) may only be carried out by competent installers who are members of a body approved by the HSE. The Council for Registered Gas Installers (CORGI) was approved for this purpose in 1991.

Under the scheme, individual gas fitting operatives are required to have their competence assessed and must possess a certificate issued by the accredited certification body, which consumers can ask to see.

Gas installation businesses are required to demonstrate to the registration body that these certificates are held by all their employees engaged in gas work.

Inspections

CORGI is empowered to appoint inspectors to undertake a continuing programme of work/site inspections. Inspections are based on:

- the nature of the business
- the number of operatives employed on gas work
- the type of work carried out
- prior history, e.g. record of complaints, enforcement, etc.

Inspections of registered businesses take place every three years as a minimum, the frequency of inspection being based on the risks assessed for the business.

Inspectors may also inspect the work of a registered business which is the subject of a complaint, carrying out a full and unbiased investigation of that complaint and promoting a fair and satisfactory resolution of it.

Inspectors may also take appropriate action to secure safety where, in the case of either pro-active inspection or as a result of complaint investigation, a dangerous or potentially dangerous installation is discovered.

THE GAS SAFETY (MANAGEMENT) REGULATIONS 1996

These regulations incorporate most of the key recommendations set out in the HSC's report *Britain's Gas Supply: A Safety Framework,* which dealt comprehensively with the safety implications of liberalising the domestic gas supply market.

The regulations are designed to maintain existing safety standards by ensuring that the management of the flow of gas through pipeline systems is properly controlled. No gas transporter is allowed to operate until its safety case has been accepted by the HSE.

The regulations also require gas transporters to appoint a Network Emergency Co-ordinator (NEC), who has powers to co-ordinate action in emergency circumstances where there may be a total or partial failure of supply, so as to ensure continued safe operation of the network. The NEC is required to prepare a safety case and have it accepted by the HSE.

Gas transporter's safety case

The key elements are:

- day-to-day management of its part of the network to ensure continuity of supply at the right pressure and composition so that gas appliances continue to burn safely
- arrangements for dealing with reports, from consumers and others, of gas leaks and suspected emissions of carbon monoxide; and
- arrangements for investigating fire and explosion incidents; the investigation of carbon monoxide poisoning incidents is the responsibility of gas suppliers.

NEC's safety case

The key elements are:

- arrangements to monitor the network to identify any potential national gas supply emergency, and to co-ordinate action to prevent it
- arrangements for directing gas transporters to secure a reduction in gas consumption where it is not possible to prevent a gas emergency developing
- procedures to restore gas supply safely to consumers, following an emergency
- arrangements for conducting emergency services.

Duty to co-operate

All other organisations in the liberalised market must co-operate with transporters and the NEC so that arrangements set out in safety cases can be delivered in practice.

National emergency telephone number

The regulations also require the establishment of a single national emergency telephone number to allow the public and consumers to report gas leaks or suspected emissions of carbon monoxide from appliances. British

Gas PLC provide this emergency reporting facility via a national 0800 free-phone number.

Gas transporters must provide, or secure the provision of, around-the-clock emergency cover to deal with reports of gas leaks, etc., in areas covered by their licences and to make situations safe.

THE HEALTH AND SAFETY (CONSULTATION WITH EMPLOYEES) REGULATIONS 1996

These regulations came into operation on 1 October 1996 and brought in changes to the law with regard to the health and safety consultation process between employers and employees.

Under the Safety Representatives and Safety Committees Regulations 1997, employers must consult safety representatives appointed by any trade unions they recognise.

Under the Health and Safety (Consultation with Employees) Regulations 1996 employers must consult any employees who are not covered by the Safety Representatives and Safety Committees Regulations. This may be by direct consultation with employees or through representatives elected by the employees they are to represent.

HSE Guidance

HSE Guidance accompanying the regulations details:

- which employees must be involved
- the information they must be provided with
- procedures for the election of representatives of employee safety
- the training, time off and facilities they must be provided with; and
- their functions in office.

THE HEALTH AND SAFETY (DISPLAY SCREEN EQUIPMENT) REGULATIONS 1992

These regulations implement the EC VDU Directive and came into force on 1 January 1993. Workstations put into service on or after that date must meet the requirements laid down in the Schedule to the regulations. Workstations in operation prior to 31 December 1992 had to comply with the Schedule no later than 31 December 1996.

Compliance with the regulations is based on an analysis of workstations by an employer, taking into account the criteria outlined in the Schedule. The regulations should be read in conjunction with the MHSWR as far as the general duties of employers are concerned.

Interpretation (regulation 1)

A number of important definitions are incorporated in regulation 1, including:

- *display screen equipment* means any alphanumeric or graphic display screen, regardless of the display process involved
- *operator* means a self-employed person who habitually uses display screen equipment as a significant part of their normal work
- *use* means use for or in connection with work
- *user* means an employee who habitually uses display screen equipment as a significant part of their normal work
- *workstation* means an assembly comprised of:
  - display screen equipment (whether provided with software determining the interface between the equipment and its operator or user, a keyboard or any other input device)
  - any optional accessories to the display screen equipment
  - any disk drive, telephone, modem, printer, document holder, work chair, work desk, worksurface or other item peripheral to the display screen equipment
  - the immediate work environment around the display screen equipment.

Exemptions from the regulations

The regulations do *not* apply to:

- drivers' cabs or control cabs for vehicles or machinery
- display screen equipment on board a means of transport
- display screen equipment mainly intended for public operation
- portable systems not in prolonged use
- calculators, cash registers or any equipment having a small data or measurement display required for direct use of the equipment
- window typewriters.

The analysis of workstations to assess and reduce risks (regulation 2)

Every employer shall perform a suitable and sufficient analysis of those workstations that:

- regardless of who has provided them, are used for the purposes of their undertaking by users
- have been provided by them and are used for the purposes of their undertaking by operators,

for the purpose of assessing the health and safety risks to which these people are exposed in consequence of such use.

Any assessment made by an employer shall be reviewed by them if:

- there is reason to suspect that it is no longer valid
- there has been a significant change in the matters to which it relates

and where, as a result of any such review, changes in an assessment are required, the employer concerned shall make them.

The employer shall reduce the risks identified in consequence of the above assessment to the lowest extent reasonably practicable.

Requirements for workstations (regulation 3)

Every employer shall ensure that any workstation first put into service on or after 1 January 1993 meets the requirements laid down in the Schedule to these regulations.

Every employer shall ensure that any workstation first put into service on or before 31 December 1992 meets the requirements laid down in the Schedule not later than 31 December 1996.

The daily work routine (regulation 4)

Every employer shall so plan the activities of users at work in their under-taking that their daily work on display screen equipment is periodically interrupted by breaks or changes of activity as will reduce their workload at that equipment.

Eyes and eyesight (regulation 5)

Where a person:

- is already a user on the date of coming into force of these regulations
- is an employee who does not habitually use display screen equipment as a significant part of their normal work, but is to become a user in the undertaking in which they are already employed,

their employer shall ensure that they are provided with an appropriate eye and eyesight test, requested by their employer, and that any such test is carried out by a competent person.

Any eye and eyesight test carried out shall:

- in any case where the person is already a user, be carried out as soon as practicable after being requested by the user concerned
- in any case where the person will become a user, be carried out before the employee concerned becomes a user.

At regular intervals after an employee has been provided with an eye and eyesight test, their employer shall ensure that they are provided with a fur-ther eye and eyesight test of an appropriate nature – any such test being car-ried out by a competent person.

Where a user experiences visual difficulties that may reasonably be con-sidered to be caused by work on display screen equipment, their employer shall ensure that they are provided, at the employer's request, with an appropriate eye and eyesight test – any such test being carried out by a competent person as soon as is practicable after it has been requested.

Every employer shall ensure that each user employed by them is provided with special corrective appliances appropriate for the work being done by the user concerned where:

- normal corrective appliances cannot be used
- the result of any eye and eyesight test that the user has been given in accordance with this regulation shows such provision to be necessary.

Nothing in the regulations regarding follow-up eye tests shall require an employer to provide any employee with an eye and eyesight test against that employee's will.

The provision of training (regulation 6)

Where a person:

- is already a user on the date of coming into force of these regulations
- is an employee who does not habitually use display screen equipment as a significant part of their normal work, but is to become a user in the undertaking in which they are already employed,

their employer shall ensure that they are provided with adequate health and safety training in the use of any workstation on which they may be required to work.

Every employer shall ensure that each user at work in their undertaking is provided with adequate health and safety training whenever the organisation of any workstation there which they may be required to work is substantially modified.

The provision of information (regulation 6)

Every employer shall ensure that operators and users at work in their undertaking are provided with adequate information about:

- all aspects of health and safety relating to their workstations
- the measures taken by the employer in order to comply with their duties under regulations 2 and 3 that relate to the operators and users and their work.

Every employer shall also ensure that users working in their undertaking are provided with adequate information about the measures taken by the employer in order to comply with their duties under regulations 4, 5, 6(1) and 6(2) that relate to them and their work.

The schedule

This sets out the minimum requirements for workstations contained in the annex to the Council Directive 90/270/EEC on the minimum safety and health requirements for work with display screen equipment.

The extent to which employers must ensure that workstations meet the requirements laid down in this Schedule

An employer shall ensure that a workstation meets the requirements laid down in this Schedule to the extent that:

- the requirements relate to a component that is present in the workstation concerned
- the requirements have an effect in terms of securing the health, safety and welfare of people at work
- the inherent characteristics of a given task do not make compliance with the requirements inappropriate for the workstation concerned.

The equipment

General comment

Using the equipment must not be a source of risk for operators or users.

4

The display screen

The characters on the screen shall be well-defined and clearly formed, of adequate size and with adequate spacing between the characters and lines.

The image on the screen should be stable, with no flickering or other form of instability.

The brightness and the contrast between the characters and the background shall be easily adjustable by the user, and also easily adjustable in response to ambient conditions.

The screen must swivel and tilt easily and freely to suit the needs of the operator or user.

It shall be possible to use a separate base for the screen or an adjustable table.

The screen shall be free of reflective glare and reflections liable to cause discomfort to the user.

The keyboard

The keyboard shall be tiltable and separate from the screen so as to allow the operator or user to find a comfortable working position, avoiding fatigue in the arms or hands.

The space in front of the keyboard shall be sufficient to provide support for the hands and arms of the operator or user.

The keyboard shall have a matt surface to avoid reflective glare.

The arrangement of the keyboard and the characteristics of the keys shall be such as to facilitate the use of the keyboard.

The symbols on the keys shall be adequately contrasted and legible from the working position.

The work desk or worksurface

The work desk or worksurface shall have a sufficiently large, low-reflective surface and allow a flexible arrangement of the screen, keyboard, documents and related equipment.

The document holder shall be stable and adjustable and shall be positioned so as to minimise the need for uncomfortable head and eye movement.

There shall be adequate space for operators or users to find a comfortable position.

The work chair

The work chair shall be stable and allow the user easy freedom of movement and a comfortable position.

The seat shall be adjustable in height.

The seat back shall be adjustable in both height and tilt and a footrest shall be made available to any user who wishes to have one.

The work environment

Space requirements

The workstation shall be dimensioned and designed so as to provide sufficient space for the user to change position and vary movements.

Lighting

Any room lighting or task lighting provided shall ensure satisfactory lighting conditions and an appropriate contrast between the screen and the background environment, taking into account the type of work and the vision requirements of the operator or user.

Possible disturbing glare and reflections on the screen or other equipment shall be prevented by co-ordinating workplace and workstation layout with the positioning and technical characteristics of the artificial light sources.

Reflections and glare

Workstations shall be so designed that sources of light, such as windows and other openings, transparent or translucid walls and brightly coloured fixtures or walls cause no direct glare and no distracting reflections on the screen.

Windows shall be fitted with a suitable system of adjustable covering to attenuate the daylight that falls on the workstation.

Noise

Noise emitted by equipment belonging to any workstation shall be taken into account when a workstation is being equipped, with a view, in particular, to ensuring that attention is not distracted and speech is not disturbed.

Heat

Equipment belonging to any workstation shall not produce excess heat that could cause discomfort to operators or users.

Radiation

All radiation, with the exception of the visible part of the electromagnetic spectrum, shall be reduced to negligible levels from the point of view of the protection of operators or users' health and safety.

Humidity

An adequate level of humidity shall be established and maintained.

The interface between computer and operator/user

In designing, selecting, commissioning and modifying software, and in designing tasks using display screen equipment, the employer shall take into account the following principles:

● software must be suitable for the task
● software must be easy to use and, where appropriate, adaptable to the user's level of knowledge or experience; no quantitative or qualitative checking facility may be used without the knowledge of the operators or users
● systems must provide feedback to operators or users on the performance of these systems
● systems must display information in a format and at a pace that are adapted to the operators or users
● the principles of software ergonomics must be applied, in particular to human data processing.

The regulations are accompanied by a comprehensive guide issued by the HSE.

THE HEALTH AND SAFETY (ENFORCING AUTHORITY) REGULATIONS 1998

These regulations provide that where the main activity carried on in any premises is specified in Schedule 1 to those regulations then, subject to specified exceptions, the local authority (LA) is the enforcing authority in respect of all activities carried on in those premises (regulation 3).

Schedule 1 to these regulations states the main activities which determine whether local authorities are the enforcing authorities, for example, the sale of goods, or the storage of goods for retail or wholesale distribution, the display or demonstration of goods at exhibitions, office activities, catering services, cosmetic or therapeutic treatments, sports, cultural or recreational activities, hiring out of pleasure craft for use on inland waters, the care, treatment, accommodation or exhibition of animals, activities of an undertaker and church worship or religious meetings.

Schedule 2 lists the activities in respect of which the HSE is the enforcing authority. These activities include, for instance, activities in mines, quarries, at fairgrounds, on board a sea-going ship, agriculture, in premises occupied by a radiography undertaking involving work with ionising radiation and those involving the operation of a railway.

Under regulation 4 the HSE is the enforcing authority for any part of a premises occupied by a body specified in the regulations, and for enforcement against such a body or its officers or servants. The HSE is also the enforcing authority for any part of the premises occupied by another person for the purpose of providing services at the premises for a specified body.

The HSE is the enforcing authority for premises where the main activity is indoor sports, if specified conditions are met. The HSE is the enforcing authority for the enforcement of section 6 of the HSWA (general duties of manufacturers and so on regarding articles and substances used at work), even though the main activity is listed in Schedule 1.

The activities set out in Schedule 2 are allocated for enforcement by the HSE, even though the main activity carried out is listed in Schedule 1. Further, these regulations provide that, regardless of the main activity, the licensing or registration authority for an explosives factory, magazine, store or premises for keeping explosives shall be the enforcing authority for such provisions of the Explosives Act 1875 and legislation made under it as a relevant statutory provision (regulation 4).

The regulations enable the responsibility for enforcement to be assigned by the HSE and LA jointly (to either of them) where they agree there is uncertainty in the particular case as to which one is the enforcing authority (regulation 6).

Schedule 1: The main activities that determine whether the local authorities will be the enforcing authorities

The following activities determine that the local authorities will be the enforcing authorities:

- the sale or storage of goods for retail or wholesale distribution except:
  - at container depots where the main activity is the storage of goods in the course of transit to or from dock premises, an airport or railway
  - where the main activity is the sale or storage for wholesale distribution of any dangerous substance

– where the main activity is the sale or storage of water or sewage or their by-products or natural or town gas

and, for the purposes of this point, where the main activity carried out on premises is the sale and fitting of motor car tyres, exhausts, windscreens or sunroofs, the main activity shall be deemed to be sale of goods

- the display or demonstration of goods at an exhibition for the purposes of offer or advertisement for sale
- office activities
- catering services
- the provision of permanent or temporary residential accommodation, including the provision of a site for caravans or campers
- consumer services provided in a shop, except dry cleaning or radio and television repairs (and, here, *consumer services* means services of a type ordinarily supplied to people who receive them otherwise than in the course of a trade, business or other undertaking carried out by them, whether for profit or not)
- cleaning (wet or dry) in coin-operated units in launderettes and similar premises
- the use of a bath, sauna or solarium, massaging, hair transplanting, skin-piercing, manicuring or other cosmetic services and therapeutic treatments, except where they are carried out under the supervision or control of a registered medical practitioner, or dentist registered under the Dentists Act 1984, a physiotherapist, an osteopath or chiropractor
- the practice or presentation of the arts, sports, games, entertainment or other cultural or recreational activities except where carried out in a museum, art gallery or theatre or where the main activity is the exhibition of a cave to the public
- the hiring out of pleasure craft for use in inland waters
- the care, treatment, accommodation or exhibition of animals, birds or other creatures, except where the main activity is horse breeding or horse training at a stable, or is an agricultural activity or veterinary surgery
- the activities of an undertaker, except where the main activity is embalming or the making of coffins
- church worship or religious meetings

- the provision of car parking facilities within the perimeter of an airport
- the provision of childcare, or playgroup or nursery facilities.

Schedule 2: Activities in respect of which the HSE is the enforcing authority

These activities include:

- any activity in a mine or quarry, other than a mine or quarry in which notice of abandonment has been given under section 139(2) of the Mines and Quarries Act 1954

- any activity in a fairground
- any activity in premises occupied by a radio, television or film undertaking in which the activity of broadcasting, recording or filming is carried out, and the activity of broadcasting, recording or filming wherever they are carried out (for this purpose, *film* includes video)
- the following activities carried out at premises by those who do not normally work on the premises:

  - construction work if:

    (i) regulation 7(1) of the Construction (Design and Management) Regulations 1994 (which requires projects which include or are intended to include construction work to be notified to the HSE) applies to the project which includes the work
    (ii) the whole or part of the work contracted to be undertaken by the contractor at the premises is to the external fabric or other external part of a building or structure
    (iii) it is carried out in a physically segregated area of the premises, the activities normally carried out in this area have been suspended for the purpose of enabling the construction work to be carried out, the contractor has authority to exclude from this area persons who are not attending it in connection with the carrying out of the work and the work is not the maintenance of insulation on pipes, boilers or other parts of heating or water systems or its removal from them

  - the installation, maintenance or repair of any gas system, or any work in relation to a gas fitting
  - the installation, maintenance or repair of electricity systems
  - work with ionising radiations, except work in one or more of the categories set out in Schedule 3 to the Ionising Radiations Regulations 1985

- the use of ionising radiations for medical exposure (within the meaning of regulation 2(1) of the Ionising Radiations Regulations 1985)
- any activity in premises occupied by a radiography undertaking in which work is carried out with ionising radiation
- agricultural activities, and any activity at an agricultural show that involves the handling of livestock or the working of agricultural equipment
- any activity on board a sea-going ship
- any activity in relation to a ski slope, ski lift, ski tow or cable car
- fish, maggot and game breeding, except in a zoo
- any activity in relation to a pipeline within the meaning of regulation 3 of the Pipelines Safety Regulations 1996
- the operation of a railway.

THE HEALTH AND SAFETY (FIRST AID) REGULATIONS 1981

These regulations apply to nearly all workplaces in the UK. Within the regulations, *first aid* means:

- in cases where a person will need help from a medical practitioner or nurse, treatment for the purpose of preserving life and minimising the consequences of injury or illness until such help is obtained
- treatment of minor injuries that would otherwise receive no treatment or do not need treatment by a medical practitioner or nurse.

The duties of employers

Under section 3(1), an employer shall provide, or ensure that there are provided, such equipment and facilities as are *adequate and appropriate* in the circumstances for enabling first aid to be rendered to their employees if they are injured or become ill at work.

Two main duties are imposed on employers by the regulations:

- to provide first aid
- to inform employees of the first aid arrangements.

4

Self-employed people must provide first aid equipment for their own use.

The Approved Code of Practice regarding first aid

This Code emphasises the duty of employers to consider a number of factors and determine for themselves what is adequate and appropriate in all the circumstances. Furthermore, where there are particular risks associated with the operation of an enterprise, the employer must ensure that first aiders receive training to deal with these specific risks.

Factors to be considered in assessing first aid provision include:

- the number of employees
- the nature of the undertaking
- the size of the establishment and the distribution of employees
- the location of the establishment and the locations employees go to in the course of their employment
- use of shift working (each shift would have the same level of first aid cover/protection)
- the distance from external medical services, such as the local casualty department.

The general guidance suggests that even in a simple office, there ought to be a first aider for every 50 people there.

First aid boxes

There should be at least one first aid box, the contents of which are listed in the ACOP.

THE HEALTH AND SAFETY (INFORMATION FOR EMPLOYEES) REGULATIONS 1998

These regulations require information relating to health, safety and welfare to be furnished to employees by means of posters or leaflets in the form approved and published for the purposes of the regulations by the HSE (copies of the form of poster or leaflets may be obtained from the HSE and the actual posters from HMSO).

The regulations also require that the name and address of the enforcing authority and the address of the Employment Medical Advisory Service be written in the appropriate space on the poster. Where the leaflet is given to employees, the same information must be specified in a written notice accompanying it.

The regulations provide for the issuing of certificates of exemption by the HSE, provide for a defence for a contravention of the regulations and repeal, revoke and modify various enactments relating to the provision of information to employees.

The regulations do not apply to the master and crew of a sea-going ship.

Modification to the 1989 regulations

The Health and Safety Information for Employees (Modifications and Repeals) Regulations 1995 amended the 1989 regulations, allowing the HSE to approve, as an alternative to the basic 'Health and Safety Law' poster, a particular form of poster or leaflet for use in relation to a particular employment or class of employment.

When applying for approval to display an alternative poster, applicants must demonstrate:

- a clearly defined industry or group of employers;
- a clear demand for the alternative poster; and
- that the poster would meet the same purposes as the basic poster and that the benefit justifies the development costs.

These regulations abolished 53 statutory requirements to display health and safety information, notably section 139 of the Factories Act 1961 and a series of regulations made under that Act covering specific industries.

THE HEALTH AND SAFETY (SAFETY SIGNS AND SIGNALS) REGULATIONS 1996

These regulations implement the European Safety Signs Directive 92/58/EEC on the provision and use of safety signs at work, and apply to all workplaces, including offshore installations.

The purpose of the Directive was to encourage the standardisation of safety signs throughout the Member States so that safety signs, wherever they are seen, have the same meaning.

The regulations cover various means of communicating health and safety information, including the use of illuminated signs, hand and acoustic signals (e.g. fire alarms), spoken communication and the marking of pipework containing dangerous substances.

These requirements are *in addition* to traditional signboards such as prohibition and warning signs. Fire safety signs, e.g. signs for fire exit and fire-fighting equipment are also covered. The signboards specified in the regulations are covered by BS 5378: Parts 1 and 3: 1980 *Safety signs and colours.*

Principal requirements

Employers must use a safety sign where a risk cannot be adequately avoided or controlled by other means. Where a safety sign would not help to reduce that risk, or where the risk is not significant, a sign is not required.

They promote a general move towards symbol-based signs.

The regulations extend the term 'safety sign' to include hand signals, pipeline marking, acoustic signals and illuminated signs.

The number of safety symbols are increased, and new colour meanings are introduced.

They require, where necessary, the use of road traffic signs within workplaces to regulate road traffic.

Employers are required to:

(a) maintain the safety signs which are provided by them; and
(b) explain unfamiliar signs to their employees and tell them what they need to do when they see a safety sign.

The regulations apply to all places and activities where people are employed, but exclude signs and labels used in connection with the supply of substances, products and equipment or the transport of dangerous goods.

Employers are required to mark pipework containing dangerous substances, for example, by identifying and marking pipework at sampling and discharge points. The same symbols or pictograms need to be shown as those commonly seen on containers of dangerous substances, but using the triangular-shaped warning signs.

Although the regulations specify a code of hand signals for mechanical handling and directing vehicles, they permit other equivalent codes to be used such as BS 6736: *Code of practice for hand signalling use in agricultural operations*, and BS 7121: Part 1: 1989 *Code of practice for safe use of cranes*.

Dangerous locations, for example, where people may slip, fall from heights, or where there is low headroom, and traffic routes, may need to be marked to meet requirements under the Workplace (Health, Safety and Welfare) Regulations 1992.

Although these regulations require stores and areas containing significant quantities of dangerous substances to be identified by the appropriate warning sign, i.e. the same signs as are used for marking pipework, they will mainly impact on smaller stores. This is because the majority of sites on which 25 tonnes or more of dangerous substances are stored can be expected to be marked in accordance with the Dangerous Substances (Notification and Marking of Sites) Regulations 1990. These have similar marking requirements for storage of most dangerous substances.

Stores need not be marked if:

- they hold very small quantities
- the labels on the containers can be seen clearly from outside the store.

Fire safety signs

In general, the regulations do not require any changes where existing fire safety signs containing symbols complying with BS 5499: Part 1: 1990 *Fire safety signs, notices and graphic symbols*, perhaps in order to comply with the requirements of a fire certificate. This is because the signs in BS 5499, although different in detail to those specified in the regulations, follow the same basic pattern and are, therefore, considered to comply with the regulations.

Fire warning systems

Where evacuation from buildings is needed the regulations require the fire alarm signal to be continuous. Fire alarms conforming to BS 5839: Part 1: 1988 *Fire detection and alarm systems for buildings* do not need changing, nor do other acceptable means such as manually operated sounders, such as rotary gongs or handbells.

THE HIGHLY FLAMMABLE LIQUIDS AND LIQUEFIED PETROLEUM GASES REGULATIONS 1972

These regulations lay down requirements for the following matters in relation to highly flammable liquids and so on used in a factory:

- manner of storage
- marking of storage accommodation and vessels
- precautions to be observed for the prevention of fire and explosion
- provision of fire fighting apparatus
- securing means of escape in the event of fire.

Highly flammable liquids means both liquefied flammable gas, although not aqueous ammonia, and liquefied petroleum gases and so it includes any liquid, liquid solution, emulsion or suspension that:

- gives off a flammable vapour at a temperature less than 32°C when tested in the manner set out in Schedule 1 of the regulations (closed cup flashpoint determination method)
- supports combustion when tested in a manner set out in Schedule 2 (combustibility test).

Liquefied flammable gas is any substance that would be a flammable gas at a temperature of 20°C and a pressure of 760 millimetres of mercury, but is in a liquid form as a result of the application of pressure or refrigeration or both.

The term *liquefied petroleum gas* covers both commercial butane and commercial propane, and any mixture of them.

The application of the regulations

The regulations apply to factories and impose duties on the occupier of the premises or, in some cases, the owner of the substances.

General storage requirements for highly flammable liquids

When not in use or being conveyed, all highly flammable liquids (HFLs) should be stored in a safe manner. All HFLs should be stored in one of the following ways:

- in suitable fixed storage tanks in a safe position
- in suitable closed vessels kept in a safe position in the open air and, where necessary, protected against direct sunlight
- in suitable closed vessels kept in a store room that either is in a safe position or is a fire-resisting structure
- in the case of a workroom where the aggregate quantity of HFL stored does not exceed 50 litres, in suitable closed vessels kept in a suitably placed cupboard or bin that is a fire-resisting structure.

Other storage precautions

Bund walls

Storage tanks should be provided with a bund wall enclosure that is capable of containing 110 per cent of the capacity of the largest tank within the bund.

Ground beneath vessels

The ground beneath storage vessels should be impervious to liquid and be so sloped that any minor spillage will not remain beneath the vessels, but will run away to the sides of the enclosure.

Bulk storage

Bulk storage tanks should not be located inside buildings or on the roof of a building. Underground tanks should not be sited under the floors of process buildings.

Drum storage

The area to be utilised for drum storage should be surrounded with a sill capable of containing the maximum spillage from the largest drum in store.

The marking of store rooms and containers

Every store room, cupboard, bin, tank and vessel used for storing HFLs should be clearly and boldly marked 'Highly flammable' or 'Flashpoint below 32°C' or 'Flashpoint in the range of 22°C to 32°C'.

Specific provisions for the storage of liquefied petroleum gas

All liquefied petroleum gas (LPG) must be stored in one of the following ways:

- in suitable underground reservoirs or in suitable fixed storage tanks located in a safe position, either underground or in the open air
- in suitable movable storage tanks/vessels kept in a safe position in the open air
- in pipelines or pumps forming part of an enclosed system
- in suitable cylinders kept in safe positions in the open air or, where this is not reasonably practicable, in a store room constructed of non-combustible material, having adequate ventilation, being in a safe position, of fire-resisting structure, and being used solely for the storage of LPG and/or acetylene cylinders.

LPG cylinders must be kept in a store until they are required for use, and any expended cylinder must be returned to the store as soon as is reasonably practicable. This should ensure that only the minimum amount of LPG is kept in any workplace.

The marking of store rooms and containers

Every tank, cylinder, store room and so on used for the storage of LPG should be clearly and boldly marked 'Highly flammable – LPG'.

Precautions against spills and leaks (all HFLs)

Where HFLs are to be conveyed within a factory, a totally enclosed piped system should be used, where reasonably practicable. Where not reasonably practicable, a system using closed, non-spill containers will be acceptable.

Portable vessels, when emptied, should be removed to a safe place without delay.

Where, in any process or operation, any HFL is liable to leak or be split, all reasonably practicable steps should be taken to ensure that any such

HFL should be contained or immediately drained off to a suitable container, or to a safe place, or rendered harmless.

Precautions against escaping vapours

No means likely to ignite vapour from any HFL should be present where there may be a dangerous concentration of vapours from HFL.

Where any HFL is being utilised in the workplace, reasonably practicable steps should be taken so as to minimise the risk of escape of HFL vapours into the general workplace atmosphere. Where such escape cannot be avoided, then the safe dispersal of HFL vapours should be effected, so far as is reasonably practicable.

The relaxation of fire resistance specifications in certain circumstances

In cases where either explosion pressure relief or adequate natural ventilation are required in a fire-resistant structure, a relaxation of the specification of a fire-resistant structure is allowable.

4

Fire escapes and fire certificates

There must be adequate and safe means of escape in case of fire from every room in which any HFL is manufactured, used or manipulated. This regulation does not apply where there is storage only.

Fire certificates are generally necessary where:

- HFLs are manufactured
- LPG is stored
- liquefied flammable gas is stored.

The prevention of a build-up of deposits

Whenever, as a result of any process or, operation involving any HFL, a deposit of any solid waste residue liable to give rise to a risk of fire may occur on any surface:

- steps must be taken to prevent the occurrence of all such deposits, so far as is reasonably practicable
- where any such deposits occur, effective steps must be taken to remove all such residues, as often as necessary, to prevent danger.

Smoking controls

No person may smoke in any place in which any HFL is present and where the circumstances are such that smoking will give rise to the risk of fire.

The provision of fire-fighting equipment

Appropriate fire-fighting equipment should be made readily available for use in all factories where HFLs are manufactured, used or manipulated.

The duties of employees

It is the duty of every employee to comply with them and co-operate in carrying them out.

If an employee discovers any defect in plant, equipment or appliance, it is their duty to report the defect without delay to the occupier, manager or other responsible person.

THE IONISING RADIATIONS REGULATIONS 1999

These very specialised regulations introduced stricter requirements relating to the use of ionising radiation in the workplace. The regulations implement most of the revised Basic Safety Standards Directive and replace the Ionising Radiations Regulations 1985.

The principal provisions include:

- a requirement for employees to be authorised before they use accelerators or X-ray sets for certain specified purposes;
- the introduction of HSE criteria of competence for individuals or organisations wishing to act as radiation protection advisers (RPAs);
- a requirement for employers to carry out a risk assessment before starting work activities with ionising radiation;
- enhanced requirements for restricting exposure as far as is reasonably practicable (including specific provisions covering the proper maintenance, examination and test of any engineering control design feature, safety feature or warning device provided to restrict exposure; and the working conditions of pregnant or breastfeeding employees); and
- greater flexibility in the designation of controlled and supervised areas.

The limits on effective dose (dose to the whole body) are:

- 20 millisieverts a year for employees aged over 18 (in special cases, employers may apply a dose limit of 100 millisieverts in five years with no more than 50 millisieverts in a single year, subject to strict conditions)
- 6 millisieverts a year for trainees
- 1 millisievert for any other person, including members of the public.

An ACOP, together with HSE guidance, accompanies the regulations.

THE LIFTING OPERATIONS AND LIFTING EQUIPMENT REGULATIONS (LOLER) 1998

These regulations replace most sector-specific legislation, e.g. that relating to the construction industry, on lifting equipment, creating a single set of regulations that apply to all sectors. LOLER applies over and above the general requirements of the Provision and Use of Work Equipment Regulations 1998 in dealing with specific hazards and risks associated with lifting equipment and lifting operations. The regulations are supported by an ACOP and HSE Guidance *Safe use of lifting equipment*.

Under LOLER there is an absolute duty on employers and others to undertake lifting operations safely.

Important definitions (regulation 2)

- the *1992 Regulations* means the Supply of Machinery (Safety) Regulations 1992
- *accessory for lifting* means lifting equipment for attaching loads to machinery for lifting
- *EC declaration of conformity* means a declaration which complies with:

  - regulation 22 of the 1992 Regulations
  - Article 12.1 of the Council Directive 89/686/EEC on the approximation of the laws of the Member States relating to personal protective equipment
  - regulation 8(2)(d) of the Lifts Regulations 1997

- *examination scheme* means a suitable scheme drawn up by a competent person for such thorough examinations of lifting equipment at such intervals as may be appropriate for the purpose described in regulation 9(3)
- *lifting equipment* means work equipment for lifting or lowering loads and includes its attachments used for anchoring, fixing or supporting it
- *lifting operation* means an operation concerned with the lifting or lowering of a load (regulation 8)
- *load* includes a person
- *thorough examination* in relation to a thorough examination under regulation 9:

  - means a thorough examination by a competent person
  - where it is appropriate to carry out testing for the purpose described in the paragraph, includes such testing by a competent person as is appropriate for the purpose

- *work equipment* means any machinery, appliance, apparatus, tool or installation for use at work (whether exclusively or not).

Strength and stability (regulation 4)

Every employer shall ensure that:

- lifting equipment is of adequate strength and stability for each load, having regard in particular to the stress induced at its mounting or fixed point
- every part of a load and anything attached to it and used in lifting it is of adequate strength.

Lifting equipment for lifting persons (regulation 5)

Every employer shall ensure that lifting equipment for lifting persons:

- subject to the next paragraph, is such as to prevent a person using it being crushed, trapped or struck or falling from the carrier;
- is such as to prevent so far as is reasonably practicable a person using it, while carrying out activities from the carrier, being crushed, trapped or struck or falling from the carrier;
- subject to the requirement to provide enhanced safety rope or chain as described below, has suitable devices to prevent the risk of a carrier falling;
- is such that a person trapped in any carrier is not thereby exposed to danger and can be freed.

Every employer shall ensure that if the risk crushing, trapping or falling described above cannot be prevented for reasons inherent in the site and height differences:

- the carrier has an enhanced safety coefficient suspension rope or chain; and
- the rope or chain is inspected by a competent person every working day.

Positioning and installation (regulation 6)

Every employer shall ensure that lifting equipment is positioned or installed in such a way as to reduce to as low as is reasonably practicable the risk:

- of the lifting equipment or a load striking a person; or
- from a load:
  - drifting
  - falling freely
  - being released unintentionally,

  and it is otherwise safe.

Every employer shall ensure that there are suitable devices to prevent a person from falling down a shaft or hoistway.

Marking of lifting equipment (regulation 7)

Every employer shall ensure that:

- subject to the next paragraph, the machinery and accessories for lifting loads are clearly marked to indicate their safe working loads (SWLs);
- where the SWL of machinery for lifting depends upon its configuration:

  - the machinery is clearly marked to indicate its SWL for each configuration; or
  - information which clearly indicates its SWL for each configuration is kept with the machinery;

- accessories for lifting are clearly marked so that it is possible to identify the characteristics necessary for their safe use;

- lifting equipment which is designed for lifting persons is appropriately and clearly marked to this effect; and

- lifting equipment which is not designed for lifting persons but which might be used for that by mistake is appropriately and clearly marked to the effect that it is not designed for lifting persons.

Organisation of lifting operations (regulation 8)

Every employer shall ensure that every lifting operation involving lifting equipment is:

- properly planned by a competent person
- appropriately supervised
- carried out in a safe manner.

Thorough examination and inspection (regulation 9)

Every employer shall ensure that before he puts lifting equipment *into service for the first time* he gives it a thorough examination for any defect unless either:

- the lifting equipment has not been used before and in the case of lifting equipment for which an EC declaration of conformity could or (in the case of a declaration under the Lifts Regulations 1997) should have been drawn up, the employer has received such declaration made not more than 12 months before the lifting equipment is put into service
- or, if obtained from the undertaking of another person, it is accompanied by physical evidence of thorough examination under these regulations.

Every employer shall ensure that, where the safety of lifting equipment depends upon the *installation conditions*, it is thoroughly examined:

- after installation and before being put into service for the first time; and
- after assembly and before being put into service at a new site or in a new location,

to ensure that it has been installed correctly and is safe to operate.

Subject to paragraph 6, every employer shall ensure that lifting equipment which is exposed to *conditions causing deterioration* which is liable to result in dangerous situations is:

- thoroughly examined:
  - in case of lifting equipment for lifting persons or an accessory for lifting, at least every 6 months
  - in the case of other lifting equipment, at least every 12 months; or
  - in either case, in accordance with an examination scheme; and
  - each time that exceptional circumstances which are liable to jeopardise the safety of the lifting equipment have occurred; and
- if appropriate for the purpose, is inspected by a competent person at suitable intervals between thorough examinations.

Every employer shall ensure that no lifting equipment:

- leaves his undertaking
- or if obtained from the undertaking of another person, is used in his undertaking, unless it is accompanied by physical evidence that the last thorough examination required to be carried out under this regulation has been carried out.

Reports and defects (regulation 10)

A person making a thorough examination for an employer under regulation 9 shall:

- notify the employer straight away of any defect in the lifting equipment which he believes is or could become dangerous;
- as soon as is practicable make a written report of the thorough examination which authenticated by him or on his behalf by signature or equally secure means and which contains the information specified in Schedule 1 to:
  - the employer; and
  - any person from whom the lifting equipment has been hired or leased;
- where there is in his opinion a defect in the lifting equipment involving an existing or imminent risk of serious personal injury, send a copy of the report as soon as is practicable to the relevant enforcing authority.

A person making an inspection for an employer under regulation 9 shall:

- tell the employer straight away about any defect in the lifting equipment which he believes is or could become dangerous;
- as soon as is practicable make a record of his inspection in writing.

Every employer who has been notified under this regulation shall ensure the lifting equipment is not used:

- before the defect is rectified; or
- where the defect is not yet dangerous, after the time specified that it will become dangerous and before the defect is rectified.

In this regulation *relevant enforcing authority* means:

- where the defective lifting equipment has been hired or leased by the employer, the HSE; and
- otherwise, the enforcing authority for the premises in which the defective lifting equipment was thoroughly examined.

Keeping of information (regulation 11)

Where an employer obtaining lifting equipment to which these regulations apply receives an EC declaration of conformity relating to it, he shall keep the declaration for so long as he operates the lifting equipment.

The employer shall ensure that the information contained in:

- every report made to him under regulation 10 is kept available for inspection:

  - in the case of a thorough examination under regulation 9 of lifting equipment other than an accessory for lifting, until he ceases to use the lifting equipment;
  - in the case of a thorough examination under regulation 9 of an accessory for lifting, for two years after the report is made;
  - in the case of thorough examination under regulation 9 where safety of the equipment depends on installation conditions, until he ceases to use the lifting equipment at the place it was installed or assembled;
  - in the case of a thorough examination under regulation 9 of equipment which is exposed to conditions which cause deterioration, until the next report is made or the expiration of two years, whichever is later;

- every record made under regulation 10 is kept available until the next such record is made.

Schedule 1 – information to be contained in a report of a thorough examination

The name and address of the employer for whom the thorough examination was made.

The address of the premises at which the thorough examination was made.

Particulars sufficient to identify the lifting equipment including, where known, its date of manufacture.

The date of the last thorough examination.

The SWL of the lifting equipment or (where its SWL depends upon the configuration of the lifting equipment) its SWL for the last configuration in which it was thoroughly examined.

In relation to the first thorough examination of lifting equipment after installation or after assembly at a new site or in a new location, a statement:

- that it is such thorough examination;
- (if such be the case) that is has been installed correctly and would be safe to operate.

In relation to a thorough examination of lifting equipment other than a thorough examination to which the paragraph above relates, a statement:

- about whether it is a thorough examination:

  - within an interval of six months under regulation 9;
  - within an interval of 12 months under regulation 9;
  - in accordance with an examination scheme under regulation 9; or
  - after the occurrence of exceptional circumstances under regulation 9

- (if such be the case) that the lifting equipment would be safe to operate.

In relation to every thorough examination of lifting equipment:

- identification of any part found to have a defect which is or could become a danger to persons, and a description of the defect
- particulars of any repair, renewal or alteration required to remedy a defect found to be a danger to persons
- in the case of a defect which is not yet but could become a danger to persons

  - the time by which it could become such a danger
  - particulars of any repair, renewal or alteration required to remedy it

- the latest date by which the next thorough examination must be carried out
- where the thorough examination included testing, particulars of any test
- the date of the thorough examination.

The name, address and qualifications of the person making the report; that he is self-employed or, if employed, the name and address of the employer.

The name and address of the person signing or authenticating the report on behalf of its author.

The date of the report.

THE LIFTS REGULATIONS 1997

These regulations implement Council Directive 95/16/EC on the approximation of the laws of the Member States relating to lifts (the Lifts Directive). As such they impose a regime for the placing on the market and putting into service of all types of lift and for ensuring the safety of lifts and safety components, including conformity assessment procedures. The regulations are accompanied by a series of Schedules, including essential health and safety requirements relating to the design and construction of lifts and safety components.

Interpretation (regulation 2)

Harmonised standard means a technical specification adopted by the European Committee for Standardisation or the European Committee for Electrotechnical Standardisation or both, upon a mandate from the Commission in accordance with Council Directive 83/189/EEC of 28 March 1983 laying down a procedure for the provision of information in the field of technical standards and regulations, and of which the reference number is published in the Official Journal of the European Communities.

Installer of a lift means the natural or legal person who takes responsibility for the design, manufacture, installation and placing on the market of the lift and who affixes the CE marking and draws up the EC declaration of conformity.

Lift means an appliance serving specific levels, having a car moving:

- along guides which are rigid; or
- along a fixed course even where it does not move along guides which are rigid (for example, a scissor lift),

and inclined at an angle of more than 15° to the horizontal and intended for the transport of:

- persons,
- persons and goods,
- goods alone if the car is accessible, that is to say, a person may enter it without difficulty, and fitted with controls situated inside the car or within reach of a person inside.

4

Manufacturer of the safety components means the natural or legal person who takes responsibility for the design and manufacture of the safety components and who affixes the CE marking and draws up the EC declaration of conformity.

Model lift means a representative lift whose technical dossier shows the way in which the essential safety requirements will be met for lifts which conform to the model lift defined by objective parameters and which uses identical safety components.

Relevant essential health and safety requirements in relation to a lift or safety component means those provisions of the essential health and safety requirements which are applicable to that particular lift or safety component, as the case may be.

Responsible person means:

(a) in the case of a lift, the installer of the lift;

(b) in the case of a safety component, the manufacturer of the safety component or his authorised representative ;

(c) where neither the installer of the lift nor the manufacturer of the safety component nor the latter's authorised representative established in the Community, as the case may be, have fulfilled the requirements of regulation 8(2) or 9(2) applicable to the lift or safety component, the person who places the lift or the safety component on the market.

Safe in relation to a lift or safety component, means that the lift or, in the case of a safety component, the lift in which it is to be installed, when properly installed and maintained and used for its intended purpose is not liable to endanger the health or safety of persons or, where appropriate, the safety of property.

Standard or *standard referred to in Article 5* means a technical specification approved by a recognised standardising body for repeated or continuous application, with which compliance is not compulsory; and, for the avoidance of doubt, this definition includes a harmonised standard or a transposed harmonised standard.

Supply includes offering to supply, agreeing to supply, exposing for supply and possessing for supply.

Transposed harmonised standard means a national standard of a Member State which transposes a harmonised standard.

Lifts, safety components and components (regulation 3)

These regulations apply to:

- lifts permanently serving buildings or constructions; and
- safety components for use in such lifts.

Nothing in these regulations shall preclude the placing on the market of any component, other than a safety component:

- which is intended to be incorporated into a lift to which these regulations apply; and
- in respect of which a declaration is made by the manufacturer of that component or his authorised representative established in the Community that the component is intended for such incorporation.

General duty relating to the placing on the market and putting into service of lifts (regulation 8)

Subject to regulation 12, no person who is a responsible person shall place on the market and put into service any lift unless the requirements of paragraph 2 below have been complied with in relation to it.

The requirements in respect of any lift are that:

- it satisfies the relevant essential health and safety requirements and for the purpose of satisfying those requirements:
  - where a transposed harmonised standard covers one or more of the relevant essential health and safety requirements, any lift constructed in accordance with that transposed harmonised standard shall be presumed to comply with that or, as the case may be, those essential health and safety requirements;

– by calculation or on the basis of design plans, it is permitted to demonstrate the similarity of a range of equipment to satisfy the essential safety requirements;

- the appropriate conformity assessment procedure in respect of the lift has been carried out in accordance with regulation 13;
- the CE marking has been affixed to it by the installer of the lift in accordance with Schedule 3;
- the declaration of conformity has been drawn up in respect of it by the installer of the lift containing the information listed in Part B of Schedule 2, taking account of the specifications given in the Schedule used for the conformity assessment procedure; and
- it is in fact safe.

Any technical documentation or other information in relation to a lift required to be retained under the conformity assessment procedure used shall be retained by the person specified in that respect in that conformity assessment procedure for any period specified in that procedure.

General duty relating to the placing on the market and putting into service of safety components (regulation 9)

4

Subject to regulation 12, no person who is a responsible person shall place on the market and put into service any safety component unless the requirements of paragraph 2 below have been complied with in relation to it.

The requirements in respect of any safety component are that:

- it satisfies the relevant essential health and safety requirements and for the purpose of satisfying those requirements where a transposed harmonised standard covers one or more of the relevant essential health and safety requirements, any safety component constructed in accordance with that transposed harmonised standard shall be presumed to be suitable to enable a lift on which it is correctly installed to comply with that or, as the case may be, those essential health and safety requirements;
- subject to paragraph 3 below, the appropriate conformity assessment procedure in respect of that safety component has been carried out in accordance with regulation 13(1);
- the CE marking has been affixed to it, or on a label inseparably attached to the safety component, by the manufacturer of that safety component or his authorised representative established in the Community in accordance with Schedule 3;
- a declaration of conformity has been drawn up in respect of it by the manufacturer of the safety component or his authorised representative established in the Community containing the information listed in Part A of Schedule 2, taking account of specifications given in the Schedule used for the conformity assessment procedure; and
- it is in fact safe.

Any technical documentation or other information in relation to a safety component required to be retained under the conformity assessment procedure used shall be retained by the person specified in that respect in that conformity assessment procedure for any period specified in that procedure.

General duty relating to the supply of a lift or safety component (regulation 10)

Subject to regulation 12, it shall be the duty of any person who supplies any lift or safety component but who is not a person to whom regulation 8 or 9 applies, to ensure that that lift or safety component, as the case may be, is safe.

Specific duties relating to the supply of information, freedom from obstruction of lift shafts and retention of documents (regulation 11)

The person responsible for work on the building or construction where a lift is to be installed and the installer of the lift shall:

● keep each other informed of the facts necessary for; and
● take the appropriate steps to ensure,

the proper operation and safe use of the lift; in particular it shall be ensured that shafts intended for lifts do not contain any piping or wiring or fittings other than that necessary for the operation and safety of that lift.

Where, in the case of a lift, for the purposes of regulation 8(2)(b) the appropriate conformity assessment procedure is one of the procedures set out in regulation 13(2)(a), (b) or (c), the person responsible for the design of the lift must supply to the person responsible for the construction, installation and testing all necessary documents and information to be able to operate in absolute security.

A copy of the declaration of conformity referred to in regulation 8(2)(d) or 9(2)(d) shall:

● in the case of a lift, be supplied to the Commission, the Member States and any other notified bodies, on request, by the installer of the lift together with a copy of the reports of the tests involved in the final inspection to be carried out as part of the appropriate conformity assessment procedure referred to in regulation 8(2)(b); and
● be retained by the person who draws up that declaration, for a period of 10 years:

  – in the case of a lift, from the date on which the lift was placed on the market; and
  – in the case of a safety component, from the date on which safety components of that type were last manufactured by that person.

Conformity assessments procedures (regulation 13)

For the purposes of regulation 8(2)(b) or 9(2)(b), the appropriate conformity assessment procedure shall be:

- in the case of a lift, one of the procedures set out in paragraph 2 below; and
- in the case of a safety component, one of the procedures set out in paragraph 3 below.

The procedures referred to in paragraph 1(a) above are:

- if the lift was designed in accordance with a lift having undergone an EC-type examination, it shall be constructed, installed and tested by implementing:

  - the final inspection referred to in Schedule 6;
  - the quality assurance system referred to in Schedule 11; or
  - the quality assurance system referred to in Schedule 13,

  and the procedures for the design and construction stages, on the one hand, and the installation and testing stages, on the other, may be carried out on the same lift;

- if the lift was designed in accordance with a model lift having undergone an EC-type examination as referred to in Schedule 5, it shall be constructed, installed and tested by implementing:

  - the final inspection referred to in Schedule 6;
  - the quality assurance system referred to in Schedule 11; or
  - the quality assurance system referred to in Schedule 13,

  and all permitted variations between a model lift and the lifts forming part of the lifts derived from that model lift must be clearly specified (with maximum and minimum values) in the technical dossier required as part of the appropriate conformity assessment procedure;

- if the lift was designed in accordance with a lift for which a quality assurance system pursuant to Schedule 12 was implemented, supplemented by an examination of the design if the latter is not wholly in accordance with the harmonised standards, it shall be installed and constructed and tested by implementing, in addition:

  - the final inspection referred to in Schedule 6;
  - the quality assurance system in accordance with Schedule 11; or
  - the quality assurance system in accordance with Schedule 13;
  - the unit verification procedure, referred to in Schedule 9, by a notified body; or
  - the quality assurance system in accordance with Schedule 12, supplemented by an examination of the design if the latter is not wholly in accordance with the transposed harmonised standards.

The procedures referred to in paragraph 1(b) are as follows:

- to submit the model of the safety component for EC-type examination in accordance with Schedule 5 and for production checks by a notified body in accordance with Schedule 10;
- to submit the model of the safety component for EC-type examination in accordance with Schedule 5 and operate a quality assurance system in accordance with Schedule 7 for checking production; or
- to operate a full quality assurance system in accordance with Schedule 8.

Requirements fulfilled by the person who places a lift or safety component on the market (regulation 14)

Where in the case of a lift or safety component, any of the requirements of regulations 8, 9, 11 and 13 to be fulfilled by the installer of the lift or the manufacturer of the safety component or, in the case of the latter, his authorised representative established in the Community, have not been so fulfilled such requirements may be fulfilled by the person who places that lift or safety component on the market.

Nothing in this regulation shall affect the power of an enforcement authority to take action in respect of the installer of the lift, the manufacturer of the safety component or, in the case of the latter, his authorised representative established in the Community in respect of a contravention of or a failure to comply with any of those requirements.

Notified bodies (regulation 15)

For the purposes of these regulations, a notified body is a body which has been appointed to carry out one or more of the conformity assessment procedures mentioned and referred to in regulation 13 which has been:

- appointed as a notified body in the UK pursuant to regulation 16; or
- appointed by a Member State other than the UK, and has been notified by the Member State concerned to the Commission and the other Member States pursuant to Article 9(1) of the Lifts Directive.

Notified bodies appointed by the Secretary of State (regulation 16)

This regulation deals with the administrative procedures, conditions of appointment, etc. in respect of notified bodies.

Fees (regulation 17)

Notified bodies may charge fees in connection with the undertaking of duties under the regulations.

Application of Schedule 15 (regulation 19)

Enforcement arrangements are outlined in Schedule 15 to the regulations, including the provisions relating to the service of notices and other action by the HSE under section 20 of the HSWA.

Offences (regulation 20)

Any person who:

- contravenes or fails to comply with regulations 8, 9, 10 or 11(1) or (2); or
- fails to supply or retain a copy of the declaration of conformity as required by regulation 11(3),

    shall be guilty of an offence.

Penalties (regulation 21)

A person guilty of an offence under regulation 20(a) shall be liable on summary conviction:

- to imprisonment for a term not exceeding 3 months; or
- to a fine not exceeding level 5 on the standard scale,

    or to both.

A person guilty of an offence under regulation 20(b) shall be liable on summary conviction to a fine not exceeding level 5 on the standard scale.

Defence of due diligence (regulation 22)

Subject to the provisions of this regulation, in proceeding against any person for an offence under regulation 20 it shall be a defence for that person to show that he took all reasonable steps and exercised all due diligence to avoid committing the offence.

Liability of persons other than the principal offender (regulation 23)

Where the commission by any person of an offence under regulation 20 is due to the act or default committed by some other person in the course of any business of his, the other person shall be guilty of the offence and may be proceeded against and punished by virtue of this paragraph whether or not proceedings are taken against the first-mentioned person.

Where a body corporate is guilty of an offence under these regulations (including where it is guilty by virtue of paragraph 1 above in respect of any act or default which is shown to have been committed with the consent or connivance of, or to be attributable to any neglect on the part of, any director, manager, secretary or other similar officer of the body corporate or any person who was purporting to act in any such capacity he, as well as the body corporate, shall be guilty of that offence and shall be liable to be proceeded against and punished accordingly.

THE MANAGEMENT OF HEALTH AND SAFETY AT WORK REGULATIONS 1999

These regulations (MHSWR) are accompanied by an ACOP issued by the HSC. The duties, because of their wide-ranging general nature, overlap with many existing regulations, such as the Control of Substances Hazardous to Health Regulations (COSHHR) 1994. Where duties overlap, compliance with the duty in the more *specific* regulation will normally be sufficient to comply with the corresponding duty in the MHSWR. However, where the duties in the MHSWR go *beyond* those in the more specific regulations, additional measures will be needed to comply fully with the MHSWR.

Because of their importance in the areas of health and safety management, the principal requirements of the regulations are dealt with below.

Interpretation (regulation 1)

The important definitions included here are as follows:

- *the 1996 Act* means the Employment Rights Act 1996
- *the assessment* means, in the case of an employer, the assessment made by him in accordance with regulation 3
- *child*
  - as respects England and Wales, means a person who is not over compulsory school age, construed in accordance with section 8 of the Education Act 1996; and
  - as respects Scotland, means a person who is not over compulsory school age, construed in accordance with section 31 of the Education (Scotland) Act 1980
- *Given birth* means delivering a live child or, after 24 weeks of pregnancy, a stillborn child
- *New or expectant mother* means an employee who is pregnant, who has given birth within the previous 6 months, or who is breastfeeding
- *employment business* – means a business (whether or not carried out with a view to profit and whether or not carried out in conjunction with any other business) that supplies people (other than seafarers) who are employed in it to work for and under the control of others in any capacity
- *fixed-term contract of employment* – means a contract of employment for a specific term that is fixed in advance or can be ascertained in advance by reference to some relevant circumstance

- *the preventive and protective measures* – means the measures that have been identified by the employer or by the self-employed person in consequence of the assessment as to the measures they need to take to comply with the requirements and prohibitions imposed on them by or under the relevant statutory provisions and Part II of the Fire Precautions (Workplace) Regulations 1997.
- *Young person* means any person who has not attained the age of 18 years.

The disapplication of these regulations (regulation 2)

These regulations shall not apply to or in relation to the master or crew of a sea-going ship or to the employer of these people in respect of the normal shipboard activities of a ship's crew under the direction of the master.

Regulations 3(4), (5), 10(2) and 19 shall not apply to occasional work or short-term work involving:

- domestic service in a private household; or
- work regulated as not being harmful, damaging or dangerous to young people in a family undertaking.

Risk assessment (regulation 3)

4

Every employer shall make a suitable and sufficient assessment of:

- the risks to the health and safety of their employees to which they are exposed while at work
- the risks to the health and safety of those not in their employ arising out of or in connection with the conduct by them of their undertaking, for the purpose of identifying the measures they need to take to comply with the requirements and prohibitions imposed on them by or under the relevant statutory provisions and by Part II of the Fire Precautions (Workplace) Regulations 1997.

Similar provisions to these apply in the case of self-employed people (excluding the Fire Precautions (Workplace) Regulations 1997).

Any assessment shall be reviewed by the employer or self-employed person who made it if:

- there is reason to suspect that it is no longer valid
- there has been a significant change in the matters to which it relates,

and where, as a result of any such review, changes to an assessment are required, the employer or self-employed person shall make them.

An employer shall not employ a *young person* unless he has, in relation to risks to the health and safety of young persons, made or reviewed an assessment as described above.

In making or reviewing the assessment, an employer who employs or is to employ a young person shall take particular account of:

- the inexperience, lack of awareness of risks and immaturity of young persons
- the fitting-out and layout of the workplace and the workstations
- the nature, degree and duration of exposure to physical, biological and chemical agents
- the form, range and use of work equipment and the way in which it is handled
- the organisation of processes and activities
- the extent of the health and safety training provided to young persons
- risks from processes, agents and work listed in the Annex to the Council Directive 94/33/EC on the protection of young people at work.

Where the employer employs five or more employees, he shall record:

- the significant findings of the assessment
- any group of his employees identified by it as being especially at risk.

Comment

Risk assessment is the principal feature of all modern protective legislation. Information on the practical application of risk assessment is provided in the ACOP.

Principles of prevention to be applied (regulation 4)

Where an employer implements any preventive and protective measures he shall do so on the basis of the principles specified in Schedule 1 to these regulations.

Health and safety arrangements (regulation 5)

Every employer shall make and give effect to such arrangements as are appropriate, having regard to the nature of his activities and the size of his undertaking, for the effective planning, organisation, control, monitoring and review of the preventive and protective measures.

Where the employer employs five or more employees, he shall record these arrangements.

Comment

There is a need here to consider the systems necessary to ensure the effective management of health and safety requirements. Such systems should be integrated with other management systems, such as those for the financial, personnel, production, engineering, purchasing and other areas of management activity. The main elements of management practice, that is, planning, organising, controlling, monitoring and reviewing, should be taken into account.

Health surveillance (regulation 6)

Every employer shall ensure that their employees are provided with such health surveillance as is appropriate, having regard to the risks to their health and safety that are identified by the assessment.

Comment

Health surveillance may already be required in order to comply with existing legislation, such as the Control of Substances Hazardous to Health Regulations (COSHHR) 1999. However, health surveillance may also be necessary where the risk assessment under these MHSWR indicates that:

- there is an identifiable disease or adverse health condition related to the work concerned
- valid techniques are available to detect indications of the disease or condition, such as certain forms of biological monitoring
- there is a reasonable likelihood that the disease or condition may occur under the particular conditions of work
- surveillance is likely to further the protection of the health of the employees to be covered.

The primary objective of any health surveillance activity is to detect adverse health effects at an early stage, thereby enabling further harm to be prevented.

Health and safety assistance (regulation 7)

All employers (except a qualified self-employed person or a partnership that includes a qualified person; see below) shall appoint one or more competent persons to assist them in undertaking the measures they need to take to comply with the requirements and prohibitions imposed on them by or under the relevant statutory provisions.

Where an employer appoints competent persons in accordance with the above, he shall make arrangements for ensuring adequate co-operation between them.

The employer shall ensure that the number of persons appointed under the first paragraph, the time available for them to fulfil their functions and the means at their disposal are adequate for the size of their undertaking, the risks to which their employees are exposed and the distribution of those risks throughout the undertaking.

The employer shall ensure that:

- any person appointed by them in accordance with the first paragraph who is not in their employ:
    - is informed of the factors known, or suspected by the employer to affect the health and safety of any other person who may be affected by the conduct of their undertaking
    - has access to the information referred to in regulation 10

- any person appointed by them in accordance with the first paragraph is given such information about any person working in their undertaking who is:

    - employed by them under a fixed-term contract of employment
    - employed in an employment business, as is necessary to enable that person to carry out properly the function specified in the paragraph.

A person shall be regarded as competent for the purposes of the first paragraph where they have sufficient training and experience or knowledge and other qualities to undertake properly the measures referred to in that paragraph.

The first paragraph shall *not* apply to a self-employed employer who is not in partnership with any other person where they have sufficient training and experience or knowledge and other qualities to undertake properly the measures referred to in that paragraph.

The first paragraph shall not apply to individuals who are employers and who are together carrying on business in partnership where at least one of the individuals concerned has sufficient training and experience or knowledge and other qualities to:

- properly undertake the measures they need to take to comply with the requirements and prohibitions imposed on them by or under the relevant statutory provisions
- properly assist their fellow partners in undertaking the measures they need to take to comply with the requirements and prohibitions imposed on them or under the relevant statutory provisions.

Where there is a competent person in the employer's employment, that person shall be appointed for the purposes of the first paragraph in preference to a competent person not in his employment.

Comment

The concept of 'competent persons' is not new to health and safety legislation, the appointment of such people being required under the Construction (Health, Safety and Welfare) Regulations 1996, Noise at Work Regulations 1989 and so on.

The degree of competence of the competent person for the purposes of these regulations will depend on the risks identified in the risk assessment. While there is no specific emphasis on the qualifications of such people, broadly, a competent person should have such skill, knowledge and experience as to enable them to identify defects and understand the implications of these defects. The depth of training, accountabilities and responsibilities, authority and level of reportability within the management system of the competent person is, therefore, important. The competent person should not be placed so far down the management system that their recommendations carry no weight.

The procedures for serious and imminent danger and for danger areas (regulation 8)

Every employer shall:

- establish and, where necessary, give effect to appropriate procedures to be followed in the event of serious and imminent danger to people at work in their undertaking
- nominate a sufficient number of competent people to implement these procedures in so far as they relate to the evacuation from the premises of employees and others at work
- ensure that none of their employees have access to any area occupied by them to which it is necessary to restrict access on grounds of health and safety unless the employee concerned has received adequate health and safety instruction.

Without prejudice to the generality of the first point above, the procedures referred to in it shall:

- so far as is practicable, require any one at work who is exposed to serious and imminent danger to be informed of the nature of the hazard and of the steps to be taken to protect them from it
- enable those concerned (if necessary by taking appropriate steps in the absence of guidance or instruction and in the light of their knowledge and the technical means at their disposal) to stop work and immediately proceed to a place of safety in the event of their being exposed to serious, imminent or unavoidable danger
- save in exceptional cases for reasons duly substantiated (which cases and reasons shall be specified in those procedures), require the people concerned to be prevented from resuming work in any situation where there is still a serious imminent danger.

A person shall be regarded as competent for the purposes of the second point given in the first paragraph where they have sufficient training and experience or knowledge and other qualities to enable them to implement the evacuation procedures referred to there properly.

Comment

The aim here is that employers establish procedures to be followed by workers if situations present serious and imminent danger, and make clear under what circumstances they should stop work and move to a place of safety. Fundamentally, the requirement is for an organisation's emergency plans/procedures to cover foreseeable high-risk situations, such as fire, explosion and so on and ensure that staff are trained in these procedures.

Contacts with external services (regulation 9)

Every employer shall ensure that any necessary contacts with external services are arranged, particularly as regards first aid, emergency medical care and rescue work.

Information for employees (regulation 10)

Every employer shall provide their employees with comprehensible and relevant information on:

- the risks to their health and safety identified by the assessment
- the preventive and protective measures
- the procedures referred to in regulation 8 and the measures referred to in regulations 4(2)(a) of the Fire Precautions (Workplace) Regulations 1997
- the identity of those persons nominated by him in accordance with regulation 8 and regulation 4(2)(b) of the Fire Precautions (Workplace) Regulations 1997
- the risks notified to him in accordance with regulation 11.

Every employer shall, before employing a *child*, provide a parent of the child with comprehensible and relevant information on:

- the risk to the child's health and safety identified by the assessment
- the preventive and protective measures
- the risks notified in accordance with regulation 11.

Comment

The significant feature of this regulation is that the information provided to workers must be 'comprehensible', that is, written in such a way as to be easily understood by the people to whom it is addressed. On this basis, the mode of presentation of such information should take account of their level of training, knowledge and experience. It may also need to consider people with language difficulties or with disabilities that may impede their receipt of information. Information can be provided in whatever form is most suitable in the circumstances, so long as it is comprehensible, such as a staff handbook, posters or other media. Specific provisions apply in the case of the employment of children.

Co-operation and co-ordination (regulation 9)

Where two or more employers share a workplace (whether on a temporary or permanent basis), each such employer shall:

- co-operate with the other employers concerned so far as is necessary to enable them to comply with the requirements and prohibitions imposed on them by or under the relevant statutory provisions and by Part II of the Fire Precautions (Workplace) Regulations 1997
- taking into account the nature of their activities, take all reasonable steps to co-ordinate the measures they take to comply with the requirements and prohibitions imposed on him by or under the relevant statutory provisions and by Part II of the Fire Precautions (Workplace) Regulations 1997 with the measures the other employers concerned are taking to comply with those that apply to them

- take all reasonable steps to inform the other employers concerned of the risks to their employees' health and safety arising from or in connection with how they conduct their undertaking.

The first paragraph applies to employers sharing a workplace with self-employed people and to self-employed people sharing a workplace with other self-employed people in the same way that it applies to employers sharing a workplace with other employers, the words 'employers' and 'employees' being construed accordingly.

Comment

This regulation makes provision for employers jointly occupying a work site, such as in cases of construction sites, trading estates or office blocks, to co-operate and co-ordinate their health and safety activities, particularly where there are risks that are common to everyone, irrespective of their work activity. Fields of co-operation and co-ordination include health and safety training, safety monitoring procedures, emergency arrangements and welfare amenity provisions. In some cases, the appointment of a health and safety co-ordinator should be considered.

People working in host employers' or self-employed people's undertakings (regulation 12)

Every employer and every self-employed person shall ensure that the employer of any employees from an outside undertaking who are working in their undertaking are provided with comprehensible information on:

- the risks to those employees' health and safety arising out of or in connection with the conduct by the host employer's or self-employed person's undertaking
- the measures taken by the host employer or self-employed person in compliance with the requirements and prohibitions imposed on them by or under the relevant statutory provisions and by Part II of the Fire Precautions (Workplace) Regulations 1997 in so far as the requirements and prohibitions relate to those employees.

The first paragraph shall apply to a self-employed person who is working in the undertaking of an employer or a self-employed person as it applies to employees from an outside undertaking who are working there, and the words 'employer', 'employees' and 'self-employed person' shall be construed accordingly. The reference to the Fire Precautions (Workplace) Regulations 1997 does not apply to a self-employed person working on someone else's premises.

Every employer shall ensure that every person working in their undertaking who is not their employee and every self-employed person (who is not themselves an employer) shall ensure that any person working in their undertaking is provided with appropriate instructions and comprehensible information regarding any risks to that person's health and safety that arise from the way they conduct their undertaking.

Every host employer shall:

- ensure that the employer of any employees from an outside undertaking working in their undertaking is provided with sufficient information to enable the employees' employer to identify any person nominated by the host employer in accordance with regulation 8 to implement evacuation procedures as far as these employees are concerned
- take all reasonable steps to ensure that the employees from an outside undertaking receive sufficient information to enable them to identify any person nominated by the host employer in accordance with regulation 8 to implement evacuation procedures as far as they are concerned.

This last paragraph above shall also apply to a self-employed person who is working in an employer's undertaking and the words 'employer' and 'employees' shall be construed accordingly.

Comment

This regulation applies to a wide range of work activities where employees undertake work in other people's premises, such as contract cleaners, maintenance staff and so on. These people must be provided with adequate information and instructions regarding relevant risks to their health and safety.

Capabilities and training (regulation 13)

Every employer shall, in entrusting tasks to their employees, take into account their capabilities as regards health and safety.

Every employer shall ensure that their employees are provided with adequate health and safety training:

- on their being recruited into the employer's undertaking
- on their being exposed to new or increased risks because of:

  - their being transferred or given a change of responsibilities within the employer's undertaking
  - the introduction of new work equipment into or a change respecting work equipment already in use within the employer's undertaking
  - the introduction of new technology into the employer's undertaking
  - the introduction of a new system of work into or a change respecting a system of work already in use within the employer's undertaking.

The training referred to in the points above shall:

- be repeated periodically where appropriate
- be adapted to take account of new or changed risks to the health and safety of the employees concerned
- take place during working hours.

Comment

This regulation introduced, for the first time in health and safety legislation, a human factors approach to ensuring appropriate levels of health and safety provision. When allocating work to employees, employers should ensure that the demands of the job do not exceed the employees' ability to carry out the work without risk to themselves and others. There is a need, therefore, for employers to consider both the physical and mental abilities of employees before allocating tasks, together with their level of knowledge, training and experience. Training, in particular, should reflect this aspect of human capability.

Employees' duties (regulation 14)

Every employee shall use any machinery, equipment, dangerous substance, transport equipment, means of production or safety device provided for them by their employer in accordance both with any training in the use of this equipment and the instructions for using it that have been provided by the employer in compliance with the requirements and prohibitions imposed on them by or under the relevant statutory provisions.

Every employee shall inform their employer or any other employee of that employer with specific responsibility for the health and safety of fellow employees:

- of any work situation that a person with the first-mentioned employee's training and instruction would reasonably consider represented a serious and immediate danger to health and safety
- of any matter that would reasonably be considered to represent a short-coming in the employer's protection arrangements for health and safety,

in so far as that situation or matter either affects the health and safety of the employee or arises out of or in connection with their own activities at work, and has not previously been reported to their employer or to any other employee of that employer in accordance with this paragraph.

Comment

This regulation reinforces and expands the duties of employers towards employees under section 7 of the HSWA. In the light of these requirements, employers should install some form of hazard reporting system, whereby employees can report hazards to their employer or appointed competent person. Such a system should ensure a prompt response where hazards are identified, with the competent person signing off the hazard report when the hazard has been eliminated or controlled.

Temporary workers (regulation 15)

Every employer shall provide any person they have employed under a fixed-term contract with comprehensible information on:

- any special occupational qualifications or skills required to be held by that employee if they are to carry out their work safely
- any health surveillance required to be provided to that employee by or under any of the relevant statutory provisions

and shall provide such information before the employee concerned commences their duties.

Every employer and every self-employed person shall provide any person employed in an employment business who is to carry out work in their undertaking with similar comprehensible information to that mentioned in the first paragraph above.

Every employer and every self-employed person shall ensure that every person running an employment business, the employees of which are to carry out work in their undertaking, is provided with comprehensible information on:

- any special occupational qualifications or skills that these employees are required to hold if they are to carry out their work safely
- the specific features of the jobs to be filled by these employees (in so far as those features are likely to affect their health and safety)

and the person running the employment business concerned shall ensure that the information so provided is given to the said employees.

Comment

This regulation supplements previous regulations as it requires the provision of information but with additional requirements for temporary workers, that is, those employed on fixed duration contracts and those employed in employment businesses, but working under the control of a user company. The use of temporary workers will also have to be notified to the competent person.

Risk assessment in respect of new or expectant mothers (regulation 16)

Where:

- the persons working in an undertaking include women of child-bearing age; and
- the work is of a kind which could involve risk, by reasons of her condition, to the health and safety of a new or expectant mother, or to that of her baby, from any processes or working conditions, or physical, biological or chemical agents, including those specified in Annexes I and II of Council Directive 92/85/EEC on the introduction of measures to encourage improvements in the safety and health at work of pregnant workers and workers who have recently given birth or are breastfeeding,

the assessment required by regulation 3 shall also include an assessment of such risk.

Where, in the case of an individual employee, any other action which the employer is required to take under the relevant statutory provisions would not avoid the risk referred to in the first paragraph, the employer shall, if it is reasonable to do so, and would avoid such risks, alter her working conditions or hours of work.

If it is not reasonable to alter the working conditions or hours of work, or if it would not avoid such risk, the employer shall, subject to section 67 of the Employment Rights Act 1996, suspend the employee from work so long as is necessary to avoid such risk.

References in the above paragraphs to risk mean, in relation to risk from any infectious or contagious disease, a level of risk at work which is above the level of risk to which a new or expectant mother may be expected to be exposed outside the workplace.

Certificate from registered medical practitioner in respect of new or expectant mothers (regulation 17)

Where:

- a new or expectant mother works at night; and
- a certificate from a registered medical practitioner or a registered midwife shows that it is necessary for her health or safety that she should not be at work for any period of such work identified in the certificate,

the employer shall, subject to section 67 of the Employment Rights Act 1996, suspend her from work for so long as is necessary for her health or safety.

Notification by new or expectant mothers (regulation 18)

The employer is not required to take any action under regulation 16 until an employee has notified the employer in writing that she is pregnant, has given birth within the previous six months, or is breastfeeding.

Nothing in regulations 16 or 17 shall require the employer to maintain action taken in relation to an employee

- where her working conditions or hours were altered, or she was suspended from work and where she has failed within a reasonable time of receiving a written request from her employer to produce for the employer's inspection a certificate from a registered medical practitioner or registered midwife showing that she is pregnant;
- once the employer knows that she is no longer a new or expectant mother; or
- if the employer cannot establish whether she remains a new or expectant mother.

Protection of young persons (regulation 19)

All employers shall ensure that young persons they employ are protected at work from any risks to their health or safety which are a consequence of their lack of experience, or absence of awareness of existing or potential risks or of the fact that young persons have not yet fully matured.

Subject to the paragraph below, no employer shall employ a young person for work:

- which is beyond his physical or psychological capacity;
- involving harmful exposure to agents which are toxic or carcinogenic, cause heritable genetic damage or harm to the unborn child or which in any other way chronically affect human health;
- involving harmful exposure to radiation;
- involving the risk of accidents which it may reasonably be assumed cannot be recognised or avoided by young persons owing to their insufficient attention to safety or lack of experience or training; or
- in which there is a risk to health from:

  - extreme cold or heat
  - noise
  - vibration,

  and in determining whether work will involve harm or risks for the purposes of this paragraph, regard shall be had to the results of the assessment.

Nothing in the paragraph above shall prevent the employment of a young person who is no longer a child for work where:

- it is necessary for his training;
- the young person will be supervised by a competent person;
- any risk will be reduced to the lowest level that is reasonably practicable.

The provisions contained within this regulation are without prejudice to:

- the provisions contained elsewhere in these regulations; and
- any prohibition or restriction on the employment of any person, arising otherwise than by this regulation.

Exemption certificates (regulation 20)

These regulations may, by certification from the Secretary of State for Defence, exempt members of the forces generally from having to comply with them.

Provisions as to liability (regulation 21)

Nothing in the relevant statutory provisions affords an employer a defence in any criminal proceedings for a contravention of those provisions by reason of any act or default of:

- any employee of his
- a person appointed by him under regulation 7.

The exclusion of civil liability (regulation 22)

Breach of a duty imposed by these regulations shall not confer a right of action in any civil proceedings. (Similar provisions apply in the case of the HSWA.) This does not apply to any duty imposed by these regulations on an employer with regard to young persons or to the extent that it relates to an employee risk to new or expectant mothers.

Schedule 1

This Schedule specifies the general principles of prevention set out in Article 6(2) of the Council Directive 89/391/EEC.

General principles of prevention

- avoiding risks
- evaluating the risks that cannot be avoided
- combating the risks at source
- adapting the work to the individual, especially as regards the design of workplaces, the choice of work equipment and the choice of working and production methods, with a view, in particular, to alleviating monotonous work and work at a pre-determined work rate and to reducing their effect on health
- adapting to technical progress
- replacing the dangerous by the non-dangerous or the less dangerous
- developing a coherent overall prevention policy which covers technology, organisation of work, working conditions, social relationships and the influence of factors relating to the working environment
- giving collective protective measures priority over individual protective measures
- giving appropriate instructions to employees.

Footnote to the MHSWR

The duties specified in these regulations are of an *absolute* nature, qualified by the word 'shall', compared with, for instance, the HSWA, where the duties are qualified by the phrase 'so far as is reasonably practicable', a lower level of duty.

The 1999 version of the regulations brought in specific provisions with regard to young persons, children and pregnant workers, first aid, emergency medical care and rescue work, together with specifying general principles of accident and ill-health prevention.

THE MANUAL HANDLING OPERATIONS REGULATIONS 1992

These regulations came into force on 1 January 1993 and implement the European 'Heavy Loads' Directive. They supplement the general duties placed on employers and others by the HSWA and the broad requirements of the MHSWR 1992. They are supported by guidance notes issued by the HSE.

A number of terms are defined as follows.

- The term *injury* does not include injury caused by any toxic or corrosive substance that:
  - has leaked or spilled from a load
  - is present on the surface of a load but has not leaked or spilled from it
  - is a constituent part of a load.

- The term *load* includes any person and any animal.
- The term *manual handling operations* means any transporting or supporting of a load (including the lifting, putting down, pushing, pulling, carrying or moving thereof) by hand or by bodily force.

The duties of employers (regulation 4)

Each employer shall so far as is reasonably practicable, avoid the need for their employees to undertake any manual handling operations at work that involve a risk of their being injured.

Where it is *not* reasonably practicable to avoid such a need, the employer shall:

- make a suitable and sufficient assessment of all such manual handling operations to be undertaken by them, having regard to the factors that are specified in the Schedule
- take appropriate steps to reduce the risk of injury to these employees to the lowest level reasonably practicable
- take appropriate steps to provide any of these employees with general indications and, where it is reasonably practicable to do so, precise information on:

  - the weight of each load
  - the heaviest side of any load, the centre of gravity of which is not positioned centrally.

In determining for the purposes of this regulation whether manual handling operations at work involve a risk of injury and in determining the appropriate steps to reduce that risk regard shall be had in particular to:

(a) the physical suitability of the employee to carry out the operations;

(b) the clothing, footwear or other personal effects he is wearing;

(c) his knowledge and training;

(d) the results of any relevant risk assessment carried out pursuant to regulation 3 of the Management of Health and Safety at Work Regulations 1999;

(e) whether the employee is within a group of employees identified by that assessment as being especially at risk; and

(f) the results of any health surveillance provided pursuant to regulation 6 of the Management of Health and Safety at Work Regulations 1999.

Any such assessment shall be reviewed by the employer if:

- there is reason to suspect that it is no longer valid
- there has been a significant change in the manual handling operations to which it relates

and where, as a result of any such review, changes to an assessment are required, the relevant employer shall make them.

The regulations further require that each employee while at work shall make full and proper use of any system of work provided for their use by their employer to reduce the risk of injury as mentioned above in compliance with these regulations.

Schedule 1

Factors the employer must bear in mind and questions they must consider when making an assessment of manual handling operations are as follows.

4

The tasks

Do they involve:

- holding or manipulating loads at distance from the trunk?
- unsatisfactory bodily movement or posture, especially:
  - twisting the trunk
  - stooping
  - reaching upwards?
- excessive movement of loads, especially:
  - excessive lifting or lowering distances
  - excessive carrying distances?
- excessive pushing or pulling of loads?
- risk of sudden movement of loads?
- frequent or prolonged physical effort?
- insufficient rest or recovery periods?
- a rate of work imposed by a process?

The loads

Are they:

- heavy?
- bulky or unwieldy?
- difficult to grasp?
- unstable or with contents likely to shift?
- sharp, hot or otherwise potentially damaging?

The working environment

Are there:

- space constraints preventing good posture?
- uneven, slippery or unstable floors?
- variations in level of floors or work surfaces?
- extremes of temperature or humidity?
- conditions causing ventilation problems or gusts of wind?
- poor lighting conditions?

Individual capability

Does the job:

- require unusual strength, height or other feature?
- create a hazard to those who might reasonably be considered to be pregnant or have a health problem?
- require special information or training for its safe performance?

Other factors

Is movement or posture hindered by personal protective equipment or clothing?

For an example of an assessment check-list, see Figure 4.3.

THE NOISE AT WORK REGULATIONS 1989

These regulations came into operation on 1 January 1990 and are accompanied by a number of Noise Guides issued by the HSE. They give effect in Great Britain to provisions of the Council Directive 86/188/EEC on the protection of workers from the risks related to exposure to noise.

The regulations bring in the concepts of 'daily personal noise exposure' and 'action levels' (see Figure 4.4).

The relevant definitions regarding these concepts are:

- *daily personal noise exposure,* which means the level of daily personal noise exposure of an employee ascertained in accordance with Part 1 of the Schedule to the regulations, but taking no account of the effect of any personal ear protector used
- *exposed* means exposed to while at work, and *exposure* shall be construed accordingly
- the *first action level* means a daily personal noise exposure of 85 decibels (A)
- the *peak action level* means a level of peak sound pressure of 200 pascals
- the *second action level* means a daily personal noise exposure of 90 decibels (A).

Manual handling of loads

EXAMPLE OF AN ASSESSMENT CHECKLIST

Note: This checklist may be copied freely. It will remind you of the main points
to think about while you:
– consider the risk of injury from manual handling operations
– identify steps that can remove or reduce the risk
– decide your priorities for action.

SUMMARY OF ASSESSMENT	Overall priority for remedial action: Nil/Low/Med /High*
Operations covered by this assessment	Remedial action to be taken: ....
Locations:	Date by which action is to be taken ....
Personnel involved:	Date for reassessment: ....
Date of assessment:	Assessor's name: ....Signature......................

circle as appropiate

Section A – Preliminary:
Q1 Do the operations involve a significant risk of injury? Yes/No*
If 'Yes' go to Q2. If 'No' the assessment need go no further.
If in doubt answer 'Yes'. You may find the guidelines in Appendix 1 helpful.

Q2 Can the operations be avoided/mechanised/automated at reasonable cost? Yes/No*
If 'No' go to Q3. If 'Yes' proceed and then check that the result is satisfactory.

Q3 Are the operations clearly within the guidelines in Appendix 1? Yes/No*
If 'No' go to Section B. If 'Yes' you may go straight to Section C if you wish.

Section C – Overall assessment of risk:
Q What is your overall assessment of the risk of injury? Insignificant/Low/Med/High*
If not 'Insignificant' go to Section D. If 'Insignificant' the assessment need go no further.

Section D – Remedial action:
Q What remedial steps should be taken, in order of priority?

i ....

ii ....

iii ....

iv ....

v ....

And finally:
– complete the SUMMARY above
– compare it with your other manual handling assessments
– decide your priorities for action

– TAKE ACTION ....AND CHECK THAT IT HAS THE DESIRED EFFECT

● **FIG 4.3 Example of an assessment check-list**

Section B – More detailed assessment, where necessary:

Questions to consider: (If the answer to a question is 'Yes' place a tick against it and then consider the level of risk)		Level of risk: (Tick as appropriate)			Possible remedial action (Make rough notes in this column in preparation for completing Section D)
	Yes	**Low**	**Med**	**High**	
The tasks – do they involve:					
● holding loads away from trunk?					
● twisting?					
● stooping?					
● reaching upwards?					
● large vertical movement?					
● long carrying distances?					
● strenuous pushing or pulling?					
● unpredictable movement of loads?					
● repetitive handling?					
● insufficient rest or recovery?					
● a workrate imposed by a process?					
The loads – are they:					
● heavy?					
● bulky/unwieldy?					
● difficult to grasp?					
● unstable/unpredictable?					
● intrinsically harmful (eg sharp/hot?)					
The working environment – are there:					
● constraints on posture?					
● poor floors?					
● variations in levels?					
● hot/cold/humid conditions?					
● strong air movements?					
● poor lighting conditions?					
Individual capability – does the job:					
● require unusual capability?					
● hazard those with a health problem?					
● hazard those who are pregnant?					
● call for special information/training?					
Other factors –					
is movement or posture hindered by clothing or personal protective equipment?					

Deciding the level of risk will inevitably call for judgement. The guidelines in Appendix 1 may provide a useful yardstick.

When you have completed Section B go to Section C.

● **FIG 4.3 continued**

The assessment of exposure (regulation 4)

Under regulation 4, every employer shall, when any of their employees is likely to be exposed to the first action level or above or to the peak action level or above, ensure that a competent person makes a noise assessment that is adequate for the purposes of:

- identifying which employees are so exposed
- providing them with information about the noise to which these employees may be exposed that will facilitate compliance with the employer's duties under regulations 7, 8, 9 and 11.

This noise assessment shall be reviewed when:

- there is reason to suspect that the assessment is no longer valid
- there has been a significant change in the work to which the assessment relates,

and where, as a result of the review, changes in the assessment are required, these changes shall be made.

Assessment records (regulation 5)

Following any noise assessment, the employer shall ensure that an adequate record (see Figure 4.5) of the assessment and of any review of it that is carried out, is kept until a further noise assessment is made.

Reducing the risk of damage to hearing (regulation 6)

This regulation places a general duty on every employer to reduce the risk of damage to the hearing of their employees due to exposure to noise to the lowest level reasonably practicable.

Reducing exposure to noise (regulation 7)

Every employer shall, when any of their employees are likely to be exposed to the second action level or above or to the peak action level or above, reduce, so far as is reasonably practicable (other than by the provision of ear protectors), the exposure to noise of that employee.

Ear protection (regulation 8)

While the emphasis under the regulations is on noise control, the regulations do recognise the need for ear protection. Under regulation 8, every employer shall ensure, so far as is practicable, that when any of their employees are likely to be exposed to the first action level or above in circumstances where

ACTION REQUIRED WHERE Lep.d IS LIKELY TO BE: (see note 1 below)	below 85dB(A)	85dB(A) 1st AL	90dB(A) 2nd AL (2)
EMPLOYER'S DUTIES **General Duty to Reduce Risk** Risk of hearing damage to be reduced to the lowest level	●	●	●
Assessment of Noise Exposure Noise assessments to be made by a Competent Person (Reg. 4) Record of assessments to be kept until a new one is made (Reg. 5)		● ●	● ●
Noise Reduction Reduce exposure to noise as far as is reasonably practicable by means other than ear protectors (Reg. 7)			●
Provision of Information To Workers Provide adequate information, instruction and training about risks to hearing, what employees should do to minimise risk, how they can obtain ear protectors if they are exposed between 85 and 90 dB(A), and their obligations under the Regulations (Reg. 11) Mark ear protection zones and notices, so far as reasonably practicable (Reg. 9)		●	● ●
Ear Protectors Ensure so far as is practicable that protectors are:- – provided to employees who ask for them (Reg.8(1)) – provided to all exposed (Reg.8(2)) – maintained and repaired (Reg.10(1)(b)) – used by all exposed (Reg.10(1)(a)) Ensure so far as is reasonably practicable that all who go into a marked area protection zone use ear protectors (Reg.9 (1)(b))		● ●	● ● ● ● (3)
Maintenance and Use of Equipment Ensure so far as is practicable that:- – all equipment provided under the Regulations is used except for the ear protectors provided between 85 and 90 dB(A) (Reg.10(1)(a)) – all equipment is maintained. (Reg.10(1)(b))		● ●	● ●
EMPLOYEE'S DUTIES **Use of Equipment** So far as practicable:- use ear protectors (Reg.10(2)) use any other protective equipment (Reg.10(2)) report any defects discovered to his/her employer (Reg. 10(2))		● ●	● ● ●
MACHINE MAKER'S AND SUPPLIER'S DUTIES **Provision of Information** Provide information on the noise likely to be generated (Reg.12)		●	●

NOTES: (1)   *The dB(A) action levels are values of daily personal exposure to noise (LEP.d)*
           (2)   *All the actions indicated at 90dB(A) are also required where the peak sound pressure above 200 Pa (140dB re 20 uPa).*
           (3)   *Lep.d is the daily personal noise exposure of an employee derived from a formula incorporated in the Schedule to the Regulations.*

● **FIG 4.4 The action levels specified in the Noise at Work Regulations 1989**

NOISE EXPOSURE RECORD

Name and address of premises, department etc. _____

_____

_____

_____

_____

Date of survey _____   Survey made by _____

Workplace Number of persons	Noise level (Leq(s) or sound level)	Daily exposure period	LEP.d dB(A)	Peak pressure (where appropriate)	Comments/ remarks

4

General comments _____

Instruments used _____

Date of last calibration _____   Signature _____

                                            Date _____

● **FIG 4.5 Example of a noise exposure record**

the daily personal noise exposure of that employee is likely to be less than 90 decibels (A), that employee is provided, at their own request, with suitable and sufficient personal ear protectors.

Furthermore, when any of their employees is likely to be exposed to the second action level or above or to the peak action level or above, every employer shall ensure, so far as is practicable, that this employee is provided with suitable ear protectors which, when properly worn, can reasonably be expected to keep the risk of damage to the employee's hearing to below that arising from exposure to the second action level or, as the case may be, to the peak action level.

Ear protection zones (regulation 8)

Regulation 8 is concerned with the demarcation and identification of ear protection zones, that is, areas where employees must wear ear protection.

Thus, every employer shall, in respect of any premises under their control, ensure, so far as is reasonably practicable, that:

- each ear protection zone is demarcated and identified by means of a sign (specified in paragraph A.3.3 of Appendix A to Part 1 of BS 5878) that includes text indicating:

  - that is an ear protection zone
  - the need for their employees to wear personal ear protectors while in any such zone

- none of their employees enters any such zone unless that employee is wearing personal ear protectors.

In this regulation, *ear protection zone* means any part of the premises referred to above where any employee is likely to be exposed to the second action level or above or to the peak action level or above, and *Part 1 of BS 5378* has the same meaning as in regulation 2(1) of the Safety Signs Regulations 1980.

The maintenance and use of equipment (regulation 10)

Every employer shall ensure, so far as is practicable, that anything provided by them:

- to or for the benefit of an employee complies with their duties under these regulations (other than personal ear protectors provided pursuant to regulation 8) and is fully and properly used
- complies with their duties under these regulations and is maintained in an efficient state, in efficient working order and in good repair.

Every employee shall, so far as is practicable, use personal ear protectors fully and properly when they are provided by their employer pursuant to regulation 7 and any other protective measures provided by their employer in compliance with the employer's duties under these regulations. If the employee discovers any defect in the protective equipment or other measures, they shall report it to their employer straight away.

The provision of information to employees (regulation 11)

Every employer shall, in respect of any premises under their control, provide each of their employees who is likely to be exposed to the first action level or above or to the peak action level or above with adequate information, instruction and training on:

- the risk of damage to that employee's hearing that such exposure may cause
- what steps that employee can take to minimise the risk
- the steps that the employee *must* take in order to obtain the personal ear protectors referred to in regulation 8
- that employee's obligations under these regulations.

The modification of the duties of manufacturers and so on regarding articles for use at work and articles of fairground equipment (regulation 12)

The duties under section 6 of the HSWA on the part of manufacturers, designers and so on were modified to include a duty to ensure that, where any such article is likely to cause any employee to be exposed to the first action level or above or to the peak action level or above, adequate information is provided concerning the noise likely to be generated by this article.

THE NOTIFICATION OF COOLING TOWERS AND EVAPORATIVE CONDENSERS REGULATIONS 1992

These regulations require the notification to local authorities of wet cooling towers and evaporative condensers, which are components of many of the air-conditioning systems found in large buildings, and of industrial cooling towers.

Knowledge of the whereabouts of such equipment is of particular help to local authorities in the investigation of outbreaks of legionnaires' disease.

Notification is done by completing a standard form available from the local authority (LA). Any changes to the information contained in the form must be notified within one month of their occurring. The LA must also be informed, as soon as reasonably practicable, when equipment ceases to be operational.

The regulations came into operation on 2 November 1992.

The information required

The information required on a notification form includes:

- the name and address of the premises where the cooling towers and evaporative condensers are situated
- the number of such devices there are on site
- the name, address and telephone number of the person in control of the premises
- brief information on the whereabouts in the premises of the equipment.

THE NOTIFICATION OF NEW SUBSTANCES REGULATIONS 1993

These regulations implement in the UK an EU-wide scheme designed to ensure that a chemical placed on the market in one Member State can be validly placed on the market in others without duplicating authorisation.

The key points of these regulations are:

- all new substances must go through a notification procedure before being placed on the market;
- each EU country has its own competent authority to whom notification must be made under the Dangerous Substances Directive as amended; in the case of the UK, this is the HSE and the Department of Environment acting jointly;
- they apply only to specialist importers and manufacturers of chemicals, the notifier being the manufacturer, importer or 'sole representative';
- notification involves the delivery to the competent authority of a technical dossier and other information set out in the regulations;
- a risk assessment is optional;
- all new substances must be notified irrespective of their potential for harm, but where a substance is notified as being a dangerous substance, then notification must include proposals for compliance with the classification and labelling requirements of CHIP 2, together with a draft CHIP 2 safety data sheet;
- a full notification is required if the amount of the new substance to be placed on the market is to be one tonne or more per year, and the technical dossier supplied must comply with Part A of Schedule 2;
- for below one tonne per year but not below 100 kilogrammes per year, the notification requirements are only for a summary of the dossier and information and the technical dossier must comply with Part B of Schedule 2; the competent authority can require a reduced notification to be in full rather than in summary form;
- following notification the competent authority is allowed a period to consider the notification documentation before the notified new substance can be placed on the market;

- there are three information levels once a substance has been placed on the market, the trigger levels being when the quantity placed reaches:

  10 tonnes per year or 50 tonnes in total
  100 tonnes per year or 500 tonnes in total
  1000 tonnes per year or 5000 tonnes in total

  in which case additional testing may be required by the competent authority;
- certain substances are treated as having been notified, including polymers containing less than 2 per cent in combined form of a new substance, small quantities for scientific research and development, and substances on the market for process-orientated research and development;
- information on various tests of the substance is required to be incorporated in the technical dossiers accompanying a new substance;
- the competent authority is required to carry out a risk assessment of the real or potential risks created by a notified substance;
- the conclusion from the risk assessment will be to allocate the substance to one of the levels of concern set out in Article 3.4 of the Risk Assessment Directive;
- pesticides or plant protection products are excluded from the scope of the regulations.

THE PERSONAL PROTECTIVE EQUIPMENT AT WORK REGULATIONS 1992–2002

People at work use a wide range of personal protective equipment (PPE), including safety boots, aprons, wellingtons, eye protectors and ear protectors, among others. Up to 1993, there was very little guidance or, indeed, legislation covering this matter, other than the duty to *provide* such equipment under, for instance, the Protection of Eyes Regulations 1974.

The PPER came into operation on 1 January 1993. As such, they revoke much of the older, more specific pieces of legislation, replacing them with one, complete set of regulations covering all forms of PPE.

The regulations:

- amend certain regulations made under the HSWA that deal with PPE, so that they fully implement the EC Directive in circumstances where they apply
- cover all aspects of the *provision, maintenance and use* of PPE at work and in other circumstances
- revoke and replace almost all pre-HSWA and some post-HSWA legislation that deals with PPE.

It should be appreciated that specific requirements of current regulations dealing with PPE, namely the Control of Lead at Work Regulations 1998, the Ionising Radiations Regulations 1999, the Control of Asbestos at Work Regulations 2002, the Control of Substances Hazardous to Health Regulations (COSHHR) 2002, the Noise at Work Regulations 1989 and the Construction (Head Protection) Regulations 1989, take precedence over the more general requirements of the PPER.

Under the regulations *personal protective equipment* means 'all equipment (including clothing affording protection against the weather) which is intended to be worn or held by a person at work and which protects him against one or more risks to his health and safety, and any addition or accessory designed to meet this objective' – a very broad definition indeed.

Regulations 4 to 12 do not apply in respect of PPE, which include:

- ordinary working clothes and uniforms that do not specifically protect the health and safety of the wearer
- an offensive weapon (within the meaning of section 1(4) of the Prevention of Crime Act 1953) used as self-defence or deterrent equipment
- portable devices for detecting and signalling risks and nuisances
- PPE used for protection while travelling on a road
- equipment used during the playing of competitive sports.

The provision of PPE (regulation 4)

Under regulation 4, every employer *shall* ensure that *suitable* PPE is provided to their employees who may be exposed to a risk to their health and safety while at work, except where and to the extent that such risk has been adequately controlled by other means that are equally or more effective. Similar provisions to these apply in the case of self-employed people.

PPE shall not be considered suitable unless:

- it is appropriate for the risk or risks involved, the conditions at the place where exposure to the risk may occur, and the period for which it is worn;
- it takes account of ergonomic requirements and the state of health of the person or persons who may wear it, and the characteristics of the workstation of each such person;
- it is capable of fitting the wearer correctly, if necessary after adjustments have been made within the range for which it is designed;
- *so far as is practicable*, it is effective in preventing or adequately controlling the risk or risks involved without increasing the overall risk.

Where it is necessary to ensure that PPE is hygienic and otherwise free of risk to health, every employer and every self-employed person shall ensure that PPE provided under this regulation is provided to a person for use only by him.

The compatibility of PPE (regulation 5)

Every employer shall ensure that, where the presence of more than one risk to health or safety makes it necessary for their employees to wear or use more than one item of PPE, such equipment is compatible and continues to be effective against the risk or risks in question. Similar provisions as above apply in the case of self-employed people.

Assessing PPE requirements (regulation 6)

As with the other regulations that came into operation at the same time, considerable emphasis is placed on assessment. Regulation 6 is quite clear on this matter, as follows:

- before choosing any PPE that they are required to provide, an employer or self-employed person shall ensure that an assessment is made to determine whether the PPE they intend to provide is suitable
- the assessment shall include:
  - an assessment of any risk or risks that have not been avoided by other means
  - the definition of the characteristics that the PPE must have in order to be effective against the risks identified in the assessment above, taking into account any risks that the equipment itself may create
  - a comparison of the characteristics of the PPE available with the characteristics referred to defined above
  - an assessment of whether the PPE is compatible with other PPE which is in use and which an employee would be required to wear simultaneously.
- The assessment shall be reviewed if:
  - there is reason to suspect that it is no longer valid
  - there has been a significant change in the matters to which it relates

and where, as a result of any such review, changes in the assessment are required, the relevant employer or self-employed person shall ensure that they are made.

The maintenance and replacement of PPE (regulation 7)

Regulation 7 places a duty on the employer to maintain PPE. The word 'maintenance' here includes the replacement or cleaning of PPE, as appropriate, and keeping it *in an efficient state, in efficient working order and in good repair*. Again, similar provisions to these apply in the case of self-employed people. From a management viewpoint , 'maintenance in an efficient state, in efficient working order and in good repair' implies the need for a formal system for checking all forms of PPE issued to staff on a regular basis and the recording of such checks.

Accommodation for PPE (regulation 8)

Employers and the self-employed must ensure that 'appropriate accommodation', such as a locker, is provided for PPE when it is not being used.

Information, instruction and training (regulation 9)

Employers must provide adequate and appropriate information to enable the employee to know:

- the risk or risks that the PPE will avoid or limit
- the purpose for which and the manner in which the PPE is to be used
- any action to be taken by the employee to ensure that the PPE remains in an efficient state, in efficient working order and in good repair and shall ensure that such information is kept available to employees.

The information and instruction provided shall not be considered adequate and appropriate unless it is comprehensible to the people to whom it is given.

Without prejudice to the generality of paragraph 1 above the employer shall, where appropriate, and at suitable intervals, organise demonstrations in the wearing of PPE.

The use of PPE (regulation 10)

Under regulation 10, there are the following duties on both employers and employees:

- every employer shall take *all reasonable steps* to ensure that any PPE provided to their employees is properly used
- every employee shall use any PPE in accordance with both any training in the use of the PPE concerned that has been received by them and the instructions regarding the use of it that have been given to them
- every self-employed person shall make full and proper use of any PPE
- every employee and self-employed person shall take all reasonable steps to ensure that the PPE is returned to the accommodation provided for it after use.

Reporting loss or defect (regulation 11)

Employees who have been provided with PPE are required to report to their employer any loss of or obvious defect in it straight away.

HSE guidance notes

These regulations are accompanied by extensive guidance on the selection, use and maintenance of PPE, including a specimen risk survey table for the use of PPE.

PRESSURE EQUIPMENT REGULATIONS 1999

These extensive regulations implement the Pressure Equipment Directive 97/23/EC of the European Parliament and of the Council on the approximation of the laws of the Member States relating to pressure equipment. They apply to the placing on the market and putting into service of a wide range of pressure equipment, such as pressure vessels, piping, safety and pressure accessories, along with fittings and ancillary equipment, including heat exchangers. They are accompanied by a series of Schedules dealing with, for instance, essential safety requirements for pressure equipment, conformity assessment procedures, CE marking and enforcement arrangements.

Interpretation (regulation 2)

Assembly means several pieces of pressure equipment assembled by a manufacturer to constitute an integrated and functional whole.

Enforcement authority means:

(a) in the case of pressure equipment and assemblies for use in the workplace:

 (i) in Great Britain, the Health and Safety Executive; and

 (ii) in Northern Ireland, the Health and Safety Executive for Northern Ireland;

(b) in the case of pressure equipment and assemblies for private use or consumption:

 (i) in Great Britain, weights and measures authorities; and

 (ii) in Northern Ireland, every district council.

Essential safety requirements means the requirements set out in Schedule 2.

European approval for materials or *European approval for pressure equipment materials* means a technical document defining the characteristics of materials intended for repeated use in the manufacture of pressure equipment which are not covered by any harmonised standard as referred to in paragraph 4.2(b) of Schedule 2.

Fluid means gases, liquids and vapours in pure phase as well as mixtures thereof; a fluid may contain a suspension of solids.

Fluid in Group 1 means a fluid being a dangerous fluid, that is to say a substance or preparation covered by the definitions of Article 2(2) of Council Directive 67/548/EEC of 27 June 1967 on the approximation of the laws, regulations and administrative provisions relating to the classification, packaging and labelling of dangerous substances as specified in that Article as:

- explosive
- extremely flammable
- highly flammable
- flammable (where the maximum allowable temperature is above flash-point)
- very toxic
- toxic
- oxidising.

Fluid in Group 2 means a fluid which is not a fluid in Group 1.

Maximum allowable pressure or *PS* means the maximum pressure for which the equipment is designed, as specified by the manufacturer, defined at a location specified by the manufacturer, being the location of connection of protective or limiting devices or the top of equipment or, if either of the foregoing are not appropriate, any point specified by the manufacturer.

Maximum allowable temperature or *TS* means the maximum or minimum temperatures, as the case may be, for which the equipment is designed, as specified by the manufacturer.

Nominal size or *DN* means a numerical designation of size which is common to all components in a piping system other than components indicated by outside diameters or by thread size; that is to say a convenient round number for reference sizes which is only loosely related to manufacturing dimensions and designated by the letters *DN* followed by a number.

Permanent joints means joints which cannot be disconnected except by destructive methods.

Piping means piping components intended for the transport of fluids when connected together for integration into a pressure system, such components include in particular a pipe or system of pipes, tubing, fittings, expansion joints, hoses, other pressure-bearing components as appropriate or heat exchangers consisting of pipes for the purpose of cooling or heating air.

Pressure means pressure relative to atmospheric pressure being gauge pressure; vacuum is designated by a negative value.

Pressure accessories means devices with an operational function and having pressure-bearing housings.

Pressure equipment means vessels, piping, safety accessories and pressure accessories; where applicable, pressure equipment includes elements attached to pressurised parts, such as flanges, nozzles, couplings, supports, lifting lugs, and similar.

Responsible person means:

(a) the manufacturer or his authorised representative established within the Community; or

(b) where neither the manufacturer nor his authorised representative is established within the Community, the person who places the pressure equipment or assembly on the market or puts it into service as the case may be.

Safe in relation to pressure equipment or an assembly, means that the pressure equipment or assembly when properly installed and maintained and used for its intended purpose is not liable to endanger the health or safety of persons and, where appropriate, domestic animals or property, and, where the context admits, cognate expressions shall be construed accordingly.

Safety accessories means devices designed to protect pressure equipment against the allowable limits being exceeded; such devices include:

- devices for directing pressure limitation, such as safety valves, bursting disc safety devices, buckling rods, controlled safety pressure relief systems; and
- limiting devices, which either activate the means for correction or provide for shutdown or shutdown and lockout, such as pressure switches or temperature switches or fluid level switches and safety related measurement control and regulation devices.

Standard means a technical specification approved by a recognised standardising body for repeated or continuous application, with which compliance is not compulsory; and, for the avoidance of doubt, includes a harmonised standard or a transposed harmonised standard.

Supply includes offering to supply, agreeing to supply, exposing for supply and possessing for supply and cognate expressions shall be construed accordingly.

Transposed harmonised standard means a national standard of a Member State which transposes a harmonised standard.

Vessel means a housing designed and built to contain fluids under pressure including its direct attachments up to the coupling point connecting it to other equipment; a vessel may be composed of more than one chamber.

Volume or *V* means the internal volume of a chamber, including the volume of nozzles to the first connection or weld and excluding the volume of permanent internal parts.

Pressure equipment and assemblies (regulation 3)

These regulations apply to pressure equipment and assemblies with a maximum allowable PS greater than 0.5 bar.

Excluded pressure equipment and assemblies (regulation 4)

These regulations shall not apply to the products listed in Schedule 1.

For the avoidance of doubt these regulations shall not apply to the assembly of pressure equipment on the site and under the responsibility of the user.

General duty relating to the placing on the market or putting into service of pressure equipment (regulation 7)

Subject to regulation 11, no person who is a responsible person shall place on the market or put into service any pressure equipment described in paragraph 2 unless the requirements of paragraph 3 have been complied with in relation to it.

For the purposes of paragraph 1, the following are pressure equipment:

(a) vessels, except those referred to in sub-paragraph (b), for:

   (i) gases, liquefied gases, gases dissolved under pressure, vapours and also those liquids whose vapour pressure at the maximum allowable temperature is greater than 0.5 bar above normal atmospheric pressure (1013 mbar) within the following limits:

   (aa) for fluids in Group 1, with a volume greater than 1L and a product of PS and V greater than 25 bar-L, or with a pressure PS greater than 200 bar;

   (bb) for fluids in Group 2, with a volume greater than 1L and a product of PS and V greater than 50 bar-L, or with a pressure PS greater than 1000 bar, and all portable extinguishers and bottles for breathing apparatus;

   (ii) liquids having a vapour pressure at the maximum allowable temperature of not more than 0.5 bar above normal atmospheric pressure (1013 mbar) within the following limits:

   (aa) for fluids in Group 1, with a volume greater than 1L and a product of PS and V greater than 200 bar-L, or with a pressure PS greater than 500 bar;

   (bb) for fluids in Group 2, with a pressure PS greater than 10 bar and a product PS and V greater than 10,000 bar-L, or with a pressure PS greater than 1000 bar;

(b) fires or otherwise heated pressure equipment with a risk of overheating intended for generation of steam or super-heated water at temperatures higher than 110°C having a volume greater than 2L, and all pressure cookers;

(c) piping intended for:

   (i) gases, liquefied gases, gases dissolved under pressure, vapours and those liquids whose vapour pressure at the maximum allowable temperature is greater than 0.5 bar above normal atmospheric pressure (1013 mbar) within the following limits:

   (aa) for fluids in Group 1, with a DN greater than 25;

   (bb) for fluids in Group 2, with a DN greater than 32 and a product of PS and DN greater than 1000 bar;

   (ii) liquids having a vapour pressure at the maximum allowable temperature of not more than 0.5 bar above normal atmospheric pressure (1013 mbar), within the following limits:

   (aa) for fluids in Group 1, with a DN greater than 25 and a product of PS and DN greater than 2000 bar;

(bb) for fluids in Group 2, with a PS greater than 10 bar, a DN greater than 200 and a product of PS and DN greater than 5000 bar;

(d) subject to regulation 9, safety and pressure accessories intended for equipment covered by sub-paragraphs (a), (b) and (c), including where such equipment is incorporated into an assembly.

The requirements in respect of pressure equipment described in paragraph 2 above are that:

(a) it satisfies the relevant essential requirements and for the purpose of satisfying those requirements where a transposed harmonised standard covers one or more of the relevant essential requirements, any pressure equipment constructed in accordance with the transposed harmonised standard shall be presumed to comply with that or, as the case may be, those essential requirements;

(b) the appropriate conformity assessment procedure in respect of the pressure equipment has been carried out in accordance with regulation 13(1);

(c) the CE marking has been affixed to it by the manufacturer of the pressure equipment or his authorised representative established within the Community in accordance with regulation 16 and Schedule 5;

(d) a declaration of conformity has been drawn up in respect of it by the manufacturer of the pressure equipment or his authorised representative established within the Community containing the information listed in Schedule 6; and

(e) it is in fact safe.

Any technical documentation or other information in relation to an item of pressure equipment required to be retained under the conformity assessment procedure used shall be retained by the person specified in that respect in that conformity assessment procedure for any period specified in that procedure.

General duty relating to the placing on the market or putting into service of assemblies (regulation 8)

Subject to regulation 11, no person who is a responsible person shall place on the market or put into service an assembly described in paragraph 2 below unless the requirements of paragraph 3 below have been complied with in relation to it.

For the purposes of paragraph 1 above, an assembly is:

(a) an assembly which includes at least one item of pressure equipment described in paragraph 2 of regulation 7 and:

   (i) is intended for generating steam or superheated water at a temperature higher than 110°C comprising at least one item of fired or other heated pressure equipment presenting a risk of overheating; or

(ii) not being a product included in sub-paragraph (i) above, is intended by the manufacturer to be placed on the market and put into service as an assembly; or

(b) an assembly which is intended for generating warm water at temperatures not greater than 110°C, which is manually fed with the solid fuels and has a PS.V greater than 50 bar-L.

(a) The requirements in respect of an assembly described in paragraph 2(a) above are that:

(i) it satisfies the relevant essential requirements and for the purpose of satisfying those requirements where a transposed harmonised standard covers one or more of the relevant essential requirements, an assembly constructed in accordance with that transposed harmonised standard shall be presumed to comply with that or, as the case may be, those essential safety requirements;

(ii) the appropriate conformity assessment procedure in respect of the assembly has been carried out in accordance with regulation 14;

(iii) the CE marking has been affixed to it by the manufacturer of that assembly or his authorised representative established within the Community in accordance with regulation 16 and Schedule 5;

(iv) a declaration of conformity has been drawn up in respect of it by the manufacturer of the assembly or his authorised representative established within the Community containing the information listed in Schedule 6; and

(v) it is in fact safe.

(b) The requirements in respect of any assembly described in paragraph 2(b) above are that:

(i) it satisfies the essential requirements referred to in paragraphs 2.10, 2.11, 3.4, 5(a) and 5(d) of Schedule 2; and

(ii) it complies with the requirements listed in sub-paragraphs (a)(ii) to (v) inclusive above.

Any technical documentation or other information in relation to an assembly required to be retained under the conformity assessment procedure used shall be retained by the person specified in that respect in that conformity assessment procedure for any period specified in that procedure.

Requirement for pressure equipment or assemblies to comply with sound engineering practice (regulation 9)

Subject to regulation 11, no person who is a responsible person shall place on the market or put into service any pressure equipment or assembly to which these regulations apply below or equal to the limits in regulations 7(2)(a), (b) or (c) or regulation 8(2) unless the requirements of paragraph 2 below have been complied with in relation to it.

The requirements referred to in paragraph 1 above are that the pressure equipment or assembly:

- is designed and manufactured in accordance with sound engineering practice in order to ensure safe use;
- must be accompanied by adequate instructions for use;
- must bear markings to permit identification of the manufacturer or his authorised representative established within the Community; and
- be safe.

For the avoidance of doubt, pressure equipment or assemblies to which this regulation applies shall not bear the CE marking.

General duty relating to the supply of pressure equipment or assemblies (regulation 10)

Subject to regulation 11, no person, who is not a responsible person, shall supply any pressure equipment or assembly unless that pressure equipment or assembly is safe.

Classification of pressure equipment (regulation 12)

For the purposes of regulation 13, pressure equipment referred to in regulation 7(2) shall be classified by category in accordance with Schedule 3 according to ascending level of hazard. (The appropriate table for pressure equipment coming within regulation 7(2) is listed in Schedule 3, tables 1 to 9 accordingly.)

Where a vessel is composed of a number of chambers, it shall be classified in the highest category applicable to the individual chambers and where a chamber contains several fluids, classification shall be on the basis of the fluid which requires the highest category.

Conformity assessment procedures for pressure equipment (regulation 13)

This regulation specifies the appropriate conformity assessment procedure for pressure equipment, which shall be one of the procedures set out in Schedule 4.

Conformity assessment procedures for assemblies (regulation 14)

This regulation specifies the relevant conformity assessment procedure to be applied by the manufacturer in respect of an assembly.

Exclusion for pressure equipment and assemblies for use for experimentation (regulation 15)

Regulations 13 and 14 shall not apply to an item of pressure equipment or an assembly the use of which is in the interests of experimentation.

Experimentation means pressure equipment or an assembly designed or adapted for the purpose only of conducting any test or measurement in relation to that pressure equipment or assembly.

CE marking (regulation 16)

This regulation lays down provisions with respect to CE marking of pressure equipment according to circumstances.

The CE marking shall be accompanied by the identification number of the notified body involved at the production control phase.

European approval of materials (regulation 17)

European approval for materials shall be issued, at the request of one or more manufacturers of materials or equipment, by one of the notified bodies referred to in regulation 18 specifically designated for that task and in accordance with the procedures set out in Schedule 7.

Notified bodies (regulation 18)

For the purposes of these regulations, a *notified body* is a body which has been appointed to carry out one or more of the conformity assessment procedures mentioned or referred to in regulation 13 or 14 or to issue a European approval for materials as referred to in regulation 17 and which has been:

- appointed as a notified body in the United Kingdom pursuant to regulation 20; or
- appointed by a Member State other than the United Kingdom, and has been notified by the Member State concerned to the Commission and the other Member States pursuant to Article 112 of the Pressure Equipment directive.

Notified bodies and recognised third-party organisations appointed by the Secretary of State (regulation 20)

The Secretary of State may from time to time appoint such persons as he thinks fit to be notified bodies or recognised third-party organisations for the purposes of these regulations.

This regulation lays down specific provisions relating to the above appointments including their functions, termination of appointments, refusal to issue an EC-type examination certificate or EC design-examination certificate, quality assurance procedures and procedures for making unexpected visits to manufacturers.

Fees (regulation 21)

This regulation outlines procedures for a notified body or recognised third-party organisation to charge fees in connection with, or incidental to, carrying out its duties in relation to the functions referred to in regulation 20(2)(c) as it may determine according to certain provisions.

User inspectorates (regulation 22)

A *user inspectorate* is a body which has been appointed to carry out one or more of the conformity assessment procedures referred to in regulations 13 and 14 and as more specifically defined in paragraph 7 and which has been:

- appointed in the United Kingdom pursuant to paragraph 2; or
- appointed by a Member State other than the United Kingdom, and has been notified by the Member State concerned to the Commission and other Member States pursuant to Article 12 of the Pressure Equipment directive.

Provisions relating to the appointment of user inspectorates, conformity assessment procedures applicable by user inspectorates, restrictions on the use of pressure equipment and assemblies and the non-use of CE marking with respect to pressure equipment and assemblies are further dealt with in this regulation.

Conditions for pressure equipment and assemblies being taken to conform with the provisions of these regulations (regulation 23)

Subject to paragraph 2 below, any pressure equipment or assembly which:

- bears the CE marking in accordance with regulation 7(3)(c) or 8(3)(a)(iii); and
- is accompanied by an EC declaration of conformity in accordance with regulation 7(3)(d) or 8(3)(a)(iv),

shall be taken to conform with all the provisions of these regulations which apply to it, including the appropriate conformity assessment procedure specified in regulation 13, unless there are reasonable grounds for suspecting that it does not conform.

Paragraph 1 above does not apply in relation to an enforcement authority where a person fails or refuses to make available to the enforcement authority the documentation which he is required, by the conformity assessment procedure which applies to that pressure equipment or assembly, to retain a copy thereof.

Schedules to the regulations

The regulations include the following Schedules which provide extensive detail on the more technical aspects of the regulations, incorporating the various Annexes to the Directive:

1. Excluded pressure equipment and assemblies
2. Essential safety requirements (Annex I to the Pressure Equipment Directive)
3. Conformity assessment tables (Annex II to the Pressure Equipment Directive)
4. Conformity assessment procedure (Annex III to the Pressure Equipment Directive)
5. CE marking (Annex VI to the Pressure Equipment Directive)
6. EC declaration of conformity (Annex VII to the Pressure Equipment Directive)
7. European approval for materials
8. Enforcement.

THE PRESSURE SYSTEMS SAFETY REGULATIONS 2000

These regulations impose safety requirements with respect to pressure systems which are used or intended to be used at work. They further impose safety requirements with a view to preventing certain pressure vessels from becoming pressurised. The regulations are accompanied by an ACOP. The majority of duties are of an absolute nature.

Interpretation (regulation 2)

Competent person means a competent individual person (other than an employee) or a competent body of persons corporate or unincorporated; and accordingly any reference in these regulations to a competent person performing a function includes a reference to his performing it through his employees.

Danger in relation to a pressure system means reasonably foreseeable danger to persons from system failure, but (except in the case of steam) it does not mean danger from the hazardous characteristics of the relevant fluid other than from its pressure.

Examination means a careful and critical scrutiny of a pressure system or part of a pressure system, in or out of service as appropriate, using suitable techniques, including testing where appropriate, to assess:

(a) its actual condition; and

(b) whether, for the period up to the next examination, it will not cause danger when properly used if normal maintenance is carried out, and for this purpose *normal maintenance* means such maintenance as it is reasonable to expect the user (in the case of an installed system) or owner (in the case of a mobile system) to ensure is carried out independently of any advice from the competent person making the examination.

Installed system means a pressure system other than a mobile system.

Mobile system means a pressure system which can be readily moved between and used in different locations but it does not include a pressure system of a locomotive.

Pressure system means:

(a)   a system comprising one or more pressure vessels of rigid construction, any associated pipework and protective devices;

(b)   the pipework with its protective devices to which a transportable pressure receptacle (TPR) is, or is intended to be, connected; or

(c)   a pipeline and its devices,

which contains or is liable to contain a relevant fluid, but does not include a TPR.

Pipeline means a pipe or system of pipes used for the conveyance of relevant fluid across the boundaries of premises, together with any apparatus for inducing or facilitating the flow of relevant fluid through, or through a part of, the pipe or system, and any valves, valve chambers, pumps, compressors or similar works which are annexed to, or incorporated in the course of, the pipe or system.

Pipework means a pipe or system of pipes together with associated valves, pumps, compressors and other pressure containing components and includes a hose or bellows, but does not include a pipeline or any protective device.

Protective devices means devices designed to protect the pressure system against system failure and devices designed to give warning that system failure might occur, and include bursting discs.

Relevant fluid means:

(a)   steam;

(b)   any fluid or mixture of fluids which is at a pressure greater than 0.5 bar above atmospheric pressure, and which fluid or mixture of fluids, is:

  (i)   a gas; or

  (ii)   a liquid which would have a vapour pressure greater than 0.5 bar above atmospheric pressure when in equilibrium with its vapour at either the actual temperature of the liquid or 17.5°C; or

(c)   a gas dissolved under pressure in a solvent contained in a porous substance at ambient temperature and which could be released from the solvent without the application of heat.

Safe operating limits means the operating limits (incorporating a suitable margin of safety) beyond which failure is liable to occur.

Scheme of examination means the written scheme referred to in regulation 8.

System failure means the unintentional release of stored energy (other than from a pressure relief system) from a pressure system.

User in relation to a pressure system, or a vessel to which regulation 15 applies, means the employer or self-employed person who has control of the operation of the pressure system or such a vessel or, in the case of a pressure system or such a vessel at or in:

(a)   a mine within the meaning of section 180 of the Mines and Quarries Act 1954 it means the manager for the time being of that mine;

(b)   a quarry within the meaning of regulation 3 of the Quarries Regulations 1999 it means the operator for the time being of that quarry.

Design and construction (regulation 4)

Any person who designs, manufactures, imports or supplies any pressure system or any article which is intended to be a component part of any pressure system shall ensure that the following paragraphs are complied with.

- The pressure system or article, as the case may be, shall be properly designed and properly constructed from suitable material, so as to prevent danger.
- The pressure system or article, as the case may be, shall be so designed and constructed that all necessary examinations for preventing danger can be carried out.
- Where the pressure system has any means of access to its interior, it shall be so designed and constructed as to ensure, so far as is practicable, that access can be gained without danger.
- The pressure system shall be provided with such protective devices as may be necessary for preventing danger; and any such device designed to release contents shall do so safely, so far as is practicable.

Provision of information and marking (regulation 5)

Any person who:

- designs for another any pressure system or any article which is intended to be a component part thereof; or
- supplies (whether as manufacturer, importer or in any other capacity) any pressure system or any such article;

shall provide sufficient written information concerning its design, construction, examination, operation and maintenance as may reasonably foreseeably be needed to enable the provisions of these regulations to be complied with.

The employer of a person who modifies or repairs any pressure system shall provide sufficient written information concerning the modification or repair as may reasonably foreseeably be needed to enable the provisions of these regulations to be complied with.

Any person who manufactures a pressure vessel shall ensure that before it is supplied by him the information specified in Schedule 3 is marked on the vessel, or on a plate attached to it, in a visible, legible and indelible form; and no person shall import a pressure vessel unless it is so marked.

No person shall remove from a pressure vessel any mark or plate containing any of the information specified in Schedule 3.

No person shall falsify any mark on a pressure system, or on a plate attached to it, relating to its design, construction, test or operation.

Installation (regulation 6)

The employer of a person who installs a pressure system at work shall ensure that nothing about the way in which it is installed gives rise to danger or otherwise impairs the operation of any protective device or inspection facility.

Safe operating limits (regulation 6)

The user of an installed system and owner of a mobile system shall not operate the system or allow it to be operated unless he has established the safe operating limits of that system.

The owner of a mobile system shall, if he is not also the user of it:

- supply the user with a written statement specifying the safe operating limits of that system; or
- ensure that the system is legibly and durably marked with such safe operating limits and that the mark is clearly visible.

Written scheme of examination (regulation 8)

The user of an installed system and owner of a mobile system shall not operate the system or allow it to be operated unless he has a *written scheme for the periodic examination*: by a competent person, of the following parts of the system, namely:

- all protective devices;
- every pressure vessel and every pipeline in which (in either case) a defect may give rise to danger; and
- those parts of the pipework in which a defect may give rise to danger;

and such parts of the system shall be identified in the scheme.

The user or owner shall:

- ensure that the scheme has been drawn up, or certified as being suitable, by a competent person;
- ensure that:
  - the content of the scheme is reviewed at appropriate intervals by a competent person for the purpose of determining whether it is suitable in current conditions of use of the system; and

– the content of the scheme is modified in accordance with any recommendations made by that competent person arising out of that review.

No person shall certify or draw up a scheme of examination as above unless the scheme is suitable and:

- specifies the nature and frequency of examination;
- specifies any measures necessary to prepare the pressure system for safe examination other than those it would be reasonable to expect the user or owner respectively to take without specialist advice; and
- where appropriate provides for an examination to be carried out before the pressure system is used for the first time.

References above to the suitability of the scheme are references to its suitability for the purposes of preventing danger from those parts of the pressure system included in the scheme.

Examination in accordance with the written scheme (regulation 9)

The user of an installed system and owner of a mobile system shall:

- ensure that those parts of the system included in the scheme are examined by a competent person within the intervals specified and, where appropriate, before the system is used for the first time; and
- before each examination take all appropriate safety measures to prepare the system for examination, including any such measures as are specified in the scheme of examination.

Where a competent person undertakes an examination for the purposes of paragraph 1 above, he shall carry out that examination properly and in accordance with the scheme of examination.

Where a competent person has carried out an examination for the purposes of paragraph 1 above, he shall, subject to paragraph 4 and Regulation 14(4), make a *written report of the examination*, sign or add his name to it, date it and send it to the user or owner respectively; and the said report shall be so sent as soon as is practicable after completing the examination (or, in the case of integrated installed systems where the examination is part of a series, as soon as is practicable after completing the last examination in that series), and in any event to arrive:

- within 28 days of the completion of the examination (or, in the case of integrated installed systems where the examination is part of a series, within 28 days of the completion of the last examination in that series); or
- before the date specified in the report under paragraph 5(b);

whichever is the sooner.

Where the competent person is the user or owner respectively, the requirement to send the report to the user or owner shall not apply, but he shall make the report by the time it would have been required to have been sent to him if he had not been the competent person.

The report specified by paragraph 3 shall:

- state which parts of the pressure system have been examined, the condition of those parts and the results of the examination;
- specify any repairs or modifications to, or changes in the established safe operating limits of, the parts examined which, in the opinion of the competent person, are necessary to prevent danger or to ensure the continued effective working of the protective devices, and specify the date by which any such repairs or modifications must be completed or any such changes to the safe operating limits must be made;
- specify the date within the limits set by the scheme of examination after which the pressure system may not be operated without a further examination under the scheme of examination; and
- state whether in the opinion of the competent person the scheme of examination is suitable (for the purpose of preventing danger from those parts of the pressure system included in it) or should be modified, and if the latter state the reasons.

The user or owner respectively of a system which has been examined under this regulation shall ensure that the system is not operated, and no person shall supply such a mobile system for operation, after (in each case): **4**

- the date specified under paragraph 5(b), unless the repairs or modifications specified under that paragraph have been completed, and the changes in the established safe operating limits so specified have been made; or
- the date specified under paragraph 5(c) (or if that date has been postponed under paragraph 7, the postponed date) unless a further examination has been carried out under the scheme of examination.

The date specified in a report under paragraph 5(c) may be postponed to a later date by agreement in writing between the competent person who made the report and the user or owner respectively if:

- any such postponement does not give rise to danger;
- only one such postponement is made for any one examination; and
- such postponement is notified by the user or owner in writing to the enforcing authority for the premises in which the pressure system is situated, before the date specified in the report under paragraph 5(c).

Where the competent person above is the user or owner respectively the reference to an agreement in writing shall not apply, but there shall be included in the notification under sub-paragraph (c) a declaration that the postponement will not give rise to danger.

The owner of a mobile system shall ensure that the date specified under paragraph 5(c) is legibly and durably marked on the mobile system and that the mark is clearly visible.

Action in case of imminent danger (regulation 10)

If the competent person carrying out an examination under the scheme of examination is of the opinion that the pressure system or part of the pressure system will give rise to imminent danger unless certain repairs or modifications have been carried out or unless suitable changes to the operating conditions have been made, then without prejudice to the requirements of regulation 9, he shall forthwith make a written report to that effect identifying the system and specifying the repairs, modifications or changes concerned and give it:

● in the case of an installed system, to the user; or
● in the case of a mobile system, to the owner and to the user, if any,

and the competent person shall within 14 days of the completion of the examination send a written report containing the same particulars to the enforcing authority for the premises at which the pressure system is situated.

Where a report is given in accordance with the above, the user or owner respectively shall ensure that the system is not operated until the repairs, modifications or changes have been carried out or made.

Operation (regulation 11)

The user or owner respectively shall provide for any person operating the system adequate and suitable instructions for:

● the safe operation of the system; and
● the action to be taken in the event of an emergency.

The user of a pressure system shall ensure that it is not operated except in accordance with the instructions provided in respect of that system.

Maintenance (regulation 12)

The user of an installed system and the owner of a mobile system shall ensure that the system is properly maintained in good repair, so as to prevent danger.

Modification and repair (regulation 13)

The employer of a person who modifies or repairs a pressure system at work shall ensure that nothing about the way in which it is modified or repaired gives rise to danger or otherwise impairs the operation of any protective device or inspection facility.

Keeping of records, etc (regulation 14)

The user of an installed system and the owner of a mobile system shall keep:

● the last report relating to the system by the competent person;

- any such previous reports if they contain information which will materially assist in assessing whether:
  - the system is safe to operate; or
  - any repairs or modifications to the system can be carried out safely;
- any:
  - information provided pursuant to regulation 5; or
  - instructions specified in section 3.4 to Annex 1 to Directive 97/23/EC of the European Parliament and of the Council on the approximation of the laws of the Member States concerning pressure equipment, which Annex is set out in Schedule 2 to the Pressure Equipment Regulations 1999, and provided pursuant to regulation 7(1) and (3) or 8(1) and 3(b)(i) of those regulations,

    which relate to those parts of the pressure system included in the scheme of examination; and
- any agreement made pursuant to regulation 9(7), and, in a case to which Regulation 9(8) applies, a copy of the notification referred to in Regulation 9(7)(c), until a further examination has been carried out since that agreement or notification under the scheme of examination.

Precautions to prevent pressurisation of certain vessels (regulation 15)

Paragraph 2 shall apply to a vessel:

- which is constructed with a permanent outlet to the atmosphere or to a space where the pressure does not exceed atmospheric pressure; and
- which could become a pressure vessel if that outlet were obstructed.

The user of a vessel to which this paragraph applies shall ensure that the outlet is at all times kept open and free from obstruction when the vessel is in use.

Regulation 15 – Precautions to prevent pressurisation

Under regulation 15 the user of a vessel to which Part IV applies shall ensure that the outlet referred to in regulation 14(a) is at all times kept open and free from obstructions when the vessel is in use.

Defence (regulation 16)

In any proceedings for an offence for a contravention of the provisions of these regulations, it shall, subject to paragraphs (2) and (3), be a defence for the person charged to prove –

- that the commission of the offence was due to the act or default of another person not being one of his employees (hereinafter called 'the other person'); and

- that he took all reasonable precautions and exercised all due diligence to avoid the commission of the offence.

The person charged shall not, without leave of the court, be entitled to rely on the above defence unless, within a period ending seven clear days:

- before the hearing to determine the mode of trial, where the proceedings are in England or Wales;
- before the intermediate diet, where the proceedings are summary proceedings in Scotland; or
- before the first diet, where the proceedings are solemn proceedings in Scotland,

he has served on the prosecutor a notice in writing giving such information identifying or assisting in the identification of the other person as was then in his possession.

Where a contravention of these regulations by any person is due to the act or default of the other person, that other person shall be guilty of the offence which would, but for any defence under this regulation available to the first-mentioned person, be constituted by the act or default.

Schedule 1

Regulation 3(2)

Part I lists 25 pressure systems or parts of same which are excepted from all regulations.

Part II outlines pressure systems excepted from certain regulations.

Schedule 2

This Schedule modifies certain duties in cases where pressure systems are supplied by way of lease, hire or other arrangements

Schedule 3 – Marking of pressure vessels

The information referred to in regulation 5(4) is as follows:

1. The manufacturer's name
2. A serial number to identify the vessel
3. The date of manufacture of the vessel
4. The standard to which the vessel was built
5. The maximum allowable pressure of the vessel
6. The maximum allowable pressure of the vessel where it is other than atmospheric
7. The design temperature.

THE PROVISION AND USE OF WORK EQUIPMENT REGULATIONS 1998

These regulations originally implemented the EC Directive of 30 November 1989 'concerning the minimum safety and health requirements for the use of work equipment by workers at work', generally known as the 'Machinery Safety' Directive 89/655/EEC. The 1992 version of the regulations was revoked by the Provision and Use of Work Equipment Regulations 1998.

These comprehensive regulations are intended:

- to implement the Machinery Safety Directive
- to simplify and clarify existing laws on the provision and use of work equipment by reforming older legislation
- to form a coherent, single set of key health and safety requirements concerning the provision and use of work equipment.

The regulations are supported by an Approved Code of Practice and Guidance Notes prepared by the HSE.

The majority of the requirements of the regulations are of an absolute nature.

Interpretation (regulation 2)

The following terms are important:

- *inspection* in relation to an inspection under regulation 6:

  - means such visual or more rigorous inspection by a competent person as is appropriate for the purpose described in that paragraph;
  - where it is appropriate to carry out testing for the purpose, includes testing the nature and extent of which are appropriate for the purpose.

- *Thorough inspection* in relation to a thorough examination under regulation 32:

  - means a thorough examination by a competent person;
  - includes testing the nature and extent of which are appopriate for the purpose described in the paragraph.

- *Work equipment* means any machinery, appliance, apparatus or tool or installation for use at work (whether exclusively or not).
- *Use* in relation to work equipment means any activity involving work equipment and includes starting, stopping, programming, setting, transporting, repairing, modifying, maintaining, servicing and cleaning.

The application of the requirements specified under these regulations (regulation 3)

The requirements imposed by these regulations on an employer shall apply in respect of work equipment provided for or used by any of their employees at work.

The requirements imposed by these regulations on an employer shall also apply to:

- a self-employed person regarding work equipment they use at work
- a person who has control to any extent, of:
  - work equipment;
  - a person at work who supervises or manages the use of work equipment; or
  - the way in which work equipment is used at work,

and to the extent of his control.

The suitability of work equipment (regulation 4)

Every employer shall ensure that work equipment is so constructed or adapted as to be suitable for the purpose for which it is used or provided.

In selecting work equipment, every employer shall bear in mind the working conditions and risks to the health and safety of workers that exist in the premises or undertaking in which the work equipment is to be used, as well as any additional risk posed using the equipment that is to be used, as well as any additional risk posed using the equipment.

Every employer shall ensure that work equipment is used only for those operations and under those conditions for which it is suitable.

In this regulation 'suitable' means suitable in any respect which it is reasonably foreseeable will affect the health or safety of any person.

Maintenance (regulation 5)

Every employer shall ensure that work equipment is *maintained in an efficient state, in efficient working order and in good repair.*

Every employer shall ensure that where any machinery has a maintenance log, the log is kept up to date.

Inspection (regulation 6)

Every employer shall ensure that, where the safety of work equipment depends on the installation conditions, it is inspected:

- after installation and before being put into service for the first time; or
- after assembly at a new site or in a new location

to ensure that it has been installed correctly and is safe to operate.

Every employer shall ensure that work equipment exposed to conditions causing deterioration which is liable to result in dangerous situations is inspected:

- at suitable intervals; and
- each time that exceptional circumstances which are liable to jeopardise the safety of the work equipment have occurred

to ensure that health and safety conditions are maintained and that any deterioration can be detected and remedied in good time.

Every employer shall ensure that the result of an inspection made under this regulation is recorded and kept until the next inspection under this regulation is recorded.

Every employer shall ensure that no work equipment:

- leaves his undertaking; or
- if obtained from the undertaking of another person, is used in his undertaking

unless it is accompanied by physical evidence that the last inspection required to be carried out under this regulation has been carried out.

This regulation does not apply to:

- a power press to which regulations 32–35 apply
- a guard or protection device for the tools of such power press
- work equipment for lifting loads including persons
- winding apparatus to which the Mines (Shafts and Winding) Regulations 1993 apply
- work equipment required to be inspected by regulation 29 of the Construction (Health, Safety and Welfare) Regulations 1996.

4

Specific risks (regulation 7)

Where the use of work equipment is likely to involve a specific risk to health or safety, every employer shall ensure that:

- the use of this equipment is restricted to those given the task of using it
- repairs, modifications, maintenance or servicing of the work equipment is restricted to those who have been specifically designated to do this (whether or not they are also authorised to perform other operations).

The employer shall ensure that those designated to do the repairs and so on mentioned above have received adequate training in the operations they will need to carry out to fulfil this role.

Information and instructions (regulation 8)

Every employer shall ensure that all those who use work equipment have available to them adequate health and safety information and, where appropriate, written instructions pertaining to the use of the equipment.

Every employer shall ensure that any of their employees who supervise or manage the use of work equipment have available to them similar adequate information.

Without prejudice to the generality of first two paragraphs above, the information and instructions required by either of these paragraphs shall include information and, where appropriate, written instructions on:

- the conditions in which and the methods by which the work equipment may be used
- foreseeable abnormal situations and the action to be taken if such a situation were to occur
- any conclusions to be drawn from experience in using the work equipment.

Information and instruction required by this regulation shall be readily comprehensible to those concerned.

Training (regulation 9)

Every employer shall ensure that all those who use work equipment have received adequate training for the purposes of health and safety, including training in the methods that may be adopted when using the work equipment, any risks that such use may entail and the precautions to be taken.

Every employer shall ensure that any of their employees who supervise or manage the use of work equipment have received similar adequate training.

Conformity with EU requirements (regulation 10)

Every employer shall ensure that an item of work equipment conforms at all times with any essential requirements, other than requirements which, at the time of its being first supplied or put into service in any place in which these regulations apply, did not apply to work equipment of its type.

In this regulation 'essential requirements', in relation to an item of work equipment, means requirements relating to the design and construction of work equipment of its type in any of the instruments listed in Schedule 1 (being instruments which give effect to community directives concerning the safety of products).

Dangerous parts of machinery (regulation 11)

Every employer shall ensure that the measures listed below are taken in order to:

- prevent access to any dangerous part of machinery or any rotating stock-bar
- stop the movement of any dangerous part of machinery or rotating stock-bar before any part of a person enters a danger zone.

The measures required shall consist of:

- the provision of fixed guards enclosing every dangerous part or rotating stock-bar where and to the extent that it is practicable to do so, but where to the extent that it is not, then
- the provision of other guards or protection devices where to the extent that it is practicable to do so, but where to the extent that it is not, then

- the provision of jigs, holders, push-sticks or similar protection appliances used in conjunction with the machinery where and to the extent that it is practicable to do so,

and the provision of such information, instruction, training and supervision as is necessary.

All guards and protection devices provided in relation to the first two points in the list above shall:

- be suitable for the purpose for which they are provided
- be of good construction, sound material and adequate strength
- be maintained in an efficient state, in efficient working order and in good repair
- not give rise to any increased risk to health or safety
- not be easily bypassed or disabled
- be situated at sufficient distance from the danger zone
- not unduly restrict the view of the operating cycle of the machinery, where such a view is necessary
- be so constructed or adapted that they allow operations necessary to fit or replace parts and for maintenance work, but restrict this access so that it is allowed only to the area where the work is to be carried out and, if possible, without having to dismantle the guard or protection device.

All jigs, holders, push-sticks or similar protection appliances provided shall comply with the requirements as to suitability, construction and so on listed above.

In this regulation:

- *danger zone* – means any zone in or around machinery in which a person is exposed to a risk to health or safety from contact with a dangerous part of machinery or a rotating stock-bar
- *stock-bar* – means any part of a stock-bar that projects beyond the head-stock of a lathe.

Protection against specified hazards (regulation 12)

Every employer shall take measures to ensure that the exposure of a person using work equipment to any risk to their health or safety from any of the hazards listed below is either prevented or, where this is not reasonably practicable, adequately controlled.

The measures required by the above paragraph shall:

- be measures other than the provision of the PPE or of information, instruction, training and supervision, so far as is reasonably practicable
- include, where appropriate, measures to minimise the effects of the hazard as well as to reduce the likelihood of the hazard occurring.

The hazards referred to above are:

- any article or substance falling or being ejected from work equipment
- rupture or disintegration of parts of work equipment
- working equipment catching fire or overheating

- the unintended or premature discharge of any article or of any gas, dust liquid, vapour or other substance that, in each case, is produced, used or stored in the work equipment
- the unintended or premature explosion of the work equipment or any article or substance produced, used or stored in it.

For the purposes of this regulation, *adequate* means adequate only in relation to the nature of the hazard and the nature and degree of exposure to the risk, and *adequately* shall be construed accordingly.

High or very low temperatures (regulation 13)

Every employer shall ensure that work equipment, parts of work equipment and any article or substance produced, used or stored in work equipment that, in each case, is at a high or very low temperature shall have protection, where appropriate, so as to prevent burns, scalds or searing.

Control systems and controls (regulations 14–17)

These regulations deal with general and specific safety requirements as follows:

- regulation 14 – Controls for starting or making a significant change in operating conditions
- regulation 15 – Stop controls
- regulation 16 – Emergency stop controls
- regulation 17 – Controls.

Control systems (regulation 18)

Every employer shall ensure, so far as is reasonably practicable, that all control systems of work equipment:

- are safe; and
- are chosen making due allowance for the failures, faults and constraints to be expected in the planned circumstances of use.

Without prejudice to the generality of this statement, a control system shall not be safe unless:

- its operation does not create an increased risk to health or safety
- it ensures, so far as is reasonably practicable, that any fault in or damage to any part of the control system or the loss of supply of any source of energy used by the work equipment cannot result in additional or increased risk to health or safety
- it does not impede the operation of any control required by regulations 15 or 16

Isolation from sources of energy (regulation 19)

Every employer shall ensure that, where appropriate, work equipment is provided with suitable means to isolate it from all its sources of energy.

Without prejudice to the generality of the first paragraph, the means mentioned in it shall not be suitable unless they are clearly identifiable and readily accessible.

Every employer shall take appropriate measures to ensure that the reconnection of any energy source to work equipment does not expose any person using the work equipment to any risk to their health or safety.

Stability (regulation 20)

Every employer shall ensure that work equipment, or any part of work equipment, is stabilised by clamping or otherwise, where necessary, for the purposes of health or safety.

Lighting (regulation 21)

Every employer shall ensure that suitable and sufficient lighting, which takes account of the operations to be carried out, is provided at any place where a person uses work equipment.

Maintenance operations (regulation 22)

Every employer shall take appropriate measures to ensure that work equipment is so constructed or adapted that, so far as is reasonably practicable, maintenance operations that involve a risk to health or safety can be carried out while the work equipment is shut down or, in other cases:

- without exposing the person carrying them out to a risk to their health or safety
- appropriate measures can be taken for the protection of any person carrying out maintenance operations that involve a risk to their health or safety.

Markings (regulation 23)

Every employer shall ensure that work equipment is marked in a clearly visible manner with any marking appropriate for reasons of health and safety.

Warnings (regulation 24)

Every employer shall ensure that work equipment incorporates any warnings or warning devices that are appropriate for reasons of health and safety.

Without prejudice to the generality of this statement, warnings given by means of warning devices on work equipment shall not be appropriate unless they are unambiguous, easily perceived and easily understood.

Employees carried on mobile work equipment (regulation 25)

Every employer shall ensure that no employee is carried by mobile work equipment unless:

- it is suitable for carrying persons
- it incorporates features for reducing to as low as is reasonably practicable risks to their safety, including risks from wheels and tracks.

Rolling over of mobile work equipment (regulation 26)

Every employer shall ensure that, where there is a risk to an employe riding on mobile work equipment from its rolling over, it is minimised by:

- stabilising the work equipment
- a structure which ensures that the work equipment does no more than fall on its side
- a structure giving sufficient clearance to anyone being carried if it over-turns further than that
- a device giving comparable protection.

Where there is a risk of anyone being carried by mobile work equipment being crushed by its rolling over, the employer shall ensure that it has a suitable restraining system for him.

This regulation shall not apply to a fork-lift truck which has a structure to ensure it only falls on its side or which gives sufficient clearance to the driver if it overturns further than that.

Compliance with this regulation is not required where:

- it would increase the overall risk to safety
- it would not be reasonably practicable to operate the mobile work equipment in consequence
- it would not be reasonably practicable in relation to an item of work equipment provided for use in the workplace before 5 December 1998.

Overturning of fork-lift trucks (regulation 27)

Every employer shall adapt or equip a fork-lift truck described above which carries an employee so as to reduce to as low as is reasonably practicable the risk to safety from its overturning.

Self-propelled work equipment (regulation 28)

Every employer shall ensure that, where self-propelled work equipment may, while in motion, involve risk to the safety of persons:

- it has facilities for preventing it being started by an unauthorised person
- it has appropriate facilities for minimising the consequences of a collision where there is more than one item of rail-mounted work equipment in motion at the same time
- it has a device for braking and stopping
- where safety constraints so require, emergency facilities operated by readily accessible controls or automatic systems are available for braking and stopping the work equipment in the event of failure of the main facility
- where the driver's direct field of vision is inadequate to ensure safety, there are adequate devices for improving his vision so far as is reasonably practicable
- if provided for use at night or in dark places:
    - it is equipped with lighting appropriate to the work being carried out
    - it is otherwise sufficiently safe for such use

- if it, or anything carried or towed by it, constitutes a fire hazard and is liable to endanger employees, it carries appropriate fire-fighting equipment unless such equipment is kept sufficiently close to it.

Remote-controlled self-propelled work equipment (regulation 29)

Every employer shall ensure that where remote-controlled self-propelled work equipment involves a risk to safety while in motion:

- it stops automatically once it leaves its control range
- where the risk is of crushing or impact
- it incorporates features to guard against such risk unless other appropriate devices are able to do so.

Drive shafts (regulation 30)

Where the seizure of the drive shaft between mobile work equipment and its accessories or anything towed is likely to involve a risk to safety every employer shall:

- ensure that the work equipment has a means of preventing such seizure or where such seizure cannot be avoided
- take every possible measure to avoid an adverse effect on the safety of an employee.

Every employer shall ensure that:

- where mobile work equipment has a shaft for the transmission of energy between it and other mobile work equipment; and
- the shaft could become soiled or damaged by contact with the ground while uncoupled,

the work equipment has a system for safeguarding the shaft.

Power presses to which Part IV does not apply (regulation 31)

Regulations 32–35 shall not apply to a power press which is described in Schedule 2.

Thorough examination of power presses, guards and protection devices (regulation 32)

Every employer shall ensure that a power press is not put into service for the first time after installation, or after assembly at a new site or in a new location unless:

- it has been thoroughly examined to ensure that it:

  - has been installed correctly; and
  - would be safe to operate; and

- any defect has been remedied.

Every employer shall ensure that a guard, other than one to which the paragraph below relates, or protection device, is not put into service for the first time on a power press unless:

- it has been thoroughly examined when in position on that power press to ensure that it is effective for its purpose; and
- any defect has been remedied.

Every employer shall ensure that a part of a closed tool which acts as a fixed guard is not used on a power press unless:

- it has been thoroughly examined when in position on any power press in the premises to ensure that it is effective for its purpose; and
- any defect has been remedied.

For the purpose of ensuring that health and safety conditions are maintained, and that any deterioration can be detected and remedied in good time, every employer shall ensure that:

- to every power press is thoroughly examined, and its guards and protection devices are thoroughly examined when in position on that power press:

  - at least every 12 months, where it has fixed guards only; or
  - at least every 6 months, in other cases; and
  - each time that exceptional circumstances have occurred which are liable to jeopardise the safety of the power press or its guards or protection devices; and

- any defect is remedied before the power press is used again.

Where, before the coming into force of these regulations, a power press, guard or protection device was required to be thoroughly examined by regulation 5(2) of the Power Presses Regulations 1965, the first thorough examination under the paragraph above shall be made before the date by which a thorough examination would have been required by regulation 5(2) had it remained in force.

The requirement to carry out thorough examinations does not apply to that part of a closed tool which acts as a fixed guard.

In this regulation, 'defect' means a defect notified under regulation 34 other than a defect which has not yet become a danger to persons.

Inspection of guards and protection devices (regulation 33)

Every employer shall ensure that a power press is not used after the setting, re-setting or adjustments of its tools, save in trying out its tools or save in die proving, unless:

- its every guard and protection device has been inspected and tested while in position on the power press by a person appointed in writing by the employer who is:

  - competent
  - undergoing training for that purpose and acting under the immediate supervision of a competent person;

  and who has signed a certificate described below; or

- the guards and protection devices have not been altered or disturbed in the course of the adjustment of its tools.

Every employer shall ensure that a power press is not used after the expiration of the fourth hour of a working period unless its every guard and protection device has been inspected and tested while in position on the power press by a person appointed and competent or in training as above and who has signed a certificate described below.

A certificate referred to in this regulation shall:

- contain sufficient particulars to identify every guard and protection device inspected and tested and the power press on which it was positioned at the time of the inspection and test
- state the date and time of the inspection and test
- state that every guard and protection device on the power press is in position and effective for its purpose.M

In this regulation *working period*, in relation to a power press, means:

- the period in which the day's or night's work is done; or
- in premises where a shift system is in operation, a shift.

Reports (regulation 34)

A person making a thorough examination for an employer under regulation 32 shall:

- notify the employer straight away of any defect in a power press or its guard or protection device which he believes is or could become a danger to persons;
- as soon as is practicable make a written report of the thorough examination to the employer in writing authenticated by him or on his behalf by signature or equally secure means and containing the information specified in Schedule 3; and
- where there is in his opinion a defect in a power press or its guard or protection device which is or could become a danger to persons, send a copy of the report as soon as is practicable to the enforcing authority for the premises in which the power press is situated.

A person making an inspection and test for an employer under regulation 33 shall straight away notify the employer of any defect in a guard or protection device which he believes is or could become a danger to persons and give the reason for his opinion.

Keeping of information (regulation 35)

Every employer shall ensure that the information in every report made pursuant to regulation 34(1) is kept available for inspection for two years after it is made.

Every employer shall ensure that a certificate under regulation 33 is kept available for inspection:

- at or near the power press to which it relates until superseded by a later certificate; and
- after that, until 6 months have passed since it was signed.

Exemption for the armed forces (regulation 36)

The Secretary of State for Defence may, in the interests of national security, by a certificate in writing exempt any of the home forces, any visiting force or any headquarters from any requirement or prohibition imposed by these regulations and any such exemption may be granted subject to conditions and to a limit of time and may be revoked by the said Secretary of State by a certificate in writing at any time.

Transitional provision (regulation 37)

The requirements in regulations 25–30 do not apply to work equipment provided for use in the undertaking or establishment before 5 December 1998 until 5 December 2002.

Schedule 2

Power presses to which regulations 32–35 do not apply

- a power press for the working of hot metal

- a power press not capable of a stroke greater than 6 millimetres
- a guillotine
- a combination punching and shearing machine, turret punch press or similar machine for punching, shearing or cropping
- a machine, other than a press brake, for bending steel sections
- a straightening machine
- an upsetting machine
- a heading machine
- a riveting machine
- an eyeletting machine
- a press-stud attaching machine
- a zip fastener bottom stop attaching machine
- a stapling machine
- a wire stitching machine
- a power press for the compacting of metal powders.

Schedule 3

Information to be contained in a report of a thorough examination of a power press, guard or protection device

The name of the employer for whom the thorough examination was made. The address of the premises at which the thorough examination was made. In relation to each item examined:

- that is a power press, interlocking guard, fixed guard or other type of guard or protection device
- where known its make, type and year of manufacture
- the identifying mark of:

  - the manufacture,
  - the employer.

In relation to the first thorough examination of a power press after installation or after assembly at a new site or in a new location:

- that it is such thorough examination
- either that it has been installed correctly or would be safe to operate or the respects in which it has not been installed correctly or would not be safe to operate
- identification of any part found to have a defect, and a description of the defect.

In relation to a thorough examination of a power press other than one to which the paragraph above relates:

- that it is such other thorough examination
- either that the power press would be safe to operate or the respects in which it would not be safe to operate

- identification of any part found to have a defect which is or could become dangerous, and a description of the defect.

In relation to a thorough examination of a guard or protection device:

- either that it is effective for its purpose or the respects in which it is not effective for its purpose
- identification of an part found to have a defect which is or could become dangerous, and a description of the defect.

Any repair, renewal or alteration required to remedy a defect found to be dangerous.

In the case of a defect which is not yet but could become dangerous:

- the time by which it could become dangerous;
- any repair, renewal or alteration required to remedy it.

Any other defect which requires remedy.

Any repair, renewal or alteration referred to above which has already been effected.

The date on which any defect which could become dangerous was notified to the employer under regulation 34(1)(a).

The qualification and address of the person making the report; that he is self-employed or if employed, the name and address of his employer.

The date of the thorough examination.

The date of the report.

The name of the person making the report and, where different, the name of the person signing or otherwise authenticating it.

Second-hand, hired and leased work equipment

Second-hand equipment

In situations where existing work equipment is sold by one company to another and brought into use by the second company, it becomes 'new equipment' and must meet the requirements for such equipment, *even though* it is second-hand. This means that the purchasing company will need to ascertain that the equipment meets the specific hardware provisions of regulations 11–24 before putting it into use.

Hired and leased equipment

Such equipment is treated in the same way as second-hand equipment, namely that it is classed as 'new equipment' at the hire/lease stage. On this basis, organisations hiring or leasing an item of work equipment will need to check that it meets the requirements of regulations 11–24 before putting it into use.

THE REPORTING OF INJURIES, DISEASES AND DANGEROUS OCCURRENCES REGULATIONS (RIDDOR) 1995

These regulations cover the requirement to report certain categories of injury and disease sustained by people at work, together with specified dangerous occurrences and gas incidents, to the relevant enforcing authority. The relevant enforcing authority in most cases is the HSE. In the case of commercial premises, injuries and diseases must be reported to the local authority, in most cases the environmental health officer.

Notification and reporting of injuries and dangerous occurrences (regulation 3)

The regulations require the 'responsible person' to notify the relevant enforcing authority 'by quickest practicable means' (i.e. telephone or fax) and subsequently make a report within 10 days on the approved form in the event of:

- the death of any person as a result of an accident arising out of or in connection with work;
- any person at work suffering a specified major injury as a result of an accident arising out of or in connection with work;
- any person who is not at work suffering an injury as a result of an accident arising out of or in connection with work and where that person is taken from the site of the accident to a hospital for treatment in respect of that injury;
- any person who is not at work suffering a major injury as a result of an accident arising out of or in connection with work at a hospital; or
- where there is a dangerous occurrence.

The responsible person must also report as soon as practicable, and in any event within 10 days of the accident, using the approved form, any situation where a person at work is incapacitated for work of a kind which he might reasonably be expected to do, either under his contract of employment, or, if there is no such contract, in the normal course of his work, for more than three consecutive days (excluding the day of the accident but including any days which would not have been working days) because of an injury resulting from an accident arising out of or in connection with work. (**Note**: The injured person may not necessarily be away from work but, perhaps, undertaking light duties.)

Reporting of the death of an employee (regulation 4)

Where an employee, as a result of an accident at work, has suffered a reportable injury which is a cause of his death within one year of the date of that accident, the employer must inform the relevant enforcing authority in writing of the death as soon as it comes to his knowledge, whether or not the accident has been reported.

Reporting of cases of disease (regulation 5)

Where:

- a person at work suffers from any of the occupational diseases specified in column 1 of Part I of Schedule 3 and his work involves one of the activities specified in the corresponding entry in column 2 of that Part; or
- a person at an offshore workplace suffers from any of the diseases specified in Part II of Schedule 3,

the responsible person shall forthwith send a report thereof to the relevant enforcing authority on the approved form, unless he forthwith makes a report thereof to the HSE by some other means so approved.

The above requirement applies only if:

- in the case of an employee, the responsible person has received a written statement prepared by a registered medical practitioner diagnosing the diseases as one of those specified in Schedule 3; or
- in the case of a self-employed person, that person has been informed, by a registered medical practitioner, that he is suffering from a disease so specified.

Reporting of gas incidents (regulation 6)

Whenever a conveyor of flammable gas through a fixed pipe distribution system, or a filler, importer or supplier (other than by means of retail trade) of a refillable container containing liquefied petroleum gas receives notification of any death or any major injury which has arisen out of or in connection with the gas distributed, filled, imported or supplied, as the case may be, by that person, he must forthwith notify the HSE of the incident and within 14 days send a report on the approved form.

Whenever an employer or self-employed person who is a member of a class of person approved by the HSE for the purposes of paragraph (3) of the Gas Safety (Installation and Use) Regulations 1994 (i.e. a CORGI registered gas installation business) has sufficient information for it to be reasonable for him to decide that a gas fitting or any flue or ventilation used in connection with that fitting, by reason of its design, construction, manner of installation, modification or servicing, is or has been likely to cause death, or any major injury by reason of:

- accidental leakage of gas;
- inadequate combustion of gas; or
- inadequate removal of the products of combustion of gas,

he must within 14 days send a report of it to the HSE on the approved form, unless he has previously reported such information.

Note

Reports must be made on:

- *Form 2508 – Report of an injury or dangerous occurrence* (see Figure 4.6)
- *Form 2508A – Report of a case of disease* (see Figure 4.7)
- *Form 2508G – Report of a gas incident.*

Records (regulation 7)

A record must be kept by the responsible person of all reportable injuries, diseases and dangerous occurrences.

In the case of reportable injuries and dangerous occurrences, the record must contain in each case:

- the date and time of the accident or dangerous occurrence
- in the case of an accident suffered by a person at work, the following particulars of that person:
  - full name
  - occupation
  - nature of injury

- in the case of an accident suffered by a person not at work, the following particulars of that person (unless they are not known and it is not reasonably practicable to ascertain them)
  - full name
  - status (for example 'passenger', 'customer', 'visitor' or 'bystander')
  - nature of injury

- place where the accident or dangerous occurrence happened
- a brief description of the circumstances in which the accident or dangerous occurrence happened
- the date on which the event was first reported to the relevant enforcement authority
- the method by which the event was reported.

In the case of diseases specified in Schedule 3 and reportable under regulation 5, the following information should be recorded:

- date of diagnosis of the disease
- name of the person affected
- occupation of the person affected
- name or nature of the disease
- the date on which the disease was first reported to the relevant enforcing authority
- the method by which the disease was reported.

Health and Safety at Work etc Act 1974
The Reporting of Injuries, Diseases and Dangerous Occurrences Regulations 1985

HSE
Health & Safety
Executive

Report of an injury or dangerous occurrence

Filling in this form
This form must be filled in by an employer or other responsible person.

Part A

About you

1 What is your full name?

2 What is your job title?

3 What is your telephone number?

About your organisation

4 What is the name of your organisation?

5 What is its address and postcode?

6 What type of work does the organisation do?

Part B

About the incident

1 On what date did the incident happen?

/ /

2 At what time did the incident happen?
(Please use the 24-hour clock eg 0600)

3 Did the incident happen at the above address?

Yes ☐ Go to question 4
No ☐ Where did the incident happen?
☐ elsewhere in your organisation – give the name, address and postcode
☐ at someone else's premises – give the name, address and postcode
☐ in a public place – give details of where it happened

If you do not know the postcode, what is the name of the local authority?

4 In which department, or where on the premises, did the incident happen?

P2508 (01/96)

Part C

About the injured person

If you are reporting a dangerous occurrence, go to Part F.
If more than one person was injured in the same incident, please attach the details asked for in Part C and Part D for each injured person.

1 What is their full name?

2 What is their home address and postcode?

3 What is their home phone number?

4 How old are they?

5 Are they
☐ male?
☐ female?

6 What is their job title?

7 Was the injured person (tick only one box)
☐ one of your employees?
☐ on a training scheme? Give details:

☐ on work experience?
☐ employed by someone else? Give details of the employer:

☐ self-employed and at work?
☐ a member of the public?

Part D

About the injury

1 What was the injury? (eg fracture, laceration)

2 What part of the body was injured?

Continued overleaf

• **FIG 4.6 Report of an injury or dangerous occurrence (Form 2508)**

3 Was the injury (tick the one box that applies)

☐ a fatality?

☐ a major injury or condition? (see accompanying notes)

☐ an injury to an employee or self-employed person which prevented them doing their normal work for more than 3 days?

☐ an injury to a member of the public which meant they had to be taken from the scene of the accident to a hospital for treatment?

4 Did the injured person (tick all the boxes that apply)

☐ become unconscious?

☐ need resuscitation?

☐ remain in hospital for more than 24 hours?

☐ none of the above.

Part E

About the kind of accident

Please tick the one box that best describes what happened, then go to Part G.

☐ Contact with moving machinery or material being machined

☐ Hit by a moving, flying or falling object

☐ Hit by a moving vehicle

☐ Hit something fixed or stationary

☐ Injured while handling, lifting or carrying

☐ Slipped, tripped or fell on the same level

☐ Fell from a height.
How high was the fall?

[_____] **metres**

☐ Trapped by something collapsing

☐ Drowned or asphyxiated

☐ Exposed to, or in contact with, a harmful substance

☐ Exposed to fire

☐ Exposed to an explosion

☐ Contact with electricity or an electrical discharge

☐ Injured by an animal

☐ Physically assaulted by a person

☐ Another kind of accident (describe it in Part G)

Part F

Dangerous occurrences

Enter the number of the dangerous occurrence you are reporting. (The numbers are given in the Regulations and in the notes which accompany this form.)

[_____]

Part G

Describing what happened

Give as much detail as you can. For instance

* the name of any substance involved
* the name and type of any machine involved
* the events that led to the incident
* the part played by any people.

If it was a personal injury, give details of what the person was doing. Describe any action that has since been taken to prevent a similar incident. Use a separate piece of paper if you need to.

Part H

Your signature

Signature

[_____]

Date

[/ /]

Where to send the form
Please send it to the Enforcing Authority for the place where it happened. If you do not know the Enforcing Authority, send it to the nearest HSE office.

For official use
Client number   Location number   Event number
[_____] [_____] [_____]   ☐ INV REP ☐ Y ☐ N

Health and Safety at Work etc Act 1974
The Reporting of Injuries, Diseases and Dangerous Occurrences Regulations 1995

Report of a case of disease

Filling in this form
This form must be filled in by an employer or other responsible person.

Part A

About you

1 What is your full name?

2 What is your job title?

3 What is your telephone number?

About your organisation

4 What is the name of your organisation?

5 What is its address and postcode?

6 Does the affected person usually work at this address?
Yes ☐ Go to question 7
No ☐ Where do they normally work?

7 What type of work does the organisation do?

Part B

About the affected person

1 What is their full name?

2 What is their date of birth?
/   /

3 What is their job title?

4 Are they
☐ male?
☐ female?

5 Is the affected person (tick one box)
☐ one of your employees?
☐ on a training scheme? Give details:

☐ on work experience?
☐ employed by someone else? Give details:

☐ other? Give details:

F2508A (01/96)

Continued overleaf

● **FIG 4.7 Report of a case of disease (Form 2508A)**

Part C

The disease you are reporting

1 Please give:

- the name of the disease, and the type of work it is associated with; or

- the name and number of the disease (from Schedule 3 of the Regulations – see the accompanying notes).

2 What is the date of the statement of the doctor who first diagnosed or confirmed the disease?

[/ /]

3 What is the name and address of the doctor?

Part D

Describing the work that led to the disease

Please describe any work done by the affected person which might have led to them getting the disease.

If the disease is thought to have been caused by exposure to an agent at work (eg a specific chemical) please say what that agent is.

Give any other information which is relevant.

Give your description here

Continue your description here

4

Part E

Your signature

Signature

Date

[/ /]

Where to send the form

Please send it to the Enforcing Authority for the place where the affected person works. If you do not know the Enforcing Authority, send it to the nearest HSE office.

For official use	
Client number	Location number
Event number	
	☐ INV REP ☐ Y ☐ N

Note

The system for recording accidents and diseases is not specified in RIDDOR. An employer can simply retain photocopies of reports in a file or, in the case of accidents, utilise the BI 510 Accident Book, identifying those accidents reportable under RIDDOR. Alternatively, details may be stored on a computer, provided that details from the computer file can be retrieved and printed out readily when required.

Notifiable and reportable major injuries

These are listed in Schedule 1 to RIDDOR as follows.

- Any fracture, other than to the fingers, thumbs or toes
- Any amputation
- Dislocation of the shoulder, hip, knee or spine
- Loss of sight (whether temporary or permanent)
- A chemical or hot metal burn to the eye or any penetrating injury to the eye
- Any injury resulting from electric shock or electrical burn (including any electrical burn caused by arcing or arcing products) leading to unconsciousness or requiring resuscitation or admittance to hospital for more than 24 hours
- Any other injury:

  - leading to hypothermia, heat-induced illness or to unconsciousness,
  - requiring resuscitation, or
  - requiring admittance to hospital for more than 24 hours

- Loss of consciousness caused by asphyxia or by exposure to a harmful substance or biological agent
- Either of the following conditions which result from the absorption of any substance by inhalation, ingestion or through the skin:

  - acute illness requiring medical treatment, or
  - loss of consciousness

- Acute illness which requires medical treatment where there is reason to believe that this resulted from exposure to a biological agent or its toxins or infected material.

Dangerous occurrences

Dangerous occurrences under RIDDOR are listed in Schedule 2. This Schedule classifies dangerous occurrences into five groups, namely:

General

This group includes those dangerous occurrences involving lifting machinery, etc., pressure systems, freight containers, overhead electric lines, electrical short circuit, explosives, biological agents, malfunction of radiation generators, etc., breathing apparatus, diving operations, collapse of scaffolding, train collisions, wells, pipelines or pipeline works, fairground equipment and carriage of dangerous substances by road.

This group also includes dangerous occurrences which are reportable except in relation to offshore workplaces, namely collapse of a building or structure, explosion or fire, escape of flammable substances and escape of substances.

Dangerous occurrences which are reportable in relation to mines

Listed in this group are dangerous occurrences involving fire or ignition of gas, escape of gas, failure of plant or equipment, breathing apparatus, injury by explosion of blasting material, use of emergency escape apparatus, inrush of gas or water, insecure tip, locomotives and falls of ground.

Dangerous occurrences which are reportable in relation to quarries

This group includes sinking of craft, explosive-related injuries, projection of substances outside a quarry, misfires, insecure tips, movement of slopes or faces and explosions or fires in vehicles or plant.

Dangerous occurrences which are reportable in respect of relevant transport systems

A 'relevant transport system' is defined as meaning a railway, tramway, trolley vehicle system or guided transport system. Reportable dangerous occurrences are accidents to passenger trains, accidents not involving passenger trains, accidents involving any kind of train, accidents and incidents at level crossings, accidents involving the permanent way and other works on or connected with a relevant transport system, accidents involving failure of the works on or connected with a relevant transport system, incidents of serious congestion and incidents of signals passed without authority.

Dangerous occurrences which are reportable in respect of an offshore workplace

This group includes release of petroleum hydrocarbon, fire or explosion, release or escape of dangerous substances, collapse of an offshore installation or part thereof, certain occurrences having the potential to cause death, e.g. the failure of equipment required to maintain a floating offshore installation on station, collisions, subsidence or collapse of seabed, loss of stability or buoyancy, evacuation situations and falls into water, i.e. where a person falls more than 2 metres into water.

Note

Dangerous occurrences are notifiable and reportable to the HSE irrespective of whether death and/or major injury has resulted from same.

Table 4.3 gives a summary of reporting requirements.

Table 4.3 Summary of reporting requirements

Reportable event (under RIDDOR 1995)		Responsible person
1. Special cases All reportable events in mines		The mine manager
All reportable events in quarries or in closed mine or quarry tips		The owner
All reportable events at offshore installations, except cases of disease reportable under regulation 5		The owner, in respect of a mobile installation, or the operator in respect of a fixed installation (under these regulations the responsibility extends to reporting incidents at subsea installations, except tied back wells and adjacent pipeline)
All reportable events at diving installations, except cases of disease reportable under regulation 5		The diving contractor
2. Injuries and disease		
Death, major injury, over 3-day injury, or case of disease (including cases of disease connected with diving operations and work at an offshore installation)	of an employee at work	That person's employer
	of a self-employed person at work in premises under the control of someone else	The person in control of the premises ● at the time of the event ● in connection with the carrying on of any trade, business or undertaking
Major injury, over 3-day injury, or case of disease	of a self-employed person at work in premises under their control	The self-employed person or someone acting on their behalf
Death, or injury requiring removal to a hospital for treatment (or major injury occurring at a hospital)	of a person who is not at work (but is affected by the work of someone else), e.g. a member of the public, a student, a resident of a nursing home	The person in control of the premises where, or in connection with the work going on at which, the accident causing the injury happened:

Table 4.3 continued

	• at the time of the event; and • in connection with their carrying on any trade, business or undertaking
3. Dangerous occurrences	
One of the dangerous occurrences listed in Schedule 2 to the Regulations, except:	The person in control of the premises where, or in connection with the work going on at which, the dangerous occurrence happened:
• where they occur at workplaces covered by Part 1 of this table (i.e., at mines, quarries, closed mine or quarry tips, offshore installations or connected with diving operations): or • those covered below	• at the time the dangerous occurrence happened; and • in connection with their carrying on trade, business or undertaking
A dangerous occurrence at a well	The concession owner (the person having the right to exploit or explore mineral resources and store and recover gas in any area, if the well is used or is to be used to exercise that right) or person appointed by the concession owner to organise or supervise any operation carried out by the well
A dangerous occurrence at a pipeline, but not dangerous occurrence connected with pipeline works	The owner of the pipeline
A dangerous occurrence involving a dangerous substance being conveyed by road	The operator of the vehicle

Reportable diseases

Reportable diseases are those listed in Schedule 3 of RIDDOR. There are 47 diseases and they are listed under three groups (examples from each of the three groups are shown in Table 4.4).

1. Conditions due to physical agents and the physical demands of work, e.g. malignant disease of the bones due to ionising radiation, decompression illness and subcutaneous cellulitis of the hand ('beat hand')
2. Infections due to biological agents, e.g. anthrax, brucellosis, leptospirosis, Q fever and tetanus
3. Conditions due to substances, e.g. poisoning by, for instance, carbon disulphide, ethylene oxide and methyl bromide, cancer of a bronchus or lung, bladder cancer, acne (through work involving exposure to mineral oil, tar, pitch or arsenic) and pneumoconiosis (excluding asbestosis).

Table 4.4 Reportable diseases (*Part 1 – Occupational diseases*)

Column 1 **Disease**	Column 2 **Work activity**
Conditions due to physical agents and the physical demands of work	
2 Malignant disease of the bones due to ionising radiation	Work with ionising radiation
5 Decompression illness	Work involving breathing gases or at increased pressure (including diving)
9 Subcutaneous cellulitis of the hand (beat hand)	Physically demanding work causing severe or prolonged friction or pressure on the hand
Infections due to biological agents	
15 Anthrax	(a) Work involving handling infected animals, their products or packaging containing infected material; or (b) work on infected sites
16 Brucellosis	Work involving contact with: (a) animals or their carcases (including any parts thereof) infected by brucella or the untreated products of same; or (b) laboratory specimens or vaccines of or containing brucella
20 Leptospirosis	(a) Work in places which are liable to be infested by rats, fieldmice, voles or other small mammals; (b) work at dog kennels or involving the care or handling of dogs; or

	(c)work involving contact with bovine animals or their meat products or pigs or their meat products
25 Tetanus	Work involving contact with soil likely to be contaminated by animals

Conditions due to substances

28 Poisoning by any of the following, for example: (f) carbon disulphide; (h) ethylene oxide; (l) methyl bromide	Any activity
29 Cancer of a bronchus or lung	(a) Work in or about a building where nickel is produced by decomposition of a gaseous nickel compound or where any industrial process which is ancillary or incidental to that process is carried on; or (b) work involving exposure to bis(chloromethyl) ether or any electrolytic chromium processes (excluding passivation) which involves hexavalent chromium compounds, chromate production or zinc chromate pigment manufacture
32 Bladder cancer	Work involving exposure to aluminium smelting using the Soderberg process
37 Acne	Work involving exposure to mineral oil, tar, pitch or arsenic
38 Skin cancer	
40 Byssinosis	The spinning or manipulation of raw or waste cotton or flax or the weaving of cotton or flax, carried out in each case in a room in a factory, together with any other work carried out in such a room

THE SAFETY REPRESENTATIVES AND SAFETY COMMITTEES REGULATIONS 1977

These regulations and the supporting ACOP and HSE Guidance Notes provide a basic framework within which each undertaking can develop effective relationships. The regulations are concerned with the appointment by recognised and independent trade unions of safety representatives, the functions of the representatives and the establishment of safety committees.

Interpretation (regulation 1)

The following definitions are important to these regulations:

- *recognised trade union* means an independent trade union, as defined in section 30(1) of the Trade Union and Labour Relations Act 1974, that the employer concerned recognises for the purpose of negotiations relating to or connected with one or more of the matters specified in section 29(1) of the Act regarding those employed by them or one that the Advisory, Conciliation and Arbitration Service has recommended be recognised under the Employment Act 1975, which is operative within the meaning of section 15 of the Act
- *safety representative* means a person appointed under regulation 3(1) of these regulations to be a safety representative
- *workplace* in relation to a safety representative means any place or places where the group or groups of employees they are appointed to represent are likely to frequent in the course of their employment or incidentally to it.

The appointment of safety representatives (regulation 3)

A recognised trade union may appoint safety representatives from among the employees in all cases where one or more employees are employed by an employer by whom it is recognised, except in the case of employees employed in a mine.

Each safety representative shall have functions as set out in regulation 4 and shall cease to be a safety representative when:

- their trade union notifies the employer in writing that their appointment has been terminated
- they cease to be employed at the workplace
- they resign.

A person appointed as a safety representative shall, so far as is reasonably practicable, either have been employed by their employer throughout the preceding two years or have had at least two years' experience in similar employment.

The functions of safety representatives (regulation 4)

In addition to their function under section (2)(4) of the HSWA to represent the employees in consultation with the employer under section 2(6) of the HSWA, each safety representative shall have the following functions:

- to investigate potential hazards and dangerous occurrences at the workplace and to examine the causes of accidents there
- to investigate complaints by an employee they represent relating to that employee's health, safety or welfare at work
- to make representations to the employer on matters arising from the investigations mentioned above
- to make representations to the employer on general matters affecting the health, safety or welfare at work of the employees at the workplace
- to carry out inspections
- to represent the employees they have been appointed to represent in consultations at the workplace with inspectors of the HSE and of any other enforcing authority
- to receive information from inspectors
- to attend meetings of safety committees in their capacity as a safety representative in connection with any of the above functions.

Any employer shall permit a safety representative to take such time off with pay during the employee's working hours as shall be necessary for the purpose of:

- performing their functions
- undergoing such training in aspects of these functions as may be reasonable in all the circumstances relating to any relevant provisions of a code of practice concerning time off for training approved, for the time being, by the HSC.

Here 'with pay' means in accordance with the Schedule to these regulations.

Inspections of the workplace (regulation 5)

Safety representatives shall be entitled to inspect the workplace or part of it if they have given the employer reasonable notice in writing of their intention to do so and have not inspected it, or that part of it, as the case may be, in the previous three months. They may, however, carry out more frequent inspections when this is agreed with the employer.

Where there has been a substantial change in the conditions of work (whether because of the introduction of new machinery or otherwise) or new information has been published by the HSC or HSE relevant to the hazards of the workplace since the last inspection, the safety representative, after consultation with the employer, shall be entitled to carry out a further inspection of the part of the workplace concerned, whether or not three months have not elapsed since the last inspection.

The employer shall provide such facilities and assistance as the safety representative may reasonably may require (including facilities for independent investigation by them and private discussion with the employees) for the purpose of carrying out an inspection, but this does not preclude the employer or their representative being present in the workplace during the inspection.

Inspections following notifiable accidents, occurrences and diseases (regulation 6)

Where there has been a notifiable accident or dangerous occurrence in a workplace or a notifiable disease has been contracted there and:

- it is safe for an inspection to be carried out
- the interests of employees in the group or groups that safety representatives are appointed to represent might be involved,

these safety representatives may carry out an inspection of the part of the workplace concerned and, so far as is necessary for the purpose of determining the cause, they may inspect any other part of the workplace. Where it is reasonably practicable to do so, they shall notify the employer or their representative of their intention to carry out the inspection.

The employer shall provide such facilities and assistance as the safety representatives may reasonably require (including facilities for independent investigation by them and private discussion with the employees) for the purpose of carrying out an inspection under this regulation, but this does not preclude the employer or their representative being present in the workplace during the inspection.

The inspection of documents and provision of information (regulation 7)

Safety representatives shall, if they have given the employer reasonable notice, be entitled to inspect and take copies of any document relevant to the workplace or to the employees the safety representatives represent that the employer is required to keep by virtue of any relevant statutory provisions except a document consisting of or relating to any health record of an identifiable individual.

An employer shall make available to a safety representative the information within the employer's knowledge necessary to enable them to fulfil their functions except:

- any information the disclosure of which would be against the interest of national security
- any information that they could not disclose without contravening a prohibition imposed by or under any enactment

- any information relating specifically to an individual, unless they have consented to its being disclosed
- any information the disclosure of which would, for reasons other than its effect on health, safety or welfare at work, cause substantial injury to the employer's undertaking or, where the information was supplied to them by some other person, to the undertaking of that other person
- any information obtained by the employer for the purpose of bringing, prosecuting or defending any legal proceedings.

Also, an employer is not required to produce or allow inspection of any document that is not related to health, safety or welfare.

Cases where safety representatives need not be employees (regulation 8)

Safety representatives need not be employees of the employer concerned in cases where the employees are members of the British Actors Equity Association or of the Musicians' Union.

Safety committees (regulation 9)

Any two safety representatives, at least, can request in writing that an employer establish a safety committee. Where an employer has received such a request they shall establish it in accordance with the following provisions:

- they shall consult with the safety representatives who made the request and with the representatives of recognised trade unions whose members work in any workplace the proposed committee is to oversee
- the employer shall post a notice stating the composition of the committee and the workplace or workplaces to be covered by it in a place where it may easily be read by the employees
- the committee shall be established not later than three months after the request for it has been made.

Provisions as to employment tribunals (regulation 11)

A safety representative may present a complaint to an employment tribunal that:

- the employer has failed to permit them to take time off in accordance with regulation 4(2)
- the employer has failed to pay them in accordance with regulation 4(2).

An employment tribunal shall not consider such a complaint unless it is presented within three months of the date when the failure occurred or within such further period as the tribunal considers reasonable when it is satisfied that it was *not* reasonably practicable for the complaint to be presented within this period of time.

Where an employment tribunal finds that a complaint regarding time off is well-founded, the tribunal shall make a declaration to this effect and may make an award of compensation to be paid by the employer to the employee. This shall be for an amount that the tribunal considers just equitable in all the circumstances, taking into account the employer's default in failing to permit time off to be taken by the employee and any loss sustained by the employee that is attributable to this.

Where, employment a complaint about pay, the employment tribunal finds that the employer has failed to pay the employee the whole or part of the amount required to be paid under regulation 4(2), the tribunal shall order the employer to pay the employee the amount that it finds is due to them.

The schedule relating to regulation 4(2)

Pay for time off allowed to safety representatives

Where a safety representative is permitted to take time off, their employer shall pay them:

- where the safety representative's renumeration for the work they would ordinarily have been doing during this time does not vary with the amount of work done, as if they had been doing this work for the whole of this time
- where the safety representative's remuneration for this work varies with the amount of work done, an amount calculated by means of the average hourly earnings for that work.

These average hourly earnings are the average hourly earnings of the safety representative concerned or, if no fair estimate can be made of these earnings, the average hourly earnings for work of this description by people in comparable employment with the same employer or, if there is no one else doing this work, a figure of average hourly earnings that is reasonable in the circumstances.

THE SAFETY SIGNS REGULATIONS 1980

These regulations require that safety signs – other than those for fire fighting, rescue equipment and emergency exists – conform to a standard system with regard to colours and shapes.

Prohibition signs

These are to be circular with a red band enclosing a crossed-out symbol on a white background, such as 'No smoking'.

Warnings signs

These signs are to be triangular in shape with a yellow background and black borders, symbols and text, such as 'LPG – highly flammable'.

Mandatory signs

These are to be rectangular, incorporating a blue mandatory symbol and/or text on a white background, such as 'Wear face shield'.

Safe condition signs

These are to be indicated by a green square or rectangle with symbols and lettering in white, such as 'Emergency stop'.

BS 5378, 1976

A safety sign giving health or safety information must comply with Part 1 of BS 5378, 1976, as must any strip of alternate colours used to identify a hazard, such as 'tiger striping'.

Part 1 of BS 5378 describes the safety colours as red, yellow, blue and green:

4

- *red* is for stop or prohibition signs
- *yellow* is for caution or risk of danger signs
- *blue* is for mandatory signs
- *green* is for safe condition signs.

THE SIMPLE PRESSURE VESSELS (SAFETY) REGULATIONS 1991–1994

These regulations apply to:

- simple pressure vessels, that is welded vessels made of certain types of steel or aluminium that are intended to contain air or nitrogen under pressure and manufactured in series
- relevant assemblies, that is any assembly incorporating a pressure vessel.

The regulations incorporate a number of definitions. In particular, note that:

- *safe* – means that, when a vessel is properly installed, maintained and used for the purpose for which it is intended, there is no risk (apart from one reduced to a minimum) of its being the cause or occasion of death, injury or damage to property (including domestic animals)
- *manufacturer's instructions* – are instructions issued by or on behalf of the manufacturer and include the following information:

– the manufacturer's name or mark
– the vessel type batch identification or other particulars identifying the vessel to which the instructions relate
– particulars of maximum working pressure in bar, maximum and minimum working temperature in °C and capacity in litres
– intended use of the vessel
– maintenance and installation requirements for vessel safety

and written in the official language of the member state accordingly

- *series manufacture* – means, where more than one vessel of the same type is manufactured during a given period by the same continuous manufacturing process, that it is in accordance with a common design
- *vessel* – means a simple pressure vessel, which is a welded vessel intended to contain air or nitrogen at a gauge pressure greater than 0.5 bar, not intended for exposure to flame, and having the following characteristics:

  – the components and assemblies contributing to the strength of the vessel under pressure are made either of non-alloy-quality steel or of non-alloy aluminium or of non-age-hardening aluminium alloy
  – the vessel consists of either:
    (i)   a cylindrical component with a circular cross-section, closed at each end, each end being outwardly dished or flat and being also coaxial with the cylindrical component
    (ii) two coaxial outwardly dished ends
  – the maximum working pressure (PS) is not more than 30 bar, and the PS.V (product of pressure and volume) not more than 10,000 bar litres
  – the minimum working temperature is not lower than –50°C and the maximum working temperature is not higher than 300°C in the case of steel vessels and 100°C in the case of aluminium or aluminium alloy vessels.

The regulations apply only to vessels manufactured in series. They do not apply to:

- vessels designed specifically for nuclear use, where vessel failure might or would result in an emission of radioactivity
- vessels specifically intended for installation in, or for use as part of the propulsive system of, a ship or aircraft
- fire extinguishers.

The principal requirements of the regulations

Vessels with a stored energy of *over 50 bar litres* when supplied in the UK must:

- meet the essential safety requirements, that is with regard to materials used in construction, vessel design, manufacturing processes and placing in service of vessels

- have safety clearance, that is, checks by an approved body
- bear the EC mark and other specified inscriptions
- be accompanied by manufacturer's instructions
- be safe (as defined above).

Vessels with a stored energy of *up to 50 bar litres* supplied in the UK must:

- be manufactured in accordance with engineering practice recognised as sound in the EU country
- bear specific inscriptions (but not the EC mark)
- be safe.

Similar requirements to these apply to such vessels when taken into service in the UK by a manufacturer or importer.

The regulations do *not* apply to exports to countries outside the EU or, for a transitional period, to the supply and taking into service in the UK of vessels that comply with existing UK safety requirements.

Failure to comply with these requirements:

- means that the vessels cannot be sold legally
- could result in penalties of a fine of up to £2000 or, in some cases, of imprisonment for up to three months, or both.

Categories of vessels

Different provisions are made for different categories of vessels depending on their stored energy expressed in terms of the product of the maximum working pressure in bar and its capacity in litres (PS.V).

Category A vessels

These are graded according to PS.V range thus:

- A.1 is 3000 to 10,000 bar litres
- A.2 is 200 to 3000 bar litres
- A.3 is 50 to 200 bar litres.

Category B vessels

These are vessels with a PS.V of 50 bar litres or less.

For the safety requirements for Category A and Category B vessels, see under the principal requirements of the regulations, vessels with stored energy of over or up to 50 bar litres, above.

Safety clearance

A vessel in Category A has safety clearance once an approved body has issued an EC verification certificate or an EC certificate of conformity for such a vessel.

Approved bodies

These are bodies designated by Member States, which, in the case of the UK, are those designated by the Secretary for Trade and Industry.

The EC mark and other specified inscriptions

Where an approved body has issued an EC verification certificate, it has responsibility for the application of the EC mark to every vessel covered by the certificate.

Where a manufacturer has obtained an EC certificate of conformity, they may apply the EC mark to any vessels covered by the certificate where they execute an EC declaration of conformity that they conform with a relevant national standard or the relevant prototype.

The EC mark must consist of the appropriate symbol, the last two digits of the year in which the mark is applied and, where appropriate, the distinguishing number assigned by the EC to the approved body responsible for EC verification or EC surveillance.

Other specified inscriptions to be applied to Category A and B vessels are:

- maximum working pressure in bar
- maximum working temperature in °C
- minimum working temperature in °C
- capacity of the vessel in litres
- name or mark of the manufacturer
- type and serial or batch identification of the vessel.

EC surveillance

This implies surveillance by the approved body that issued a certificate. The approved body has the following powers with respect to surveillance:

- powers of entry:
  - to take samples
  - to acquire information
  - to require additional information

- to compile reports on surveillance operations
- to report to the Secretary of State cases of wrongful application, and failures by manufacturers to:
  - carry out their undertakings
  - authorise access
  - provide other facilities.

THE SOCIAL SECURITY (INDUSTRIAL INJURIES) (PRESCRIBED DISEASES) REGULATIONS 1985–2003

These regulations list those diseases that are prescribed for the purpose of payment of disablement benefit. They were subject to minor amendments in 1987 and 1989.

A *prescribed disease* is defined in the Social Security Act 1975 as:

- a disease that ought to be treated, with regard to its causes, incidence and other relevant considerations, as a risk of occupation and not a risk common to everyone
- such that, in the absence of special circumstances, the attribution of particular cases to the nature of the employment can be established with reasonable certainty.

Schedule 1 to the regulations classifies prescribed diseases or injuries thus:

- conditions due to physical agents
- conditions due to biological agents
- conditions due to chemical agents
- miscellaneous conditions.

Within these four classifications, prescribed diseases or injuries are related to specific occupations. Each disease is numbered within the particular classification. The Schedule is very comprehensive and some examples of those listed in the four classifications are given in Table 4.5.

Under the regulations, pneumoconiosis, and the various causes of the disease, are treated separately.

Table 4.5 Prescribed diseases

Prescribed disease or injury	Occupation
Conditions due to physical agents	Any occupation involving:
2 Heat cataract	frequent or prolonged exposure to rays from molten or red-hot material
5 Subcutaneous cellulitis of the hand (Beat hand)	manual labour causing severe or prolonged friction or pressure on the hand
9 Miner's nystagmus	work in or about a mine
Conditions due to biological agents	
1 Anthrax	contact with animals infected with anthrax or the handling (including the loading or unloading or transport) of animal products or residues
5 Tuberculosis	work in or about a mine
9 Infection by Streptococcus suis	contact with pigs infected with Streptococcus suis, or with the carcases, products or residues of pigs so infected
Conditions due to chemical agents	
5 Poisoning by mercury or compound of mercury	the use or handling of, or exposure to, the fumes, dust or vapour of mercury or a compound of mercury or a substance containing mercury
21 (a) Localised new growth of the skin, papillomatous or keratotic (b) squamous-celled carcinoma of the skin	the use or handling of, or exposure to, arsenic, tar, pitch, bitumen, mineral oil (including paraffin), soot or any compound, product or residue of any of these substances, except quinone or hydroquinone
26 Damage to the liver or kidneys due to exposure to carbon tetrachloride	the use, or handling, of or exposure to the fumes of, or vapour containing, carbon tetrachloride
Miscellaneous conditions	
2 Byssinosis	work in any room where any process up to and including the weaving process is performed in a factory in which the spinning or manipulation of raw or waste cotton, or of flax, or the weaving of cotton or flax, is carried out

Prescribed disease or injury	Occupation
8 Primary carcinoma of the lung where there is accompanying evidence of one or both of the following (a) asbestosis (b) bilateral diffuse pleural thickening	(a) the working or handling of asbestos or any admixture of asbestos (b) the manufacture or repair of asbestos textiles or other articles containing or composed of asbestos (c) the cleaning of any machinery or plant used in any of the foregoing operations and of any chambers, fixtures or appliance for the collection of asbestos dust (d) substantial exposure to the dust arising from any of the foregoing operations
10 Lung cancer	(a) work underground in a tin mine (b) exposure to bis (chloromethyl) ether produced during the manufacture of chloromethyl methyl ether (c) exposure to pure zinc chromate, calcium chromate or strontium chromate.

THE SUPPLY OF MACHINERY (SAFETY) REGULATIONS 1992–1994

These regulations came into operation on 1 January 1993, and apply to:

- all types of machinery not specifically covered by legislation implementing other EC Directives
- all machinery which is powered and has moving parts
- assemblies of several machines
- parts that modify the function of machines
- components that can operate on their own.

They do not apply to:

- manually powered machinery
- machinery for transporting people or goods on the road
- machinery for lifting people
- machinery where the main hazard is electrical or specific hazards dealt with by other Directives
- machinery for export outside the EU
- second-hand machinery, other than that imported from outside the EU.

The regulations require the suppliers of most machinery to certify that machinery satisfies the relevant essential safety requirements (ESRs) set out in Schedule 3 to the regulations and be prepared, on request, to justify this certification.

Dangerous machinery

In the case of the more dangerous classes of machinery, such as saws, presses and underground equipment, the requirements are more onerous.

Suppliers must send to an approved test house a technical file detailing how the machine complies with European harmonised standards, or, if it does not, a sample of the machine itself for type approval.

The regulations also prescribe that the supplier must provide operating instructions in an EU language that the operator can understand, and installation and maintenance instructions in a language, agreed with the purchasers, that the installers will understand.

The CE mark must be affixed by the supplier to demonstrate that the machine complies with the requirements of the Directive or of any other Directive prescribing the CE mark.

Enforcement

The regulations are enforced by trading standards officers for domestic use and HSE inspectors for machinery intended for use in the workplace and elsewhere.

Implications of the regulations

New machinery must be marked with the CE mark declaring that the machine conforms to the *essential safety and health requirements* of the regulations.

THE TRANSPORTABLE PRESSURE VESSELS REGULATIONS 2001

These regulations are concerned with the safety requirements for transportable pressure vessels at both the stage of placing on the market and in subsequent use at work. Such vessels are subject to a strict regime with respect to reassessment of conformity and of periodic inspection by approved bodies. The regulations incorporate a series of Schedules covering standards for transportable pressure vessels (TPVs), conformity assessment and reassessment procedures, periodic inspection procedures and requirements for conformity marking.

Important definitions

ADR means the European Agreement concerning the International Carriage of Dangerous Goods by road signed at Geneva on 30 September 1957, as revised or re-issued from time to time.

Aerosol means a non-refillable vessel made of metal, glass or plastic and containing a gas which is compressed, liquefied or dissolved under pressure, with or without a liquid, paste or powder, and fitted with a release device allowing the contents to be ejected as solid or liquid particles in suspension in a gas, as a foam, paste or powder or in a liquid state or a gaseous state.

Bundle of cylinders means a transportable assembly of cylinders held firmly together and interconnected by a manifold.

COTIF means the Convention concerning International Carriage by Rail, as revised or re-issued from time to time.

EEC-type cylinder means a TPV:

(a) in respect of which there is an EEC Verification Certificate in force issued by an inspection body which, under the law of any Member State, is authorised to grant such a certificate for the purposes of the Pressure Vessels Framework Directive and the separate Directive relating to that type of cylinder, or, in the case of a cylinder not subject to EEC verification under any of the separate Directives, which conforms to the requirements of the Pressure Vessels Framework Directive and the separate Directive relating to that type of cylinder; and

(b) which bears all the marks and inscriptions required by the Pressure Vessels Framework Directive and the separate Directive relating to that type of cylinder.

Periodic inspection procedures means the procedures set out in Schedule 6, and *periodic inspection* shall be construed accordingly.

Placing on the market means supplying a TPV, or making it available to another person in the communities and includes importation of such a vessel.

RID means the Regulations concerning the International Carriage of Dangerous Goods by Rail which form Annex 1 to Appendix B to COTIF.

Transportable pressure vessel means a cylinder, tube or cryogenic receptacle, including:

(a) any valve or other accessory fitted to the vessel and having a direct safety function; and

(b) any permanent fitting of such a vessel,

which is used or intended to be used for the storage or transport of a gas, UN 1051 STABILISED HYDROGEN CYANIDE, UN 1052 ANHYDROUS

HYDROGEN FLUORIDE or UN 1790 HYDROFLUORIC ACID, solution, with more than 85% hydrofluoric acid, but does not include a bundle of cylinders or an aerosol.

Tube means a seamless vessel with a capacity greater than 150 litres but not more than 5000 litres.

UN number means United Nations Serial Number, that is to say, one of the four digit numbers devised by the United Nations and specified in the Approved Carriage List as a means of identification for dangerous goods, and any reference to the letter 'UN' followed by a number, in relation to certain dangerous goods, means the particular UN number for those goods specified in the Approved Carriage List.

Vehicle means a conveyance which is used for transporting TPVs by road or on a railway.

In these regulations, the words used *at work*, when referring to a TPV, include the filling, emptying, refilling, storage and transport of that vessel at work and an intention to conduct any of those activities.

Regulation 4 – Requirements relating to the placing on the market and use at work of TPVs

Subject to regulation 5, no person shall place on the market or use at work any TPV unless the requirements of paragraphs 2 to 4 below have been complied with.

- The TPV must:
  - (a) be safe and suitable for purpose;
  - (b) have been designed, manufactured and tested according to:
    - (i)   any standard listed in Schedule 2 which is relevant to that vessel; or
    - (ii) any other standard approved by the HSE for the purpose of these regulations as meeting the relevant requirements of Chapter 6.2 of the ADR;
  - (c) have been assessed by a notified body, in accordance with the relevant conformity assessment procedures specified in Schedule 4, to be in conformity with the standards identified pursuant to sub-paragraph (b); and
  - (d) bear the conformity marking, the identification number of the notified body and any marking required by the standards identified pursuant to sub-paragraph (b).
- Where the TPV includes valves (in particular, safety valves, valves for filling and emptying and cylinder valves), those valves must:
  - (a) have been designed, manufactured and tested according to:
    - (i)   any standard listed in Schedule 2 which is relevant to those valves; or
    - (ii) where there is no such standard, the requirements of the Pressure Equipment Regulations 1999; and

(b)   meet the requirements specified in paragraph 5.

• The requirements referred to in paragraphs 3 and 4 above are that the valves and accessories, as the case may be, must:

(a)   be safe and suitable for purpose;

(b)   have been assessed by a notified body, in accordance with the relevant conformity assessment procedure specified in Schedule 4, to be in conformity with the standards or requirements identified pursuant to paragraph 3 or 4 above, as the case may be, and

(c)   bear the conformity marking or the CE marking as appropriate.

• Any technical documentation or other information required to be retained under a conformity assessment procedure shall be retained by the person specified in that procedure for any period specified in that procedure.

Regulation 5 – TPVs placed on the market or used at work exclusively in Great Britain

(a)   Regulation 4 shall not apply in respect of any TPV which is placed on the market or used at work exclusively in Great Britain provided that the requirements of paragraphs 2 to 4 below are complied with.

• The TPV must:

(a)   be safe and suitable for purpose;

(b)   have been designed, manufactured and tested according to:
   (i)   any standard listed in Schedule 2 which is relevant to that vessel; or
   (ii) any other standard approved by the HSE for the purposes of these regulations as meeting the relevant requirements of Chapter 6.2 of the ADR;

(c)   where relevant, have been assessed by an approved body, in accordance with conformity assessment procedures A1, C1, F and G (and for this purpose references in those procedures to 'notified body' shall be read as if they are references to 'approved body') to be in conformity with the standards identified pursuant to sub-paragraph (b);

(d)   bear the identification number of the relevant approved body and any marking required by the standards identified pursuant to sub-paragraph (b); and

(e)   not bear the conformity marking or the identification number of any notified body.

• Where the TPV includes valves (in particular safety valves, valves for filling and emptying and cylinder valves) those valves must:

(a) have been designed, tested and manufactured according to:

   (i) a standard listed in Schedule 2 which is relevant to those valves; or

   (ii) meet the requirements of the Pressure Equipment Regulations 1999; and

(b) meet the requirements specified in paragraph 5.

- Where the TPV includes accessories, other than valves, which have a direct safety function, those accessories must:

(a) have been designed, manufactured and tested in accordance with the requirements set out in Chapter 6.2 of the ADR; and

(b) meet the requirements specified in paragraph 5 below.

- The requirements referred to in paragraphs 3 and 4 above are that the valves or the accessories, as the case may be, must:

(a) be safe and suitable for purpose;

(b) where relevant, have been assessed by an approved body, in accordance with conformity assessment procedures A1, C1, F and G to be in conformity with the standards of requirements identified pursuant to paragraph 3 or 4 above, as the case may be;

(c) bear the identification mark of the relevant approved body; and

(d) not bear the conformity marking or the identification number of any notified body.

Any technical documentation or other information required to be retained under a conformity assessment procedure shall be retained by the person specified in that procedure for any period specified in that procedure.

Regulation 7 – Periodic inspection and repeated use

The owner of a TPV shall ensure that the vessel is periodically inspected in accordance with all relevant procedures in Schedule 6 to ensure:

(a) the continued safety of that vessel; and

(b) that the vessel continues to meet the standards and requirements to which it was designed and manufactured.

No person shall fill, refill or transport a TPV unless a valid certificate issued in accordance with Schedule 6 demonstrates that the vessel is not overdue for any periodic inspection to which it is subject under these regulations.

Any person who carries out a periodic inspection of a TPV shall ensure that:

(a) that inspection complies with the relevant requirements of Schedule 6;

(b) at the end of the inspection, that vessel is marked with:

   (i) the identification number of the relevant notified body or approved body;

(ii) where required by Schedule 6, the date of the inspection; and

(iii) in the case of EEC-type cylinders undergoing their first periodic inspection, the conformity marking.

Any technical documentation or other information required to be retained under a periodic inspection procedure shall be retained by the person specified in that procedure and for any period specified in that procedure.

Regulation 8 – Notified bodies

For the purposes of these regulations, a notified body is a body which has been appointed:

(a) to carry out one or more of the following, namely:

    (i) one or more conformity assessment procedures,

    (ii) the conformity reassessment procedure,

    (iii) either or both periodic inspection procedures; and

(b) as a notified body:

    (i) pursuant to regulation 10, or

    (ii) by another Member State, and whose appointment has been communicated by that Member State to the commission and the other Member States pursuant to Article 8 of the Transportable Pressure Equipment Directive.

Regulation 9 – Approved bodies

For the purposes of these regulations, an approved body is a body which has been appointed:

(a) to carry out one or more of the following, namely:

    (i) one or more conformity assessment procedures,

    (ii) the conformity reassessment procedure,

    (iii) periodic inspection procedure module 1; and

(b) as an approved body:

    (i) pursuant to regulation 10, or

    (ii) by another Member State, and whose appointment has been communicated by that Member State to the commission and the other Member States pursuant to Article 9 of the Transportable Pressure Equipment Directive.

In respect of any conformity assessment procedure undertaken pursuant to regulation 5, an approved body shall work exclusively for the group of which it is a member.

Regulation 10 – Appointment of notified bodies and approved bodies by the HSE

The HSE may appoint such persons as it thinks fit to be notified bodies or approved bodies for the purposes of these regulations.

Regulation 10 lays down the procedure for application to the HSE and conditions of appointment by the HSE.

Regulation 11 – Conformity marking

Where a notified body or an approved body:

(a) has carried out a conformity assessment, reassessment of conformity or a periodic inspection; and

(b) requires the conformity marking to be affixed to a TPV, valve or accessory which is complete or in a state permitting final assessment,

that body shall ensure that the conformity marking is affixed in a visible, easily legible and indelible fashion.

No person shall affix any marking to a TPV, valve or accessory which is likely to render the meaning or form of the conformity marking misleading.

Any other marking may be affixed to a TPV, valve or accessory provided that the visibility and legibility of the conformity marking is not thereby reduced.

Regulation 12 – Fees

This regulation makes provision for the payment of a fee to the HSE with respect to the appointment, etc. of a notified body or an approved body.

Regulation 13 – Defence

In any proceedings for an offence under any of the provisions of these regulations it shall be a defence for the person charged to prove:

(a) that the commission of the offence was due to the act or default of another person not being one of his employees; and

(b) that he took all reasonable precautions and exercised all due diligence to avoid the commission of the offence.

WASTE ON LAND

Part II of the EPA deals with waste on land. It was enacted to provide a more comprehensive regime for dealing with waste on land and introduced a number of very significant changes in order to tighten up the control of the whole of the waste chain, not just the final deposit of waste. The underlying philosophy is that control of waste be exercised from the point at which it is produced and stored pending removal to the point at which it is finally deposited.

As such, the EPA:

- introduced a much stricter licensing system, particularly in relation to the qualifications of licence-holders, licence conditions and surrender of licences
- established a statutory duty of care over anyone who imports, produces, carries, keeps, treats or disposes of controlled waste
- reorganised the functions of the regulatory authorities to avoid the 'poacher and gamekeeper' scenario that existed under the Control of Pollution Act 1984.

'Waste' and 'controlled waste'

Waste is defined in section 75 of the EPA as:

(a) any substance which constitutes a scrap material or effluent or other unwanted surplus substance arising from the application of any process
(b) any substance and article which requires to be disposed of as being broken, worn out, contaminated, or otherwise spoiled.

Anything which is discarded or other otherwise dealt with as if it were waste shall be presumed to be waste unless the contrary is proved.

The question of what constitutes waste has been considered by the European Court of Justice in *Vessoso and Zanetti* (1990). The ECJ had to consider whether the term 'waste' in the *Framework Directive on Waste* included substances that could be recycled. The Court was clear in its judgment that waste under the Directive included materials intended for recycling, since it was waste as far as the disposer was concerned, irrespective of the intentions of the recipient. This case is important in view of the fact that the Waste Management Licensing Regulations 1994 have incorporated the EC definition of 'waste' into national law. National courts interpreting the UK Regulations are required to be consistent with the ECJ's decision in this case.

Controlled wastes are those wastes that are controlled by the EPA, namely household, industrial and commercial waste or any such waste.

THE CONTROLLED WASTE REGULATIONS 1992

These regulations classified the forms of waste in terms of the provisions relating to collection. Most household waste is collected free of charge whereas charges are made for the collection of industrial and commercial waste.

- *Household waste* includes waste from domestic properties, caravans, residential homes, educational establishments, hospitals and nursing homes

- *Industrial waste* (trade waste) means waste from any of the following premises – factories, public transport premises, premises used to supply gas, water and electricity, sewerage, postal or telecom services; waste from construction or demolition operations is also included in this definition as is waste from contaminated land
- *Commercial waste* includes waste from premises used wholly or mainly for the purposes of a trade or business, or for the purposes of sport, recreation or entertainment, except household, industrial, mining, quarrying and agricultural waste, or any other waste specified in the Waste Management Licence Regulations 1992.

Not all types of waste are 'controlled'. For example, waste from agricultural premises, waste from mines and quarries, explosive or most radioactive waste are not defined as controlled wastes. Litter is, however, classified as controlled waste.

WASTE MANAGEMENT LICENCES

Section 35 of the EPA defines a Waste Management Licence as 'a licence granted by a waste regulation authority (WRA) authorising the treatment, keeping or disposal of any specified description of controlled waste in or on specified land or the treatment or disposal of any specified description of controlled waste by means of a specified mobile plant'.

Where a WRA grants a licence, it will do so subject to any conditions that it deems appropriate. There are two types of Waste Management Licence, namely the site licence, which is granted to the occupier of land, and the mobile plant licence.

An application for a Waste Management Licence must be made to the WRA in writing accompanied by the appropriate fee. No licence can be granted in relation to land unless there is planning permission or an established use certificate which authorises the particular use. A WRA may not reflect an application that has been properly made unless:

- the applicant is not a fit and proper person;
- the rejection is necessary to prevent pollution to the environment, harm to human health or serious detriment to the amenities of the locality.

THE WASTE MANAGEMENT LICENSING REGULATIONS 1994

These regulations established a new system of waste management licensing and a redefinition of the meaning of 'waste' as previously defined in the EPA.

The regulations give effect to the definition of 'waste' provided in EC Directive 91/156 which amended the Waste Framework Directive 75/442. As a consequence, waste which is not *'directive waste'* is not household, industrial or commercial waste.

Directive waste is defined in the Directive and regulations by reference to the list of categories of waste. Directive waste is any substance or object in the categories laid down in Schedule 4 to the regulations which the producer or the person in possession of it discards, or intends to discard, or is required to discard. These categories are as follows:

- production or consumption residues
- off-specification products
- products whose date for appropriate use has expired
- materials spilled, lost or having undergone other mishap, including any materials, equipment, etc., contaminated as a result of the mishap
- materials contaminated or soiled as a result of planned actions, e.g. residues from cleaning operations, packing materials, containers
- unusable parts, e.g. reject batteries, exhausted catalysts
- substances which no longer perform satisfactorily, e.g. contaminated acids and solvents
- residues of industrial processes, e.g. slags
- residues from pollution abatement processes, e.g. scrubber sludges, spent filters
- machining or finishing residues, e.g. lathe turnings
- residues from raw materials extraction and processing, e.g. mining residues
- adulterated material, e.g. oils contaminated with PCBs
- any materials, substances or products whose use has been banned by law
- products for which the holder has no further use, e.g. agricultural, household, office, commercial and shops discards
- contaminated materials, substances or products resulting from remedial action with respect to land
- any materials, substances or products which are not contained in the above categories.

THE WASTE REGULATORY AUTHORITIES

The EPA created three different levels of waste authority, each with a different role in relation to waste management.

Waste Regulation Authorities (WRAs)

In England, the WRAs' function in non-metropolitan counties is given to county councils. In the metropolitan areas, the WRAs are the district councils, with special authorities established for Greater London, Greater Manchester and Merseyside.

The main functions of the WRA are:

- preparation of waste disposal plans
- control over the waste management licensing system
- supervision of licensed activities
- inspection of licensed land and landfill sites
- maintaining the public registers
- reporting to the Secretary of State.

Waste Disposal Authorities (WDAs)

The WDA is normally the county council in non-metropolitan areas and the district council in metropolitan areas. The functions of the WDA are:

- making arrangements for the disposal of controlled wastes collected in the area by the *waste collection authorities*
- formation of waste disposal companies
- provision of municipal waste sites for household waste to be deposited by residents
- provision of transfer stations
- waste recycling.

Waste Collection Authorities (WCAs)

These are the district councils or London boroughs. They:

- arrange for the collection of household waste in their area
- arrange for the collection of commercial or industrial waste on request
- provide bins/receptacles
- collect waste and deliver for disposal as directed by the waste disposal authority
- investigate and make arrangements for recycling.

WORK AWAY FROM BASE

A substantial number of people work away from their normal base of operations, such as employees of building contractors, people involved in the installation and servicing of equipment, drivers and company representatives. As such, they are exposed to a wide range of hazards through, in most cases, their unfamiliarity with premises, processes and working practices. In some organisations around 20 to 25 per cent of accidents to staff take place on other people's premises.

The legal requirements relating to work on other people's premises are dealt with in the Occupiers' Liability Act 1957, Health and Safety at Work etc. Act 1974 and the Management of Health and Safety at Work Regulations 1999.

Occupiers' Liability Act 1957

An occupier of premises owes a common duty of care to all lawful visitors in respect of dangers due to the state of the premises or things done or omitted to be done on them.

Health and Safety at Work, etc. Act 1974

An employer must conduct his undertaking in such a way as to ensure, so far as reasonably practicable, that persons not in his employment who may be affected thereby are not thereby exposed to risks to their health or safety (section 3).

Every person who has, to any extent, control of premises must ensure, so far as is reasonably practicable, that the premises, all means of access thereto and egress therefrom, and any plant or substances in the premises or provided for use there, is or are safe and without risks to health (section 4).

Management of Health and Safety at Work Regulations 1999

Regulation 12 deals with persons working in host employers' undertakings. Host employers must provide such persons with appropriate instructions and comprehensible information on the risks arising out of or in connection with their undertakings and, secondly, the measures taken in compliance with the requirements and prohibitions imposed upon them by or under the relevant statutory provisions in so far as the said requirements and prohibitions relate to those employees.

Summary

The duties of employers and occupiers of premises towards non-employees working in their premises or undertaking can be summarised as follows:

- a general duty of care to all lawful visitors
- a general duty not to expose non-employees to risks to their health or safety
- a general duty on controllers of premises to provide safe premises, safe access and egress, safe plant and substances
- a specific duty to provide instructions and information on risks and precautionary measures necessary by the non-employees or self-employed persons concerned.

Practical procedures to implement these requirements

1. Provision of written instructions to non-employees with regard to safe working practices generally.

2. Provision of comprehensible written information to non-employees on the hazards and precautions necessary.
3. Formal health and safety training sessions for all non-employees prior to commencing work on the host employer's premises.
4. Specification of the health and safety competence necessary for non-employees at the tender stage of contracts.
5. Operation of formal hazard reporting systems by non-employees.
6. Disciplinary procedures against non-employees for failure to comply with written instructions and safety signs, including dismissal from the site or premises in serious cases of non-compliance.
7. General supervision of non-employees and regular meetings to reinforce the safety requirements for such persons.
8. Liaison with the external employers, and certification where necessary, to ensure the employees concerned have received the appropriate information, instruction and training necessary prior to commencing work in the host employer's undertaking.
9. Pre-tender and on-going site inspections by the external employer to ensure his employees are not exposed to risks to their health or safety.

THE WORK IN COMPRESSED AIR REGULATIONS 1996

These regulations lay down requirements and prohibitions with respect to the health, safety and welfare of persons who work in compressed air.

The term *work in compressed air* means work within any working chamber, airlock or decompression chamber which (in each case) is used for the compression or decompression of persons, including a medical lock used solely for treatment purposes, the pressure of which exceeds 0.15 bar.

The regulations apply to construction work (within the meaning of the Construction (Design and Management) Regulations 1994) and have effect in addition to any applicable provisions of the Construction (Health, Safety and Welfare) Regulations 1996. They do not apply to diving operations within the meaning of the Diving Operations at Work Regulations 1981.

The regulations:

- provide for the appointment of a competent contractor (the compressed air contractor) to execute or supervise the work in compressed air included in any project
- require specified information to be notified in writing to the HSE and to specified hospitals and other bodies before work in compressed air is commenced and for further notification of the termination or suspension of such work
- require work in compressed air to be carried out only in accordance with a safe system of work and under adequate supervision
- impose requirements with regard to the provision, use and maintenance of adequate and suitable plant and equipment

- provide that a contract medical adviser be appointed to advise the compressed air contractor on matters relating to the health of persons who work in compressed air
- impose a requirement on employers for adequate medical surveillance to be carried out in respect of such employees who work in compressed air
- require compression and decompression to be carried out safely and in accordance with any procedures approved by the HSE.

THE WORKING TIME REGULATIONS 1998–2002

These regulations implemented the EC Directive on Working Practices (the 'Working Time' Directive) and apply to all areas of employment with the exception of air, rail, road, sea, inland waterway and lake transport, sea fishing, other work at sea and the activities of doctors in training.

Fundamentally, the regulations apply to atypical workers, that is those not in normal daytime employment, together with shift workers, part-time workers and night workers.

These regulations:

- entitle workers to a minimum daily rest period of 11 consecutive hours in each 24-hour period;
- provide for workers to receive a rest break where the working day is longer than 6 hours, the duration of which is to be determined by 'collective agreements or agreements between the two sides of industry' or, failing that, by national legislation
- require workers to receive a minimum weekly rest period of 35 hours in each 7-day period, although this period may be reduced to 24 hours 'if objective, technical or work organisation considerations so justify'
- limit the average working time during each 7-day period to 48 hours
- require that workers receive four weeks' annual paid leave (Article 7)
- restrict the normal hours of work of night workers to an average of 8 hours in any 24 hour period, and impose an absolute limit of 8 hours where the work involves special hazards or heavy physical or mental strain
- state that night workers should be entitled to a free health assessment, which may be provided through the national health system, prior to their assignment and at regular intervals thereafter
- require that night workers, suffering from ill-health recognised as being connected with the fact that they perform such work are, whenever possible, transferred to day work to which they are suited
- state that night and shift workers must be provided with safety and health protection arrangements that are equivalent to those applying to other workers, and that are available at all times
- require employers to take account of the general principles of adapting work to the worker, with a view in particular to alleviating monotonous work and work at a pre-determined work rate, depending upon the type

of activity, and of safety and health requirements, especially as regards to breaks during working time.

THE WORKPLACE (HEALTH, SAFETY AND WELFARE) REGULATIONS 1992

Citation and commencement (regulation 1)

These regulations came into force on 1 January 1993.

Comment

These regulations came into effect in two stages and apply to all workplaces after 1 January 1996. The majority of the requirements are of an absolute nature.

Interpretation (regulation 2)

The following definitions are important to the interpretation of the regulations:

- *new workplace* means a workplace used for the first time after 31 December 1992
- *traffic route* means a route for pedestrian traffic, vehicles or both and includes any stairs, staircases, fixed ladder, doorway, gateway, loading bay or ramp
- *workplace* means any premises or part of them that are not domestic premises and are made available to anyone as a place of work, including:
  - any place within the premises to which such a person has access while at work
  - any room, lobby, corridor, staircase, road or other place used as a means of access to or egress from the workplace or where facilities are provided for use in connection with the workplace other than a public road.

The application of these regulations (regulation 3)

The regulations apply to every workplace other than:

- a workplace that is, or is in or on, a ship
- a workplace where the only activities being undertaken are building operations or works of engineering construction
- a workplace where the only activities being undertaken are the exploration for or extraction of mineral resources
- a workplace that is situated in the immediate vicinity of another workplace or intended workplace where exploration for or extraction of mineral resources is being or will be undertaken, or where the only activities being undertaken are activities preparatory to, for the purposes of, or in connection with such exploration for or extraction of mineral resources at this other workplace.

When applied to *temporary work sites*, any requirement to ensure that a workplace complies with any of regulations 20–25 shall apply so far as is *reasonably practicable*.

The requirements of these regulations (regulation 4)

Every employer shall ensure that every workplace, modification, extension or conversion under their control and where any of their employees' work complies with any of the requirements of these regulations that:

- apply to these places
- are in force concerning them.

Everyone who has, to any extent, control of such places shall ensure that the places comply with any requirements of these regulations that:

- apply to these places
- are in force concerning them
- relate to matters within that person's control.

Control here implies the control the person has in connection with the carrying out of a trade, business or other undertaking (whether for profit or not).

Stability and solidity (regulation 4a)

Where a workplace is in a building, the building shall have a stability and solidity appropriate to the nature of the use of the workplace.

The maintenance of the workplace and of equipment, devices and systems (regulation 5)

The workplace and the equipment, devices and systems to which this regulation applies shall be maintained (which includes them being cleaned as appropriate) in *an efficient state, in efficient working order and in good repair*. They shall be subject to a suitable system of maintenance.

The equipment, devices and systems to which this regulation applies are:

- equipment and devices that if a fault occurred in them, would then be likely to fail to comply with any of these regulations; and
- mechanical ventilation systems provided pursuant to regulation 6; and
- equipment and devices intended to prevent or reduce hazards.

Ventilation (regulation 6)

Effective and suitable provision shall be made to ensure that every enclosed workplace is ventilated by a sufficient quantity of fresh or purified air.

Any plant used in order to comply with this requirement shall include an effective device to give visible or audible warning of any failure of the plant, where necessary, for reasons of health or safety.

This regulation shall not apply to any confined space in a workplace.

The temperature in indoor workplaces (regulation 7)

During working hours, the temperature in all workplaces inside buildings shall be reasonable.

Without prejudice to the generality of the above,

- a workplace shall be adequately thermally insulated where it is necessary, having regard to the type of work carried out and the physical activity of the persons carrying out the work; and
- excessive effects of sunlight on temperature shall be avoided.

A method of heating or cooling shall not be used that results in the escape into the workplace of fumes, gas or vapour of such character and to such an extent that they are likely to be injurious or offensive to anyone.

A sufficient number of thermometers shall be provided to enable people at work to determine the temperature in any workplace inside a building.

Lighting (regulation 8)

Every workplace shall have suitable and sufficient lighting, which shall, so far as is reasonably practicable, be by natural light.

Suitable and sufficient emergency lighting shall be provided and maintained in any room in circumstances in which people at work are specially exposed to danger in the event of the failure of artificial lighting.

Cleanliness and waste materials (regulation 9)

Every workplace and the furniture, furnishings and fittings therein shall be kept sufficiently clean.

The surfaces of the floor, wall and ceiling of all workplaces inside buildings shall be capable of being kept sufficiently cleaned.

So far as is reasonably practicable, waste materials shall not be allowed to accumulate, except in suitable receptacles.

Room dimensions and space (regulation 10)

Every room where people work shall have sufficient floor area, height and unoccupied space for the purposes of health, safety and welfare.

This regulation shall be complied with sufficiently if any existing workplace was subject to the provisions of the Factories Act 1961 and it does not contravene Part 1 of Schedule 1 of the Act.

Workstations and seating (regulation 11)

Every workstation shall be so arranged that it is suitable both for anyone at work in the workplace who is likely to work at it and for any work of the undertaking that is likely to be done there.

Every workstation outdoors shall be so arranged that:

- so far as is reasonably practicable, it provides protection from adverse weather
- it enables anyone at the workstation to leave it swiftly or, as appropriate, to be assisted in the event of emergency
- it ensures that any person at the workstation cannot slip or fall.

A suitable seat shall be provided for everyone at work in the workplace whose work includes operations of a kind that the work (or a substantial part of it) can or must be done sitting.

A seat shall not be considered suitable unless:

- it is suitable for the person for whom it is provided as well as for the operations to be performed
- a suitable footrest is also provided, where necessary.

The condition of floors and traffic routes (regulation 12)

Every floor in a workplace and the surface of every traffic route in a workplace shall be of a construction that is suitable for the purpose for which it is used. In particular:

- the floor or surface of the traffic route shall have no hole or slope or be uneven or slippery so as, in each case, to expose anyone to a risk to their health or safety
- every floor shall have effective means of drainage, where necessary.

So far as is reasonably practicable, every floor in a workplace and the surface of every traffic route in it shall be kept free from obstructions and from any article or substance that may cause a person to slip, trip or fall.

In considering whether a hole or slope exposes anyone to a risk to their health or safety:

- no account shall be taken of a hole where adequate measures have been taken to prevent a person falling
- account shall be taken of any handrail provided in connection with any slope.

Suitable and sufficient handrails and, if appropriate, guards shall be provided on all traffic routes that are staircases except in circumstances in which a handrail cannot be provided without obstructing the traffic route.

Falls or falling objects (regulation 13)

So far as is reasonably practicable, suitable and effective measures shall be taken to prevent any of the events specified below. Also, so far as is reasonably practicable, these measures shall be other than the provision of PPE information, instruction, training or supervision.

The events mentioned are:

- anyone falling a distance likely to cause personal injury
- anyone being struck by a falling object likely to cause personal injury.

Any area where there is a risk to health or safety from any of these events shall be clearly indicated where appropriate.

So far as is reasonably practicable, every tank, pit or structure where there is a risk of a person in the workplace falling into a dangerous substance in the tank, pit or structure, shall be securely covered or fenced.

Every traffic route over, across or in an uncovered tank, pit or structure shall be securely fenced. *Dangerous substance* here means:

- any substance likely to scald or burn
- any poisonous substance
- any corrosive substance
- any fume, gas or vapour likely to overcome a person
- any granular or free-flowing solid substance or any viscous substance that, in any case, is of a nature or quantity that is likely to cause danger to anyone.

Windows and transparent or translucent doors, gates and walls (regulation 14)

Every window or other transparent or translucent surface in a wall or partition and every transparent or translucent surface in a door or gate shall, where necessary for reasons of health or safety:

- be of safety material or be protected against breakage of the material
- be appropriately marked or incorporate features that, in either case, make it apparent.

Windows, skylights and ventilators (regulation 15)

No window, skylight or ventilator capable of being opened shall be likely to be opened, closed or adjusted in a manner that exposes any person performing such an operation to a risk to their health or safety.

No window, skylight or ventilator shall be in a position that is likely to expose anyone in the workplace to a risk to their health or safety.

The ability to clean windows and so on safely (regulation 16)

All windows and skylights in a workplace shall be of a design or so constructed that they may be cleaned safely.

In considering whether a window or skylight is safe, account may be taken of equipment used in conjunction with the window or skylight or of devices fitted to the building.

The organisation and control of traffic routes (regulation 17)

Every workplace shall be organised in such a way that pedestrians and vehicles can circulate in a safe manner.

Traffic routes shall be suitable for the people or vehicles using them, sufficient in number, in suitable positions and of sufficient size. They shall not satisfy these requirements unless suitable measures are taken to ensure that:

- pedestrians or, as the case may be, vehicles may use the traffic route without causing danger to the health or safety of people near it
- there is sufficient separation of traffic routes for vehicles from doors, gates and traffic routes of pedestrians that lead on to it
- where vehicles and pedestrians use the *same* traffic routes, there is sufficient separation between them.

All traffic routes shall be suitably indicated, where this is necessary for reasons of health or safety.

These requirements shall apply, so far as is reasonably practicable, to existing workplaces.

Doors and gates (regulation 18)

Doors and gates shall be suitably constructed (including being fitted with any necessary safety devices). They shall not comply with this requirement unless:

- any sliding door or gate has a device to prevent it coming off its track during use
- any upward-opening door or gate has a device to prevent it falling back
- any powered door or gate has suitable and effective features to prevent it causing injury by trapping somebody
- where necessary for reasons of health or safety, any powered door or gate can be operated manually, unless it opens automatically if the power fails
- any door or gate that is capable of opening by being pushed from either side is of such a construction as to provide, when closed, a clear view of the space close to both sides.

Escalators and moving walkways (regulation 19)

Escalators and moving walkways shall:

- function safely
- be equipped with any necessary safety devices
- be fitted with one or more emergency stop controls that are easily identifiable and readily accessible.

Sanitary conveniences (regulation 20)

Suitable and sufficient sanitary conveniences shall be provided at readily accessible places and shall not be suitable unless:

- the rooms containing them are adequately ventilated and lit
- they, and the rooms containing them, are kept in a clean and orderly condition
- separate rooms containing conveniences are provided for men and women, except where and so far as each convenience is in a separate room, the door of which can be secured from inside.

In existing factories, compliance with Part II of Schedule 1 shall be considered as sufficient compliance.

Washing facilities (regulation 21)

Suitable and sufficient washing facilities, including showers (if these are required by the nature of the work or for health reasons), shall be provided at readily accessible places.

Washing facilities shall not be considered suitable unless they:

- are provided in the immediate vicinity of every sanitary convenience, whether or not they are provided elsewhere as well
- are provided in the vicinity of any changing rooms required by these regulations, whether or not they are provided elsewhere as well
- they provide a supply of clean, hot and cold, or warm water (which shall be running water so far as is practicable)
- they include soap or other suitable means of cleaning
- they include towels or other suitable means of drying
- the rooms containing them are sufficiently ventilated and lit
- they and the rooms containing them are kept in a clean and orderly condition and are properly maintained
- separate facilities are provided for men and women, except where and so far as they are provided in a room the door of which is capable of being secured from inside and the facilities in each such room are intended to be used by only one person at a time.

This last point shall not apply to facilities that are used for washing only the hands, forearms and face.

Drinking water (regulation 22)

An adequate supply of wholesome drinking water shall be provided for all people at work in the workplace.

Every supply of drinking water shall:

- be readily accessible at suitable places

- be conspicuously marked by a suitable sign, where necessary, for reasons of health and safety.

Where a supply of drinking water is required, there shall also be provided a sufficient number of suitable cups or other drinking vessels, unless the supply of water is in a jet from which people can drink easily.

Accommodation for clothing (regulation 23)

Suitable and sufficient accommodation shall be provided:

- for anyone at work's own clothing that is not worn during working hours
- for special clothing that is worn by any person at work, but not taken home.

The accommodation shall not be considered suitable unless:

- where facilities to change clothing are required, it provides suitable security for clothes not worn
- where necessary to avoid risks to health or damage to the clothing, it includes separate accommodation for clothing worn at work and for other clothing
- so far as is reasonably practicable, it allows or includes facilities for drying clothing
- it is in a suitable location.

Facilities for changing clothing (regulation 24)

Suitable and sufficient facilities shall be provided for anyone at work in the workplace to change their clothing in all cases where:

- they have to wear special clothing for the purposes of their work
- they cannot, for reasons of health or propriety, be expected to change elsewhere and the facilities are easily accessible, of sufficient capacity and provided with seating.

These facilities shall not be considered suitable unless they include separate facilities for, or separate use of facilities by, men and women where necessary for reasons of propriety.

Facilities for rest and to eat meals (regulation 25)

Suitable and sufficient rest facilities shall be provided at readily accessible places. They shall:

- where necessary for reasons of health or safety include, in the case of a new workplace, extension or conversion, rest facilities provided in one or more rest rooms or, in other cases, in rest rooms or rest areas
- include suitable facilities to eat meals where food that is eaten in the workplace would otherwise be likely to become contaminated.

Rest rooms and rest areas shall:

- include suitable arrangements to protect non-smokers from discomfort caused by tobacco smoke: and
- be equipped with:
    - (i) an adequate number of tables and adequate seating with backs for the number of persons at work likely to use them at one time; and
    - (ii) seating which is adequate for the number of disabled persons at work and suitable for them.

Suitable facilities shall be provided for anyone at work who is a pregnant woman or nursing mother to rest.

Suitable and sufficient facilities shall be provided for people at work to eat meals where meals are regularly eaten in the workplace.

Where necessary, those parts of the workplace (including in particular doors, passageways, stairs, showers, washbasins, lavatories and workstations) used or occupied directly by disabled persons at work shall be organised to take account of such persons.

Schedule 1, relating to regulations 10 and 20

The provisions applicable to factories that are not new workplaces, extensions or conversions

Part I: Space

No room in the workplace shall be so overcrowded as to risk the health or safety of people at work in it.

Without prejudice to the generality of this statement, the number of people employed at a time in any workroom shall not be such that the amount of cubic space allowed for each is less than 11 cubic metres.

In calculating, for the purposes of this last provision, the amount of cubic space in any room, no space more than 4.2 metres from the floor shall be taken into account and, where a room contains a gallery, the gallery shall be treated for the purposes of this Schedule as if it were partitioned off from the remainder of the room and formed a separate room.

Part II: Number of sanitary conveniences

In workplaces where *females* work, there shall be at least one suitable water closet for use by females only for every 25 females.

In workplaces where *males* work, there shall be at least one suitable water closet for use by males only for every 25 males.

In calculating the number of males or females who work in any workplace for the purposes of this Part of the Schedule, any number not itself divisible by 25 without resulting in a fraction or remainder shall be treated as the next number higher than it which is so divisible.

Schedule 2, relating to regulation 27

Part I: Repeals

These regulations repeal:

- sections 1–7, 18, 28, 29, 57–60 and 69 of the Factories Act 1961
- sections 4–16 of the Offices, Shops and Railway Premises Act 1963
- sections 3 and 5 and, in section 25, subsections (3) and (6) of the Agriculture (Safety, Health and Welfare Provisions) Act 1956.

Part II: Revocations

These regulations revoke, wholly or partly, a substantial number of Orders and regulations, in particular:

- the Sanitary Accommodation Regulations 1938
- the Washing Facilities Regulations 1964
- the Sanitary Conveniences Regulations 1964

and other regulations dealing with safety, health and welfare in specific industries.

The ACOP and guidance notes

A very comprehensive ACOP and Guidance Notes accompany these regulations.

CURRENT TRENDS IN HEALTH AND SAFETY LEGISLATION

All modern health and safety legislation is largely driven by European Directives, for instance:

- the Directive 'on the health and safety of workers at work' was implemented in the UK as the Management of Health and Safety at Work Regulations 1992 (now replaced by the 1999 ones); and
- the Directive 'on temporary and mobile construction sites' was implemented in the UK as the Construction (Design and Management) Regulations 1994.

Regulations produced since 1992 do not, in most cases, stand on their own. They must be read in conjunction with the general duties imposed on employers under the Management of Health and Safety at Work Regulations 1999, in particular the duties relating to:

- the risk assessment
- the operation and maintenance of safety management systems
- the appointment of competent persons
- establishment and implementation of emergency procedures
- provision of information to employees which is comprehensible and relevant

- co-operation, communication and co-ordination between employers in shared workplaces, e.g. construction sites, office blocks
- provision of comprehensible health and safety information to employees from an outside undertaking
- assessment of human capability prior to allocating tasks
- provision of health and safety information, training and instruction.

Duties imposed on employers tend to be largely of an absolute nature, as opposed to qualified duties, such as 'so far as is reasonably practicable', as with the Health and Safety at Work, etc. Act 1974.

Risk assessment, taking into account the 'relevant statutory provisions', is the starting point of all health and safety management systems.

Most modern health and safety legislation requires some form of documentation, such as risk assessments and planned preventive maintenance systems, and the maintenance of records.

The principal cases

INTRODUCTION

The common law is based on cases recorded, in England and Wales, in the *Law Reports* and *All England Reports* (AER) and, in Scotland, the *Sessions Cases* (SC) and *Scots Law Times* (SLT). It is based on the doctrine of *precedent*, under which, in the vast majority of cases, a court must follow the earlier decisions of the courts at its own level and of superior courts – known as *binding precedent*. The doctrine of binding precedent is called *stare decisis*, which means 'keep to the decisions of past cases'. Other precedents established are of a *persuasive* nature, and are not binding. For the doctrine of *judicial precedent* to operate, it is necessary to know, first, the legal principle involved in a particular judgment (*ratio decidendi*) and, second, when a decision is binding and when it is persuasive.

The lower courts are bound by the higher courts and superior courts generally follow their own former decisions. In the lower courts, however, judgments are mainly concerned with questions of fact and so are not strict precedents.

THE FEATURES OF A JUDGMENT

Ratio decidendi

This means 'the reason for deciding', and is the legal principle behind a particular judgment, the actual finding on a particular fact. It is a proposition of law that decides the case in the light, or in the context, of the material facts. Whether it is binding will depend on the position in the hierarchy of the court that decided the case and of the court that is currently considering it. Thus, the *ratio decidendi* of a decision of the House of Lords binds all lower courts, whereas that of a Court of Appeal case binds the Court of Appeal and all the lower courts, but not the House of Lords.

Obiter dicta

These are 'comments by the way', such as a reference to what *could* have happened had the facts been different or a reference to an aspect of law that may not be specifically relevant to the case. *Obiter dicta* are persuasive, but not binding. However, the *obiter dicta* can be so persuasive that they may be integrated or incorporated into subsequent judgments and become part of the *ratio decidendi*.

THE PRINCIPAL CASES

The following cases form the basis for much of the current law on occupational health and safety and for its interpretation in the courts.

Donoghue v. *Stevenson* (1932) AC 562 – the duty to take reasonable care, or, the neighbour principle

The claimant purchased a bottle of ginger beer that had been manufactured by the defendant. The bottle was of dark glass so that its contents could not have been seen before they were poured out. The bottle was also sealed so that it could not have been tampered with until it reached the ultimate consumer. When the claimant poured out the ginger beer, the remains of a decomposed snail was seen floating in the ginger beer. Not unnaturally, she was taken ill, having already consumed part of the contents of the bottle.

This case established the fact that manufacturers can be liable for negligence if they fail to take reasonable care in the manufacture or preparation of their products. Lord Atkin said 'a manufacturer of products, which he sells in such a form as to show that 'he intends them to reach the ultimate consumer in the form in which they left him with no reasonable possibility of intermediate inspection, and with the knowledge that the absence of reasonable care in the preparation or putting up of the products will result in an injury to the consumer's life or property, owes a duty to the consumer to take reasonable care'.

Lord Atkin's judgment went further:

> You must take reasonable care to avoid acts or omissions which you reasonably foresee would be likely to injure your neighbour, i.e. persons who are so closely and directly affected by your act so that you ought reasonably to have them in contemplating as being so affected ...

This has subsequently come to be known as *The Neighbour Principle*. It means that there must be a close and direct relationship between the defendant and claimant, such as that between employer and worker, occupier and visitor, shopkeeper and customer, and, second, the defendant must be able to foresee a real risk of injury to the claimant if they, the defendant, do not conduct their operations or manage their property with due care.

This case established that manufacturers can be liable for negligence for a defective product if it can be shown that they have failed to take reasonable care in the preparation and marketing of it and that a consumer of the product has suffered injury, damage or loss as a result.

Wilsons & Clyde Coal Co. Ltd v. *English* (1938) AC57, 2 AER 628 – the duties of employers at common law

At common law, employers owe a general duty towards their employees to take reasonable care so as to avoid injuries, disease and death at work. More specifically, employers must:

- provide a safe place of work with safe means of access to and egress from it
- provide and maintain safe appliances, equipment and plant for doing the work
- provide and maintain a safe system for doing work
- provide competent co-employees to carry out the work.

In this case, Lord Wright said that 'the whole course of authority consistently recognises a duty which rests on the employer, and which is personal to the employer, to take reasonable care for the safety of his workmen, whether the employer be an individual, a firm or company and whether or not the employer takes any share in the conduct of the operations'.

Edwards v. *National Coal Board* (1949) 1 AER 743 – 'reasonably practicable'

This case established the legal definition of the term 'reasonably practicable', a term that qualifies the duties of employers and others under the HSWA and many regulations. A duty qualified by 'so far as is reasonably practicable', as we saw in Chapter 4 implies a lesser level of duty than that qualified by the phrase 'so far as is practicable' (see *Adsett* v. *K. & L. Steelfounders and Engineers Ltd*).

In this case, Asquith, L.J., said:

"reasonably practicable" is a narrower term than "physically possible", and implies that a computation must be made in which the quantum of risk is placed in one scale and the sacrifice involved in the measures necessary for averting the risk (whether in money, time or trouble) is placed in the other, and that, if it be shown that there is a gross disproportion between them – the risk being insignificant in relation to the sacrifice – the defendants discharge the onus upon them.

The burden of proof rests with the defendant to prove that the measures required were not 'reasonably practicable' in *all* the circumstances of the case.

Schwalb v. *Fass (H) & Son* (1946) 175 LT 345 – 'practicable' in the light of current knowledge and invention

Where a statutory duty or obligation is qualified solely by the word 'practicable', this implies a higher level of duty than one qualified by 'reasonably practicable'. 'Practicable' means something other than physically possible, it means that the measures must be possible in the light of current knowledge and invention.

Paris v. *Stepney Borough Council* (1951) AC 367 – the duty of care and the standard of care owed to vulnerable persons

This case involved a one-eyed workman employed in vehicle maintenance. The claimant was endeavouring to remove a bolt from the chassis of a vehicle and was using a hammer for this purpose. While hammering, a metal particle hit his good eye, resulting in total blindness. The claimant claimed damages from his employers for negligence on the basis that he had not been supplied with eye protection. The defendants showed in evidence that it was not standard practice to provide eye protection for activities of this nature, at least where employees had two good eyes.

The judge found for the claimant, but the Court of Appeal reversed the decisions on the grounds that the claimant's disability could be relevant *only* if it increased the risk, that is, that a one-eyed man was more likely to sustain eye injury than a two-eyed man. Having found that the risk was *not* increased, they allowed the appeal. However, the House of Lords reversed the decision and ruled that eye protection *should* have been provided for Mr Paris while undertaking work where there was a risk to his sound eye, even though the risk for a person with two sound eyes was relatively low.

Thus, an employer may owe a *higher* duty of care to an incapacitated employee than he does to an employee who is fit and has all his faculties.

Latimer v. *AEC Ltd* (1952) 2 AER 449 – the practicability of precautions and unreasonable precautions

The appellant was a milling machine operator employed by AEC Ltd. An area of a factory was flooded following a heavy storm and part of the floor was contaminated by an oily, slippery film. The floor was treated with sawdust following this event, but, owing to the very large area of floor involved, there was insufficient sawdust to cover the whole floor. The occupier of the factory was held not to be liable to an employee for injury sustained after he slipped on an untreated part of the floor.

It was held by the House of Lords that the respondents were not in breach of their statutory duty under section 25(1) of the Factories Act 1961 (for floors, steps and so on to be kept free of substances likely to cause people to slip and to be properly 'maintained'). This section referred to the general condition and soundness of construction of the floor, but did not include a 'transient and exceptional condition'. Second, it was held that the respondents had taken every reasonable step to obviate danger to the appellant and they were not liable for negligence at common law.

In this case, the practicability of the precautions was seen as appropriate in the circumstances. In the Court of Appeal, the point was put by Denning, L. J., thus: 'It is a matter of balancing the risk against the measures necessary to eliminate it'. In this situation, the cost to the defendant was too great in relation to the degree of harm that could arise.

Adsett v. *K & L Steelfounders & Engineers Ltd* (1953) 1 AER 97 and 2 AER 320 – the duties of employers with regard to dust and fume control and 'practicable' precautions

This case is one of a number of cases concerned with the interpretation of section 63(1) of the FA. This section states that, 'in every factory in which, in connection with the process carried on, there is given off any dust or fume or other impurity of such a character and to such an extent as to be likely to be injurious or offensive to the persons employed, or any substantial quantity of dust of any kind, all *practicable* measures shall be taken to protect persons employed against inhalation of dust or fumes or other impurity and to prevent it accumulating in any workroom and, in particular, where the nature of the process makes it practicable, to provide and maintain exhaust appliances as near as possible to the point of origin of the dust, fume or other impurity, so as to prevent it entering the air of any workroom'.

The various cases have hinged on two conditions of this section, namely:

● 'dust or fumes or other impurity . . . as to be likely to be injurious or offensive to persons employed'
● 'any substantial quantity of dust of any kind'.

Where either of these two conditions is fulfilled, the occupier is under a duty to take the practicable measures detailed in the latter part of the section. The meaning of the word 'practicable' means more than 'physically possible' and imposes a stricter standard than that which is 'reasonably practicable'. The measures must be possible in the light of current knowledge and invention at the time.

Marshall v. *Gotham Co. Ltd* (1954) 1 AER 937 – 'practicable' and 'reasonably practicable'

In this case, the two levels of duty, that is those of a 'practicable' and 'reasonably practicable' nature were compared. In *Marshall* v. *Gotham Co. Ltd*, Lord Reid said, 'If a precaution is practicable it must be taken unless in the whole circumstances that would be unreasonable. And as men's lives may be at stake it should not lightly be held that to take a practicable precaution is unreasonable . . . .'

Richard Thomas & Baldwin v. *Cummings* (1955) AC 321 and 1 AER 285 – 'in motion' and 'in use' in relation to machinery

This case involved the machinery fencing requirements under the FA and the actual interpretation of machinery 'in motion or use'. The phrase includes motion *or* use in the mode or manner in which the machine operated and for the purpose for which it was used, not merely movement of any kind, and does not include movement of the parts by hand in the case of a power-driven machine.

Davie v. *New Merton Board Mills and Others* (1958) 1 AER 67 – defective work equipment

The claimant was employed by the defendants on a machine and, in the ordinary course of his employment, occasionally had to use a drift, that is a tapered bar about 30 centimetres long. When in use, the drift had to be struck with a hammer. The claimant had obtained a new drift from his employers and, although it appeared to be nearly new, at the second stroke of the hammer, the drift shattered and a piece of metal struck him in the eye, causing him to lose the sight of that eye.

The defendant established that he obtained his supply of drifts from another company who were well-established and reputable suppliers and who had, in turn, obtained them from toolmakers in Sheffield.

It was held that the employers had used reasonable care and skill in providing the tool and were not liable to the claimant for the injury.

Uddin v. *Associated Portland Cement Manufacturers Ltd* (1965) 1 AER 213 – dangerous machinery, disregard of danger and contributory negligence

The claimant was employed as a machine minder by the defendants in the packing plant in a cement factory. One day, in an attempt to catch a pigeon, he leaned across a revolving shaft in a place in which he was not authorised to be, losing an arm as a result. The shaft was part of some dust extraction equipment. However, his duties were concerned solely with the cement packing plant.

He brought an action for damages against his employer, alleging that the shaft was a dangerous part of machinery that should be fenced in accordance with section 14(1) of the FA 1937.

It was held by the Court of Appeal that the defendants were in breach of their obligations under the above section and that the claimant, who, at the time of the injury,

- was performing an act wholly outside the scope of his employment
- for his own benefit
- at a place to which he knew he was not authorised to go,

was not totally debarred from recovering damages. The responsibility, therefore, was apportioned on the basis of 20 per cent to the defendants and 80 per cent to the claimant.

John Summers & Sons Ltd v. *Frost* (1955) 1 AER 870 – the absolute duty to ensure secure fencing of machinery, reasonable forseeability and breach of statutory duty

The respondent was employed by the appellants at their steelworks as a maintenance fitter. While grinding a piece of metal on a power-driven grinder, he was injured when his thumb came into contact with the revolving grindstone. The machine was fitted with an efficient guard and the only part of the revolving grindstone exposed was an arc of approximately 18 centimetres long. There was a gap of approximately 1 centimetre between the guard and the grindstone.

It was held by the House of Lords that, as the grindstone was a dangerous part of machinery within section 14(1) of the FA 1937, there was an absolute obligation in that subsection that it should be securely fenced to prevent such injury as is *reasonably foreseeable*, regardless of whether the operator using the machine is careless or inattentive. In this case, the machine was not 'securely fenced' and the appellants were in breach of their statutory duty under the FA, even though the outcome of securely fencing such a machine would render it unusable.

Kilgollan v. *Cooke & Co. Ltd* (1956) 1 AER 294 – the breach of statutory duty and the 'double-barrelled' action

This case also involved the fencing requirements for machinery under the FA 1937. Here, an employee was injured as a result of his employer's failure to comply with the fencing requirements of the FA.

In this 'double-barrelled' action, the employee sued the employer for damages in two specific actions: first, for negligence, and, second, for breach of a statutory duty.

It should be noted that this legal right of an injured party to take a 'double-barrelled' action was, to some extent, excluded as a result of section 47 of the HSWA. Thus:

- a breach of any of the general duties in sections 2–8 *will not* give rise to civil liability
- breach of any duty contained in health and safety regulations made under the HSWA *will* give rise to civil liability, unless the regulations state otherwise.

Close v. *Steel Company of Wales Ltd* (1961) 2 AER 953 – the breach of statutory duty

The appellant was operating an electric drilling machine when the drill bit shattered and a piece struck him in the eye. He claimed damages for breach of statutory duty under section 14(1) of the FA. There was no indication that this type of accident had happened previously, although bits had shattered occasionally in the past.

It was held by the House of Lords that:

- the respondents were not in breach of duty under the section because danger from the use of the bit in the drill was not a reasonably foreseeable danger as the section required
- the duty to fence dangerous parts of machinery as required by section 14(1) was a requirement that the dangerous parts be so fenced only so as to prevent the operator's body coming into contact with the machinery
- the obligation was to fence the machinery in and not to prevent fragments

of it, or the material being worked, from flying out of the machine

- if, in a factory, there is a machine that it is known from experience has a tendency to emit parts of the machine itself or of the material being worked, so as to be a danger to the operator, the absence of a guard to protect them may well give the injured person a cause for action at common law.

Dewhurst v. *Coventry Corporation* (*The Times*, 22 April 1969) – due diligence under the OSRPA 1963 and cleaning of machinery by young people

A bacon slicer was used by staff at one of Dewhurst's shops in Coventry. There were only two employees: a manager and a boy aged 16. The machine was manually operated by turning a handle that caused the cutting blade to rotate. The blade was guarded when in use and had to be removed in order to clean the machine properly.

It transpired that there were two ways of cleaning the blade, either by:

- removing the blade from the machine and wiping it
- by cleaning it in position by drawing a cloth across half the blade and then turning the handle to expose the other half and wiping it.

No locking device was fitted to prevent the blade turning. When the boy was first employed, he was instructed to use the second of these two methods, but not to turn the handle after getting the blade into position for cleaning.

The boy was, on one occasion, told to clean the machine. He removed the gate holding the bacon and the guard to the blade. While cleaning the blade, he turned the handle and the blade rotated, cutting the tip off his left index finger. He had used this method on several occasions previously as it was quicker than the first method, but not in the presence of his manager.

The company was prosecuted and convicted for a breach of section 18(1) of the OSRPA, which requires that 'no young person employed to work in premises to which this Act applies shall clean any machinery used as, or forming part of the equipment of the premises if doing so exposes him to risk of injury from a moving part of that or any adjacent machinery'.

Dewhurst appealed to the Divisional Court of the Queen's Bench, but the appeal was dismissed on the basis that section 18(1) of the Act imposed an absolute prohibition, subject to the defence in section 67, which provided that there was no liability if *all due diligence* had been used to prevent this event. The only escape from liability was to show that no reasonable care could have prevented the injury and the company had failed in their duty because they must have foreseen that, using their method, it was intended that the handle should be used and was used, at a time, when, clearly, the boy was near the blade.

Lister v. *Romford Ice & Cold Storage Ltd* (1957) 1 AER 125 – vicarious liability, the duty to non-employees and breach of contract

This case involved a lorry driver, employed by the defendants, who negligently reversed the company's vehicle into another employee (actually his father). This employee was awarded damages to be paid by the company under the doctrine of *vicarious liability*. The defendants held insurance cover for this form of liability and the insurance company paid the damages. The insurance company subsequently sued the lorry driver in the name of the company to recover what they had paid out.

It was held unanimously in the House of Lords that the lorry driver, as a servant of the company, owed them a duty to perform his tasks with reasonable care and skill, and that a servant who involves his master in vicarious liability by reason of negligence is liable in damages to the master for breach of contract. The damages in such a case amount to a complete indemnity with regard to the amount that the employer has been held vicariously liable to pay the injured claimant.

White v. Pressed Steel Fisher Ltd (1980) IRLR 176 – the training of trade union safety representatives

Under section 2(2)(c) of the HSWA, an employer must provide information, instruction and training for their employees. This case drew the distinction, however, between the health and safety training that an employer must provide, so far as is reasonably practicable, for all their employees, which includes trade union-appointed safety representatives, and the specialist training required for a safety representative in trade union aspects of their work as a representative of the union's members.

Thus, the training of safety representatives must be viewed by employers as being different to that required for other trade union officials who, by virtue of section 27 of the EPCA, must be approved by the TUC.

Thompson, Gray, Nicholson v. Smiths Ship Repairers (North Shields) Ltd (1984) IRLR 93–116 – occupational deafness and the apportionment of liability

This was one of several cases involving a claim for occupational deafness by an employee who had been exposed to noise in a number of former employments. Former cases had established the view that the last employer was specifically liable for damages. However, it was recognised that much occupational deafness occurs during the early years of exposure, and that previous employers could have been subject to a similar claim.

The outcome of this case is the tendency of the courts to apportion liability between the various employers, all of whom, through exposing the individual to various levels of noise, have contributed to the hearing loss and subsequent occupational deafness.

R. v. Swan Hunter Shipbuilders Ltd (1982) 1 AER 264 – the provision of safety instructions to both employees and non-employees

This case was brought by the HSE following the deaths of eight workmen on board HMS Glasgow in 1976, which was being fitted out in Swan Hunter's shipyard in Newcastle-upon-Tyne at that time.

Oxygen had leaked for several hours from a hose on a higher deck that passed down through a hatch and a cable duct into a compartment on the lowest deck. The last known user of the oxygen hose most likely to have been the source of the leak was an employee of Telemeter Installations, a firm of subcontractors.

The fire burned extremely rapidly and fiercely because the original flame was ignited in a confined space in an atmosphere heavily enriched with oxygen. Minutes before, four workers had tried to smoke cigarettes, but had found that these had rapidly burned away as soon as they were lit, a classic indication of oxygen enrichment.

Swan Hunter Shipbuilders Ltd were charged with the following breaches of the HSWA:

- under section 2(2)(a), the failure to provide and maintain a safe system of work
- under section 2(2)(c), the failure to provide information and instruction for their own employees
- under section 3(1), the failure to conduct their undertaking in such a way as to ensure that those *not* in their employ were not exposed to risks.

Swan Hunter submitted that it was not their responsibility to inform and instruct employees working on the ship who were not their own employees. It was this last submission that the trial judge turned down and whose decision the Court of Appeal agreed with.

Lord Justice Dunn said that Swan Hunter's had a strict duty 'to ensure, so far as was reasonably practicable', the health, safety and welfare of all their employees. If the provision of a safe system of work for the benefit of an employer's own employees involves informing and instructing those other than their own employees concerning the potential dangers, then the employer is under a duty to provide such information and instruction. Only if this was outside the bounds of reasonable practicability could there be any answer to a charge.

This case fundamentally emphasises the duty of an employer to train others. In this case, it included a duty to provide subcontractors' employees with information about the dangers of oxygen enrichment in confined spaces and such instruction as was necessary to ensure the safety of *all* the workers on the vessel.

The outcome of this case was that the subcontractors, Telemeters Ltd, was fined £15,000 and the main contractor, Swan Hunter Ltd, £3000.

British Railways Board v. *Herrington* (1972) AC 877 – injuries to trespassing children

In this case, the House of Lords adopted a more humane approach and softened its ruling, dating back to 1929, that an occupier owed virtually no duty of care to a trespasser even when that person was a child. The reason was based on a change in the climate of opinion as to the acceptable distribution of risks between occupiers and those injured on their premises.

Thus, an occupier may even owe a duty to trespassers in certain circumstances and must act with humane consideration. Public bodies, of course, take out public liability insurance to cover these types of claim.

Ebbs v. *James Whitson & Co. Ltd* (1952) 2 AER 192 – dermatitis from wood dust

The claimant was employed by the defendants as a coach builder. His job involved him in scraping and sanding monsonia wood, a species of West African walnut. The claimant contracted dermatitis as a result of exposure to the wood dust and claimed damages for breach of statutory duty under section 4(1) and section 47(1) of the FA 1937 (now sections 4 and 63 of the FA 1961).

It was held by the Court of Appeal that:

- there had been no breach of section 4(1) as this section related only to ventilation, not to other methods of rendering dust harmless
- the dust was not of such a character and given off to such an extent as to be injurious or offensive to those employed and, accordingly, there had been no breach of section 47(1) either.

Smith v. *Baker & Sons* (1891) AC 305 – *volenti non fit injuria*

In this case a worker who was involved in drilling stone in the course of his employment was injured by a stone which fell from a crane. The employer submitted the defence that the worker had voluntarily assumed this risk as a part of his normal work activities and that he was not liable. The House of Lords found that the employer had been liable on the basis that employees should not be put at risk as a result of prevailing work operations and should be entitled to protection against the risk of death and injury at work.

McWilliams (or Cummings) v. *Sir William Arrol & Co. Ltd* (1962) 1 AER 623 – breach of statutory duty; chain of causation

In this case the widow of a steel erector who had been killed by a fall in a shipbuilding yard sued, *inter alia*, the occupiers of the yard alleging a breach of section 26(2) of the Factories Act 1937 in that they had failed to provide the deceased with a safety belt. The occupiers admitted that no safety belt had been provided, but contended that, since there was evidence that the deceased would not have worn a safety belt had one been provided, their failure to provide the belt was not the cause of his death.

This view was upheld by the House of Lords where Lord Kilmuir stated that there were four steps of causation, namely:

- a duty to supply a safety belt
- a breach
- that if there had been a safety belt the deceased would have used it
- that if there had been a safety belt the deceased would not have been killed.

If the irresistible inference was that the deceased would not have worn a safety belt had it been provided, the first two steps in the chain of causation ceased to operate.

Rylands v. *Fletcher* (1868) LR 3 HL 330 – strict liability; liability without proof of negligence

There are three cases where the common law recognises an absolute or strict liability, namely:

- for the escape of fire from a defendant's premises
- for injuries caused by wild animals or ferocious domestic animals
- for damage caused by the escape of things that the defendant has brought on to his land, which are likely to do mischief if they escape, either by reason of their inherent nature or because they have been accumulated in quantity.

All three liabilities depend upon the 'escape' of something, i.e. fire, animal or dangerous substance or accumulation, from the premises where it is kept. Lord Porter said that the 'escape' which is necessary under the principle of *Rylands* v. *Fletcher* 'must be escape from a place over which a defendant has some measure of control to a place where he does not'.

Here the defendant employed an independent contractor to construct a reservoir on his land. The contractor was deemed to be competent and used competent workmen. After completion, water seeped from the reservoir and eventually flooded the claimant's mine. It was held that, though the defendant was not personally negligent, he was liable for the negligence of the contractors.

Where fire spreads and causes damage liability is strict. In high risk situations where fire may be caused by, for instance, the storage of petrol, or where activities undertaken on land could lead to explosion with the resulting potential for injury and/or damage, there is a duty on the occupier to take such measures to contain the risk and prevent its 'escape' from the premises.

In the course of judgment the House of Lords laid down the following rule:

> *Where a person for his own purposes brings and keeps on land in his occupation anything likely to do mischief if it escapes, he must keep it in at his peril, and if he fails to do so he is liable for all damage naturally accruing from the escape.*

Corn v. *Weir's Glass (Hanley) Ltd* (1960) 2 AER 300 – negligence; breach of duty of care

Mr Corn, a glazier, was carrying a large sheet of glass which necessitated use of both hands to hold same. Whilst ascending a staircase, he overbalanced and fell, causing serious injury. No handrail was provided to the staircase contrary to the requirements of the former Building (Health, Safety and Welfare) Regulations 1948.

His claim failed due to the fact that he was using both hands to carry the glass and, in any case, would not have been able to make use of a handrail.

In this case a distinction was made between a handrail and a guard rail.

A guard rail is one of such a character as will provide a physical barrier against a person falling over the side which is guarded. A handrail is a rail that can be gripped by the hand; it need not necessarily act as a physical barrier; it need only be such a rail as to enable any person, by gripping it, to steady himself against falling. This regulation does not require the handrail prescribed to be fixed on the open side of the stairs, and it may, in suitable cases, be fixed on the wall side, or in the middle.

Mersey Docks and Harbour Board v. *Coggins and Griffiths (Liverpool) Ltd* (1974) ACI – temporary employment

Here a skilled crane driver was hired out by the Mersey Docks and Harbour Board to a firm of stevedores. Whilst working for the firm of stevedores he was injured. His employers were found to be in breach of their duties as employers on the basis that, whilst he had been hired out to another employer, control over the job he performed as a crane driver rested with his permanent employers.

Thus the test of whether an employee has been temporarily employed by another employer is that of the control influenced by that temporary employer in the employee's working practice.

5

Rose v. *Plenty* (1976) 1 AER 97 – vicarious liability; unofficial employment of children

A practice employed by milk roundsmen in the past was to unofficially employ children to assist in the delivery of milk to houses. In this case a boy was injured on a milk float in circumstances where the driver, an employee of a dairy company, had actually been forbidden to 'employ' children to deliver milk and collect empty milk bottles.

In this case the milk roundsman's employers were deemed to be vicariously liable for the action of their employees, even though such persons had been expressly forbidden to engage in such practices.

Armour v. *Skeen* (1977) IRLR 310 – corporate liability

This case involved section 37 of the HSWA in terms of the duties of directors and others. In this case a council employee fell to his death from scaffolding.

It was held that the regional council's Director of Roads came within the ambit of section 37.

Tesco Supermarkets Ltd v. *Nattrass* (1971) 2 AER 127 – offences due to the act of another person; due diligence

In this case, taken under the Trade Descriptions Act 1968, an offence was committed due to the neglect of a store manager. The company submitted the defence of all due diligence, based on the wording of the defence in the Act, thus:

In any proceedings for an offence under this Act it shall, ... be a defence for the person charged to prove:

(a) *that the commission of the offence was due to a mistake or to reliance on information supplied to him or to the act or default of another person, an accident or some other cause beyond his control; and*

(b) *that he took all reasonable precautions and exercised all due diligence to avoid the commission of such an offence by himself or by any person under his control.*

It was decided that the store manager was 'another person' and that as the company had taken all reasonable precautions and exercised all due diligence to avoid the offence being committed, the actions of its employee in failing to follow instructions did not mean that the company could not rely on the defence.

R. v. Associated Octel Ltd (1996) 1 WLR 1543, 4AER – breach of duties under sections 2 and 3 of HSWA

Associated Octel Ltd, a chemicals manufacturer with its major manufacturing site at Ellesmere Port, suffered a major chemical release and fire which threatened not only employees but members of the emergency services. Because of the large inventories of chemicals held on site, the company was subject to the Control of Industrial Major Accident Hazard Regulations (CIMAH) which regulated activities on major hazard sites.

The HSE report of their investigation into this major chemical release and fire concluded that the incident might have been prevented if a more detailed assessment of the hazards and risks at the plant had been carried out by the company beforehand. The cause of the incident was a faulty pump connection to a reactor vessel.

The incident occurred on the evening of 1 February 1994 when there was a release of reactor solution from a circulating pump at the factory. The solution was highly flammable, corrosive and toxic and, in spite of the efforts of emergency services, a major fire occurred.

Associated Octel was prosecuted by the HSE in Chester Crown Court on 2 February 1996 for failing to comply with its duties under sections 2 and 3 of the Health and Safety at Work Act 1974, to ensure the safety of employees and others, such as fire-fighters. The company was fined a total of £150,000.

Although no serious injuries, ill-health or environmental effects resulted from the release of reactor solution and the fire, this was classed by the HSE as a serious incident at a major hazards site.

When the incident occurred a dense white cloud of flammable and toxic gas was released which enveloped the plant and began to move off-site. The on-site and emergency services tried an hour and a half to isolate the leak, suppress the further release of vapour and stop the cloud spreading. In spite of their efforts, however, a pool of liquid continued to collect and the flammable vapours of ethyl chloride ignited. This caused a major fire which was most intense at the base of the reactor.

Ethyl chloride is a highly flammable liquefied gas. The release also contained hydrogen chloride which is toxic and corrosive. Hydrogen chloride was also a significant combustion product during the fire.

As the fire developed there were jet flames at the top of two large process vessels on the plant and, although they and the reactor were protected by a fire resistant coating, there was concern at one stage that the vessel might explode.

Conclusions arising from the enquiry were that the release arose from a pump connection where a corroded securing flange on the pump had worked loose or a flexible joint had failed. The company inspection and maintenance procedures were found to be inadequate and although there were many measures to limit the size and effect of the release, these were not fully satisfactory because the assessment of risks was insufficient.

In particular, the assessment did not identify the risk of substantial leaks from some pumps and pipework and the need, therefore, to provide emergency shut off valves to isolate such leaks.

The plant itself was extensively damaged requiring a complete rebuild. Investigation of the immediate technical cause of the leak was hampered because plant and equipment had been subject to an intense fire and some critical components destroyed.

There were a number of important lessons to be learned from this incident, in particular the need for companies to:

- periodically review the risks posed by their plants
- choose the most appropriate techniques for assessment
- take into account the current state of scientific and technical knowledge.

The principal recommendations arising from the investigation were as follows.

1. Risk assessment
Chemical companies should make a thorough and detailed assessment of risks to prevent the loss of dangerous chemicals. This assessment should be routinely reviewed and kept up to date using the latest techniques.

2. Maintenance
A good system of inspection, examination and maintenance of plant and equipment is essential. This should include the keeping of formal records of plant history and take into account the consequences of failure as well as known reliability.

3. Mitigation measures
To limit the size and effects of any release, chemical companies should critically review whether remotely operated valves are required as part of their emergency shutdown arrangements. The HSE is to consider what guidance is required for the provision of remotely operated valves and the use of thermal insulation cladding on vessels.

4. Emergency arrangements

Companies and fire brigades should give particular attention, beforehand, to arrangements for ensuring there is clear communication about risks between staff at all levels during an emergency.

Alsop v Sheffield City Council (2002) CA – manual handling risk assessment

The claimant refuse collector was moving a wheelie bin up a steep ramp. Whilst doing so, he slipped and fell, suffering injury. In this case he had chosen to pull the bin up the ramp and did not use some adjacent steps to go to the end of the ramp which was only a short distance away. The judge initially dismissed the claim as did the Court of Appeal.

The case was based on the fact that the defendants had failed to give sufficient instruction to the claimant on how to manoeuvre the bins when confronted with this sort of situation. Moreover, the local authority had failed to undertake a risk assessment under regulation 4(1) of the Manual Handling Operations Regulations 1992.

Lord Justice Mantell adopted the approach of Lady Justice Hale in *Koonjul* v. *Thames Link Health Services* [2000] PIQR 123 and her 'element of realism' in considering risk assessment. He found that, if a risk assessment had been undertaken, the defendants would have simply informed their refuse to 'use their common sense' when moving the bins. Lord Justice Kennedy found that the employer is entitled to have regard to the 'experience of the employee said to be at risk'.

However, the regulations are clear that a risk assessment shall be undertaken. The principle of 'reasonable practicability' only applies in relation to avoiding the need to undertake manual handling operations. This was clearly set out in *Swain* v. *Denso Marston* [2000] CA. In the *Koonjul* case, Lady Justice Hale adopted a non-European approach to risk assessment and applied the common law approach, first stated by Sir Christopher Staughton in *Hawkes* v. *London Borough of Southwark* [1998] (unreported), that the requirement to carry out a risk assessment is 'merely an exhortation with no sanction attached'.

This 'realism' test, therefore, may be used to defeat claims on the basis that even if a risk assessment has been undertaken, it would not have had an effect on the cause of the accident. On this basis, the burden of proof would appear to shift from the employer to the claimant to prove that the task he was undertaking was not an everyday task or routine and that he suffered injury as a result.

STRESS AT WORK CASES

The question of stress at work has achieved considerable significance in the civil courts in the last decade.

In 2002, after hearing four appeal cases, namely:

- *Terence Sutherland (Chairman of the Governors of St Thomas Becket RC High School)* v. *Penelope Hatton*
- *Somerset County Council* v. *Leon Alan Barber*
- *Sandwell Metropolitan Borough Council* v. *Olwen Jones*
- *Baker Refractories Ltd* v. *Melvyn Edward Bishop*

Lady Justice Hale, sitting with Lords Justice Brooke and Kay, outlined a number of 'practical propositions' to assist courts in dealing with future claims for psychiatric injury arising from stress at work.

Practical propositions

1. There are no special control mechanisms applying to claims for psychiatric (or physical) illness or injury arising from the stress of doing the work the employee is required to do. The ordinary principles of employers' liability apply.
2. The threshold question is whether this kind of harm to an employee was reasonably foreseeable. This has two components:

   (a) an injury to health (as distinct from occupational stress); and

   (b) which is attributable to stress at work (as distinct from other factors).

5

3. Foreseeability depends upon what the employer knows (or ought reasonably to know) about an employee. Because of the nature of a mental disorder, it is harder to foresee than physical injury, but may be easier to foresee in a known individual than in the population at large. An employer is usually entitled to assume that the employee can withstand the normal pressures of the job unless they know of some particular problem or vulnerability.
4. The test is the same whatever the employment. There are no occupations that should be regarded as intrinsically dangerous to mental health.
5. Factors likely to be relevant in answering the threshold question include:

   (a) *The nature and extent of the work done by the employee*

   Is the workload much more than is normal for the particular job?

   Is the work particularly intellectually or emotionally demanding for this employee?

   Are demands being made of this employee unreasonable when compared with the demands made of others in the same or comparable jobs?

   Or are there signs that others doing this job are suffering harmful levels of stress?

   Is there an abnormal level of sickness or absenteeism in the same job or in the same department?

(b) *Signs from the employee of impending harm to health*

Have they a particular problem or vulnerability?

Have they already suffered from illness attributable to stress at work?

Have there recently been frequent or prolonged absences which are uncharacteristic?

Is there reason to think that these are attributable to stress at work, for example because of complaints or warnings from them or others?

6. The employer is generally entitled to take what they are told by their employee at face value, unless they have good reason to think to the contrary. They do not generally have to make searching enquiries of the employee or seek permission to make further enquiries of their medical advisers.

7. To trigger a duty to take steps, the indications of impending harm to health arising from stress at work must be plain enough for any reasonable employer to realise that they should do something about it.

8. The employer is only in breach of duty if they have failed to take the steps which are reasonable in the circumstances, bearing in mind the magnitude of the risk of harm occurring, the gravity of the harm which may occur, the costs and practicability of preventing it, and the justification for running the risk.

9. The size and scope of the employer's operation, its resources and the demands it faces are relevant in deciding what is reasonable; these include the interests of other employees and the need to treat them fairly, for example, in any redistribution of duties.

10. An employer can only reasonably be expected to take steps that are likely to do some good. The court is likely to need expert evidence on this.

11. An employer who offers a confidential counselling service, with referral to appropriate counselling or treatment services, is unlikely to be found in breach of duty.

12. If the only reasonable and effective step would have been to dismiss or demote the employee, the employer will not be in breach of its duty in allowing a willing employee to continue in the job.

13. In all cases it is necessary to identify the steps that the employer both could and should have taken before finding them in breach of their duty of care.

14. The claimant must show that the breach of duty has caused or materially contributed to the harm suffered. It is not enough to show that occupational stress has caused the harm.

15. Where the harm suffered has more than one cause, the employer should only pay for that proportion of the harm suffered which is attributable to their wrongdoing, unless the harm is truly indivisible. It is for the defendant to raise the question of apportionment.

16. The assessment of damages will take account of any pre-existing disor-
    der or vulnerability and of the chance that the claimant would have suc-
    cumbed to a stress-related disorder in any event.

CONCLUSION

Case law is an important source of law and goes back hundreds of years. It
contains the principles and rules of common law based on the decisions of
courts and reported in the various law reports. As such, common law is accumulated case law and supported by the doctrine of precedent.

   In many cases, the decisions of courts based on the common law are written into statutes, to become the written law of the land.

5

6

Recommendations for employers, managers and those in control of undertakings

Health and safety law imposes important duties on employers, managers, occupiers of premises and those controlling undertakings. An 'undertaking' can be any form of organisation, whether run for profit or not. On this basis, health and safety law applies to any type of undertaking that manages people including, for example, volunteers at a local hospital, people running a car boot sale and the organisers of a local carnival.

This chapter reviews some of the more important aspects of health and safety management necessary to ensure compliance with, in particular, the Health and Safety at Work etc. Act and the Management of Health and Safety at Work Regulations. Documentation of policies and procedures, together with the keeping of certain records, features strongly in the management process along with:

- regular safety monitoring activities, such as safety inspections;
- risk assessment;
  - workplaces, work activities and work groups
  - work equipment
  - personal protective equipment
  - manual handling operations
  - display screen equipment workstations
  - substances hazardous to health;
- the training and appointment of competent persons;
- the thorough investigation of accidents and occupational ill-health; and
- the provision of information, instruction and training for employees and others in certain cases.

All modern protective legislation – health and safety, food safety and environmental protection – is risk assessment driven. Having completed the risk assessment process, an employer must install 'preventive and protective measures' aimed at reducing or controlling hazards. These measures may include health surveillance of specific groups of employees in certain cases.

These important areas of health and safety management are dealt with below.

HEALTH AND SAFETY MANAGEMENT SYSTEMS

What is 'management'?

'Management' is defined as 'the effective use of resources in the pursuit of organisational goals'. 'Effective' implies achieving a balance between the risk of being in business and the cost of eliminating or reducing those risks.

Management entails leadership, authority and co-ordination of resources, together with planning and organisation, co-ordination and control, communication, selection and placement of subordinates, training and development of subordinates, accountability and responsibility.

Management resources include people, land, buildings and capital, together with time and the appropriate management skills in co-ordinating the use of these resources.

Key elements of successful health and safety management

The successful management of occupational health and safety includes the establishment of policy and subsequent development of policy, establishing and developing the organisation and arrangements for implementing the policy, planning strategies and implementation of those strategies to eliminate or control risks, measuring performance of managers against agreed health and safety objectives and reviewing management performance. Various forms of health and safety monitoring, such as safety audits, should result in feedback for policy revision and review.

These elements are summarised in Figure 6.

The legal duty to manage health and safety at work

In addition to the general and more specific duties placed on employers under the Health and Safety at Work Act, the Management of Health and Safety at Work Regulations place an absolute duty on employers to actually manage health and safety in the workplace, thus:

Regulation 4 – Health and safety arrangements

1. Every employer shall make and give effect to such arrangements as are appropriate, having regard to the nature of his activities and the size of his undertaking, for the effective *planning, organisation, control, monitoring and review* of the preventive and protective measures.
2. Where the employer employs five or more employees, he shall record the arrangements referred to in paragraph 1.

HSC Approved Code of Practice

The Health and Safety Commission Approved Code of Practice accompanying the Regulations enlarges on this duty thus:

'This regulation requires employers to have arrangements in place to cover health and safety. Effective management of health and safety will depend, amongst other things, on a suitable and sufficient risk assessment being carried out and the findings used effectively. The health and safety arrangements can be integrated into the management system for all other aspects of the organisation's activities. The management system adopted will need to reflect the complexity of the organisation's activities and working environment. Where the work process is straightforward and the risks generated are relatively simple to control, then very straightforward management systems may be appropriate. For large complicated organisations more complex systems may be appropriate; although the principles of the management arrangements are the same irrespective of the size of an organisation. The key elements of such effective systems can be found in *Successful health and safety management [HS(G)65 HSE Books]* or the British Standard for health and safety management systems BS8800. A successful health and safety management system will include the following elements:

- *Planning*

Employers should set up an effective health and safety management system to implement their health and safety policy which is proportionate to the hazards and risks. Adequate planning includes:

(a) adopting a systematic approach to the completion of a risk assessment. Risk assessment methods should be used to decide on priorities and to set objectives for eliminating hazards and reducing risks. This should include a programme, with deadlines for the completion of the risk assessment process, together with suitable deadlines for the design and implementation of the preventive and protective measures which are necessary;

(b) selecting appropriate methods of risk control to minimise risks;

(c) establishing priorities and developing performance standards both for the completion of risk assessment(s) and the implementation of preventive and protective measures, which at each stage minimises the risk of harm to people. Wherever possible, risks are eliminated through selection and design of facilities, equipment and processes.

- *Organisation*

This includes:

(a) involving employees and their representatives in carrying out risk assessments, deciding on preventive and protective measures and implementing those requirements in the workplace. This may be achieved by the use of a formal health and safety committee where they exist, and by the use of teamworking, where employees are involved in deciding on the preventive and protective measures and written procedures, etc.;

(b) establishing effective means of communication and consultation in which a positive approach to health and safety is visible and clear. The employer should have adequate health and safety information and make sure it is communicated to employees and their representatives, so informed decisions can be made about the choice of preventive and protective measures. Effective communication will ensure that employees are provided with sufficient information so that control measures can be implemented effectively;

(c) securing competence by the provision of adequate information, instruction and training and its evaluation, particularly for those who carry out risk assessments and make decisions about preventive and protective measures. Where necessary, this will need to be supported by the provision of adequate health and safety assistance or advice.

- *Control*

Establishing control includes:

(a) clarifying health and safety responsibilities and ensuring that the activities of everyone are well-co-ordinated;

(b) ensuring everyone with responsibilities understands clearly what they have to do to discharge their responsibilities, and ensure that they have the time and resources to discharge them effectively;

(c) setting standards to judge the performance of those with responsibilities and ensure they meet them. It is important to reward good performance as well as to take action to improve poor performance; and

(d) ensuring adequate and appropriate supervision, particularly for those who are learning and who are new to a job.

- *Monitoring*

Employers should measure what they are doing to implement their health and safety policy, to assess how effectively they are controlling risks, and how well they are developing a positive health and safety culture. Monitoring includes:

(a) having a plan and making adequate routine inspections and checks to ensure that preventive and protective measures are in place and effective. Active monitoring reveals how effectively the health and safety management system is functioning;

(b) adequately investigating the immediate and underlying causes of incidents and accidents to ensure that remedial action is taken, lessons are learnt and longer term objectives are introduced.

In both cases it may be appropriate to record and analyse the results of the monitoring activity, to identify any underlying themes or trends which may not be apparent from looking at events in isolation.

- *Review*

Review includes:

(a) establishing priorities for necessary remedial action that were discovered as a result of monitoring to ensure that suitable action is taken in good time and is completed;

(b) periodically reviewing the whole of the health and safety management system including the elements of planning, organisation, control and monitoring to ensure that the whole system remains effective.

Recording the arrangements

The regulation also requires that employers with five or more employees should record their arrangements for health and safety. The arrangements recorded should include, for example, a list of those competent persons appointed under regulation 6. As with the risk assessment, this record should form part of the same document containing the health and safety policy required under section 2(3) of the Health and Safety at Work Act.

HS(G) 65: SUCCESSFUL HEALTH AND SAFETY MANAGEMENT

The key elements of successful health and safety management are set out below, and the relationship between them is outlined in Figure 6.1.

Policy

Policy is used is used in relation to health and safety and other functional management areas (e.g. manufacturing and human resources) to convey the general intentions, approach and objectives of an organisation and the criteria and principles on which actions and responses are based. The term 'written policy statement' is used to describe those documents which record the policy of the organisation.

Organisations which are successful in achieving high standards of health and safety have safety policies which contribute to their business performance, while meeting their responsibilities to people and the environment in a way which fulfils both the spirit and the letter of the law. In this way they satisfy the expectations of shareholders, employees, customers and society at large. Their policies are cost effective and aimed at achieving the preservation and development of physical and human resources and reductions in financial losses and liabilities. Their health and safety policies influence all their activities and decisions, including those to do with the selection of resources and information, the design and operation of working systems, the design and delivery of products and services, and the control and disposal of waste.

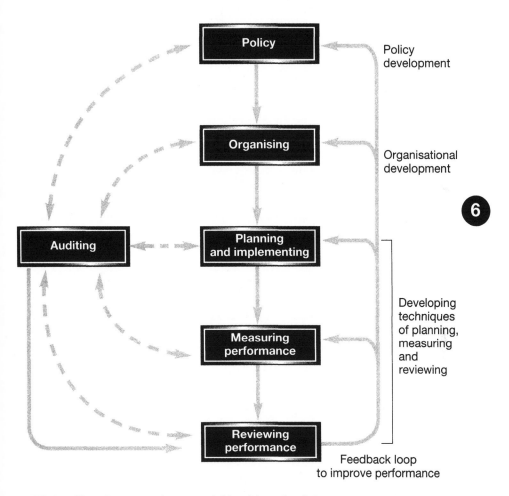

● **FIG 6.1 Key elements of successful health and safety management**

Organising

Organisation is used as a general term to describe the responsibilities and relationships between individuals which forms the social environment in which work takes place. Organising is regarded as the process of designing and establishing those responsibilities and relationships.

Organisations which achieve high health and safety standards are structured and operated so as to put their health and safety policies into effective practice. This is helped by the creation of a positive culture which secures involvement and participation at all levels. It is sustained by effective communications and the promotion of competence which enables all employees to make a responsible and informed contribution to the health and safety effort. The visible and active leadership of senior managers is necessary to develop and maintain a culture supportive of health and safety management. Their aim is not simply to avoid accidents, but to motivate and empower people to work safely. The vision, values and beliefs of leaders become the shared 'common knowledge' of all.

Planning

Planning is used to describe the process by which the objectives and methods of implementing the health and safety policy are decided. It is concerned with allocating resources (e.g. money, time or effort) to achieve objectives and decide priorities. It ranges from general topics dealing with the direction of the whole organisation to detailed issues concerned with standard setting and the control of specific risks.

These successful organisations adopt a planned and systematic approach to policy implementation. Their aim is to minimise the risks, created by work activities, products and services. They use risk assessment methods to decide priorities and set objectives for hazard elimination and risk reduction. Performance standards are established and performance is measured against them. Specific actions needed to promote a positive health and safety culture and to eliminate and control risks are identified. Wherever possible, risks are eliminated by the careful selection and design of facilities, equipment and processes or minimised by the use of physical control measures. Where this is not possible systems of work and personal protective equipment are used to control risks.

Measuring performance

Measuring means the collection of information about the implementation and effectiveness of plans and standards. This involves a variety of checking or 'monitoring' activities.

Health and safety performance in organisations which manage health and safety successfully is measured against pre-determined standards. This reveals when and where action is needed to improve performance. The success of action taken to control risks is assessed through active self-monitoring involving a range of techniques. This includes an examination of both hardware (premises, plant and substances) and software (people, procedures and systems), including individual behaviour. Failures of control are assessed through reactive monitoring which requires the thorough investigation of any accidents, ill-health or incidents with the potential to cause harm or loss. In both active and reactive monitoring the objectives are not only to determine the immediate causes of sub-standard performance but, more importantly, to identify the underlying causes and the implications for the design and operation of the health and safety management system.

Auditing and reviewing performance

Auditing is the structured process of collecting independent information on the efficiency, effectiveness and reliability of the total safety management system and drawing up plans for corrective action.

Reviewing is used to describe activities involving judgements about performance, and decisions about improving performance. Reviewing is based on information from 'measuring' and 'auditing' activities.

Learning from all relevant experience, and applying the lessons learned, are important elements in effective health and safety management. This needs to be done systematically through regular reviews of performance based on data both from monitoring activities and from independent audits of the whole health and safety management system. These form the basis for self-regulation and for securing compliance with sections 2 to 6 of the Health and Safety at Work, etc. Act 1974. Commitment to continuous improvement involves the constant development of policies, approaches to implementation and techniques of risk control. Organisations which achieve high standards of health and safety assess their health and safety performance by internal reference to key performance indicators and by external comparison with the performance of business competitors. They often also record and account for their performance in the annual reports.

DOCUMENTATION AND RECORD KEEPING REQUIREMENTS

Current health and safety legislation places considerable emphasis on the documentation of policies, procedures and systems of work and the maintenance of certain records.

The following are some of the documents and records that are required to be produced and maintained:

- Statement of Health and Safety Policy (Health and Safety at Work etc. Act 1974)
- Risk assessments in respect of:

  - workplaces (Management of Health and Safety at Work Regulations 1999 and Workplace (Health, Safety and Welfare) Regulations 1992)
  - work activities (Management of Health and Safety at Work Regulations 1999 and Workplace (Health, Safety and Welfare) Regulations 1992)
  - work groups (Management of Health and Safety at Work Regulations)
  - new or expectant mothers (Management of Health and Safety at Work Regulations 1999)

- young persons (Management of Health and Safety at Work Regulations 1999)
- work equipment (Provision and Use of Work Equipment Regulations 1998)
- personal protective equipment (Personal Protective Equipment Regulations 1992)
- manual handling operations (Manual Handling Operations Regulations 1992)
- display screen equipment (Health and Safety (Display Screen Equipment) Regulations 1992)
- substances hazardous to health (Control of Substances Hazardous to Health Regulations 2002)
- significant exposure to lead (Control of Lead at Work Regulations 1998)
- noise levels in excess of 85 dBA (Noise at Work Regulations 1989)
- before a radiation employer commences a new activity involving work with ionising radiation (Ionising Radiations Regulations 1999)
- the presence or otherwise of asbestos in non-domestic premises (Control of Asbestos at Work Regulations 2002)
- where a dangerous substance is or is liable to be present at the workplace (Dangerous Substances and Explosive Atmospheres Regulations 2002)

- Safe systems of work, including permits to work and method statements
- Pre-tender stage health and safety plan and construction phase health and safety plan (Construction (Design and Management) Regulations 1994)
- Planned preventive maintenance schedules (Workplace (Health, Safety and Welfare) Regulations 1992 and Provision and Use of Work Equipment Regulations 1998)
- Cleaning schedules (Workplace (Health, Safety and Welfare) Regulations 1992)
- Written scheme of examination for specific parts of an installed pressure system or of a mobile system and the last report relating to a system by a competent person (Pressure Systems Safety Regulations 2000)
- Written plan of work identifying those parts of a premises where asbestos is or is liable to be present in a premises and detailing how that work is to be carried out safely and without risk to health (Control of Asbestos at Work Regulations 2002)
- Records of examinations and tests of exhaust ventilation equipment and respiratory protective equipment and of repairs carried out as a result of those examinations and tests (Control of Lead at Work Regulations 1999, Control of Substances Hazardous to Health Regulations 2002 and Control of Asbestos at Work Regulations 2002)
- Record of air monitoring carried out in respect of:
  - specified substances or processes
  - lead
  - asbestos

(Control of Substances Hazardous to Health Regulations 2002, Control of Lead at Work Regulations 1998 and Control of Asbestos at Work Regulations 2002)

- Record of examination of respiratory protective equipment (Ionising Radiations Regulations 1999)
- Records of air monitoring in cases where exposure to asbestos is such that a health record is required to be kept (Control of Asbestos at Work Regulations 2002)
- Personal health records (Control of Lead at Work Regulations 1999, Ionising Radiations Regulations 1999, Control of Substances Hazardous to Health Regulations 2002 and Control of Asbestos at Work Regulations 2002)
- Personal dose records (Ionising Radiations Regulations 1999)
- Record of quantity and location of radioactive substances (Ionising Radiations Regulations 1999)
- Record of investigation of certain notifiable occurrences involving release or spillage of a radioactive substance (Ionising Radiations Regulations 1999)
- Record of suspected overexposure to ionising radiation during medical exposure (Ionising Radiations Regulations 1999)
- Major Accident Prevention Policy (Control of Major Accident Hazards Regulations 1999)
- Off-Site Emergency Plan (Control of Major Accident Hazards Regulations 1999)
- Declaration of conformity by the installer of a lift and the manufacturer of a safety component for a lift together with any technical documentation or other information in relation to a lift or safety component required to be retained under the conformity assessment procedure (Lifts Regulations 1997)
- Declaration of conformity by the manufacturer of pressure equipment and assemblies (as defined) together with technical documentation or other information in relation to an item of pressure equipment and assemblies required to be retained under the conformity assessment procedure used (Pressure Equipment Regulations 1999)
- Any technical documentation or other information required to be retained under a conformity assessment procedure and a periodic inspection procedure (Transportable Pressure Vessels Regulations 2001)
- Procedures for serious and imminent danger and for danger areas (Management of Health and Safety at Work Regulations 1999)
- Emergency procedure to protect the safety of employees from an accident, incident or emergency related to the presence of a dangerous substance at the workplace (Dangerous Substances and Explosive Atmospheres Regulations 2001)
- Contingency plan in the event of a radiation accident (Ionising Radiations Regulations 1999)

6

- Local rules in respect of controlled areas and supervised areas (Ionising Radiations Regulations 1999)
- Written arrangements for non-classified persons (Ionising Radiations Regulations 1999).

STATEMENTS OF HEALTH AND SAFETY POLICY

Under section 2(3) of the HSWA every employer has a duty to prepare, and as often as appropriate, revise a written statement of his general policy with respect to the health and safety at work of his employees and the 'organisation and arrangements' for the time being in force for carrying out that policy, and to bring the statement and any revision of it to the notice of his employees.

It is a technical offence under the HSWA for an employer not to have a written Statement of Health and Safety Policy (except in the case where less than five employees are employed).

Importance of the Statement

The Statement of Health and Safety Policy is the key document for detailing the management systems and procedures to ensure sound levels of health and safety performance. It should be revised at regular intervals, prior to, particularly, changes in the structure of the organisation, the introduction of new articles and substances, and changes to legal requirements affecting the organisation. Fundamentally, the Statement must be seen as a 'living document' which reflects the current organisational requirements, the hazards and precautions necessary, the individual responsibilities of people and systems for monitoring performance.

There is no standard format for a Statement of Health and Safety Policy. Much will depend upon the structure of the organisation, the nature of the organisation's activities, the hazards to which employees are exposed and the legislation currently applying to the organisation.

Objectives of the Statement of Health and Safety Policy

There are a number of important objectives for a Statement.

- It should affirm long range purpose.
- It should commit management at all levels and reinforce this purpose in the decision-making process.
- It should indicate the scope left for decision-making by junior managers.

Scope of the Statement of Health and Safety Policy

A well-written Statement of Health and Safety Policy should cover the following aspects:

- Management intent
- The 'arrangements' for implementing the Policy
- Individual accountabilities of directors, line managers, employees and other groups, e.g. contractors
- Details of the organisation with respect to both line and staff functions
- The role and function of health and safety specialists, e.g. health and safety advisers, occupational health nursing advisers, occupational hygienists, occupational health nurses, local safety officers, occupational physicians, company doctors and trade union-appointed safety representatives

Principal Features of a Statement of Health and Safety Policy

A Statement of Health and Safety Policy should incorporate the following features:

- A general statement of intent which states the basic objectives, supplemented by details of the 'organisation and arrangements' (rules and procedures).
- Definition of both the duties and extent of responsibility at specified line management levels for health and safety, with identification made at the highest level of the individual with overall responsibility for health and safety.
- Definition of the functions of the safety adviser / officer / consultant and his relationship to senior and line management made clear.
- The system for monitoring safety performance and publishing of information about that performance.
- An identification and analysis of the hazards that may arise and the precautions necessary on the part of employees, contractors and visitors.
- An information system which will be sufficient to produce an identification of needs and which can be used as an indicator of the effectiveness of the policy.
- A policy on the provision of health and safety information, instruction and training for all levels of the organisation.
- A commitment to consultation on health and safety and to a positive form of employee involvement.
- The Statement should bear the signature of the person with ultimate responsibility and accountability for health and safety at work, e.g. Chief Executive, Managing Director.
- It is common practice for Statements of Health and Safety Policy to incorporate a series of Appendices, such as:

  - Current health and safety legislation (the 'relevant statutory provisions') applying to the organisation
  - Duties and responsibilities for health and safety of individual levels of management
  - Role and function of the health and safety adviser and other health and safety specialists

- Hazards and precautions necessary
- Sub-policies covering:
    (i)   Smoking at work
    (ii) Alcohol at work
    (ii) Stress at work
    (iv) Bullying and harassment at work
    (v)   The provision of personal protective equipment
    (vi) Joint consultation arrangements
    (vii) Risk assessment.

SAFETY MONITORING

Safety monitoring is a pro-active strategy aimed at assessing and evaluating safety performance. It may take a number of forms.

Safety inspections

This is generally taken to mean a scheduled inspection of a workplace or part of a workplace, such as a factory, office, workshop, construction site or a public building. It can be undertaken by a manager, safety adviser or representative of employee safety.

Whilst the principal objective of a safety inspection is to identify hazards and assess the remedial action necessary, this form of monitoring may also examine maintenance standards, working practices, environmental conditions and compliance with written safety procedures.

A safety inspection is a general examination of the situation at a specific point in time rather than the in-depth approach taken by a safety survey. As with any form of safety monitoring, it is vital that the objectives are clearly defined and the outcome of the inspection in terms of recommendations for action are acted upon by management.

Safety surveys

This can take the form of an examination of a number of critical areas of operation, such as materials handling operations, or an in-depth study of all health and safety related activities in a workplace or within an organisation.

Safety surveys examine a range of issues, such as the effectiveness of the management of health and safety, environmental working conditions, health hazards, the very broad field of accident prevention and the current system for health and safety training.

At the completion of the survey management are presented with a safety survey report incorporating immediate, short-term and long-term recommendations. Implementation of the recommendations is monitored on a regular basis by the safety surveyor and progress reports prepared and issued to management.

Safety audits

The HSE publication *Successful Health and Safety Management* (HS(G)65) defines the term 'safety audit' as 'the structured process of collecting independent information on the efficiency, effectiveness and reliability of the total safety management system and drawing up plans for corrective action'.

A safety audit is the systematic measurement and validation of an organisation's management of its health and safety programme against a series of specific and attainable standards. (RoSPA). This type of monitoring subjects each area of an organisation's activities to a systematic critical examination with the objective of minimising injury and other forms of loss.

A safety audit examines, for instance, the quality of documentation from the Statement of Health and Safety Policy to other documents, such as risk assessments, permits-to-work, contractors' regulations and employee health and safety information. It will also examine health and safety management systems, management and employee attitudes to health and safety, prevention and control procedures and training arrangements. Safety audits can be designed to cover areas of particular significance to an organisation, e.g. management of hazardous substances, safety procedures for employees working away from base and specific health protection arrangements, in addition to those more general issues which apply to all workplaces.

A Specimen Safety Audit is shown in Figure 6.2.

6

Safety tours

This is an unscheduled examination of a workplace or work area, undertaken by a manager, possibly accompanied by safety representatives and/or safety committee members, to ensure, for instance, standards of housekeeping are at an acceptable level, fire protection measures are being observed or personal protective equipment is being correctly used or worn. A safety tour tends to be spontaneous as opposed to, for instance, an audit or survey, which may be planned some time ahead.

Safety sampling exercises

This technique is designed to measure, by random sampling, the accident potential in a specific workplace, or at a particular process or work activity, by identifying safety defects or omissions. Safety sampling entails the use of a safety sampling sheet which lists specific aspects to be observed and assessed, e.g. machinery safety, electrical safety, housekeeping and personal protective equipment. Each of the areas for consideration is numerically graded according to its significance and a maximum number of points are awardable for each aspect. Such a system monitors the effectiveness of the overall programme.

Safety sampling exercises are also used for comparing safety performance in comparable workplaces where, for instance, an organisation has offices, workshops, manufacturing or service units at different locations.

A typical safety sampling sheet is shown later.

Check list systems

Where there is a need to monitor safety standards on a regular basis, many organisations operate a simple check list system. In this case, a trained person, such as a supervisor, goes through the workplace or his department on a weekly basis using the check list. Such as system relies heavily on conscientious use of the check list, regular revision of same, and on the effectiveness of action taken where there is a deviation from the requirements shown in the check list.

HEALTH AND SAFETY AUDIT

Organisation

Location   **Date**

	YES/NO
DOCUMENTATION	
Legislation	
1. Are managers aware of all health and safety legislation applying to the organisation's activities and workplaces?	
2. Is this legislation available to management and employees?	
3. Have all Approved Codes of Practice, HSE Guidance Notes and internal codes of practice been studied by management with a view to ensuring compliance?	
Formal documentation	
4. Is there a formally established Statement of Health and Safety Policy?	
5. Does the existing Statement of Health and Safety Policy meet current conditions in the workplace and/or work activities?	
6. Are the 'organisation and arrangements' to implement the Health and Safety Policy still adequate?	
7. Is there a named manager with overall responsibility for health and safety?	
8. Have the hazards and precautions necessary on the part of employees and other persons been identified and recorded?	
9. Are individual responsibilities and accountabilities for health and safety clearly detailed in the Statement?	
10. Do all job descriptions adequately describe individual responsibilities and accountabilities for health and safety?	

YES/NO

11. Do written safe systems of work exist for all hazardous operations?

12 Is Permit to Work documentation available and used?

13. Has a suitable and sufficient assessment of the risks to employees and other persons been undertaken, recorded and brought to the attention of staff and other persons?

14. Have other risk assessments in respect of:

    (a) substances hazardous to health;

    (b) risks to hearing;

    (c) workplaces and/or activities and/or work groups;

    (d) work equipment;

    (e) personal protective equipment;

    (f) manual handling operations;

    (g) display screen equipment;

    been made, recorded and brought to the attention of employees and other persons where appropriate?

15. Is the fire certificate available and up-to-date?

16. Is there a record of inspections of the means of escape in the event of fire, fire appliances, fire alarms,
warning notices, fire and smoke detection equipment?

17. Is there a record of inspections and maintenance of the workplace, equipment, including guards and safety devices, systems and devices?

18. Are all examination and test certificates available, e.g. lifting appliances, pressure systems?

19. Are all necessary licences available, e.g. to store petroleum spirit?

20. Are workplace health and safety rules and procedures available, promoted and enforced?

21. Have these rules and procedures been documented in a way which is comprehensible and relevant to staff and others, e.g. Health and Safety Handbook?

22. Are disciplinary procedures for unsafe behaviour clearly documented and known to staff and other persons?

23. Is a formally written emergency procedure available?

24. Have competent persons been trained in procedures for implementing the emergency procedure?

YES/NO

25. Is documentation available for the recording of injuries, near misses, damage only accidents, diseases, ill-health and dangerous occurrences?

26. Are health and safety instruction and training records maintained?

27. Are there documented procedures for regulating the activities of contractors, visitors and other persons working on the site?

28. Is hazard reporting documentation available to employees and other persons?

29. Is there a documented planned preventive maintenance system to cover the workplace, equipment, systems and devices?

30. Are there written cleaning schedules available?

HEALTH AND SAFETY SYSTEMS

31. Have competent persons been appointed to:

(a) co-ordinate health and safety measures;

(b) implement the emergency procedure?

32. Have competent persons been adequately trained on the basis of identified and assessed risks?

33. Are the roles, functions, responsibilities and accountabilities of competent persons clearly identified?

34. Are there arrangements for specific forms of safety monitoring, e.g. safety inspections, safety sampling?

35. Is a system in operation for measuring and monitoring individual management performance on health and safety issues?

36. Are systems established for the formal investigation of accidents, ill-health, near misses and dangerous occurrences?

37. Do investigation procedures produce results which can be used to prevent future incidents?

38. Are the causes of accidents, ill-health, near misses and dangerous occurrences analysed in terms of failure of established systems of work?

39. Is a hazard reporting system in operation?

40. Are employees and other persons on site aware of the need to report hazards to management?

YES/NO

41 Is a system for preventing or controlling damage to structural items, machinery, vehicles, etc. in operation?

42. Is the system for joint consultation with employees clearly established?

43. Are the role, constitution and objectives of the Health and Safety Committee clearly identified?

44. Are the procedures for appointing or electing Committee members clearly identified?

45. Are the available facilities, including training arrangements, known to Committee members?

46. Are the capabilities of employees as regards health and safety taken into account when entrusting them with tasks?

47. Is the provision of first aid arrangements adequate?

48. Are all first aid personnel adequately trained and retrained?

49. Are the procedures covering sickness absence known to employees?

50. Is there a procedure for regulating sickness absence?

51. Are managers aware of the current sickness absence rate?

52. Do current arrangements ensure that the health and safety implications are considered at the design stage of projects?

53. Is there a formally established health and safety budget?

6

PREVENTION AND CONTROL PROCEDURES

54. Are formal inspections of workplaces, machinery, plant, tools, access equipment, electrical equipment and appliances, storage equipment, warning systems, first aid boxes, resuscitation equipment, welfare amenity areas, etc. undertaken regularly?

55. Are machinery guards and safety devices examined on a regular basis?

56. Is a Permit to Work system operated where there is a high degree of foreseeable risk?

57. Are fire and emergency procedures practised on a regular basis?

58. Where specific fire hazards are identified, are they dealt with under the current fire protection arrangements?

59. Are all items of fire protection equipment and alarms tested, examined and maintained on a regular basis?

60. Are all fire exits and escape routes marked, kept free from obstruction and operational?

61. Are all fire appliances correctly labelled, sited and maintained?

62. Are the requirements of cleaning schedules monitored?

63. Is housekeeping of a high standard, e.g. material storage, waste disposal, removal of spillages?

64. Are all gangways, stairways, fire exits, access and egress points to the workplace maintained and kept clear?

65. Is environmental monitoring of temperature, lighting, ventilation, humidity, radiation, noise and vibration undertaken on a regular basis where appropriate?

66. Is health surveillance of persons exposed to assessed risks undertaken on a regular basis?

67. Is monitoring of personal exposure to assessed health risks undertaken on a regular basis?

68. Are local exhaust ventilation systems examined, tested and maintained on a regular basis?

69. Are arrangements for the storage and handling of substances hazardous to health adequate?

70. Are all hazardous substances identified and correctly labelled, including transfer containers?

71. Is the appropriate personal protective equipment available?

72. Is the personal protective equipment worn or used by employees consistently when exposed to risks?

73. Are storage facilities for items of personal protective equipment provided?

74. Are welfare amenity provisions, i.e. sanitation, hand washing, shower and clothing storage arrangements adequate?

75. Do welfare amenity provisions promote appropriate levels of personal hygiene?

YES/NO

INFORMATION, INSTRUCTION, TRAINING AND SUPERVISION

76. Is the information provided by manufacturers and suppliers of articles and substances for use at work adequate?

77. Do employees and other persons have access to this information?

78. Is the means of promoting health and safety adequate?

79. Is effective use made of safety propaganda, e.g. posters?

80. Do current safety signs meet the requirements of the Safety Signs Regulations 1980 and the Health and Safety (Safety Signs and Signals) Regulations 1996?

81. Are safety signs adequate in terms of the assessed risks?

82. Are fire instructions prominently displayed?

83. Are hazard warning systems adequate?

84. Are the individual training needs of employees and other persons assessed on a regular basis?

85. Is health and safety training for employees undertaken:

    (a) at the induction stage;

    (b) on their being exposed to new or increased risks because of:
        (i) transfer or change in responsibilities;
        (ii) the introduction of new work equipment or a change respecting existing work equipment;
        (iii) the introduction of new technology;
        (iv) the introduction of a new system of work or change in an existing system of work?

86. Is the above training:

    (a) repeated periodically;

    (b) adapted to take account of new or changed risks;

    (c) carried out during working hours?

87. Is specific training undertaken regularly for first aid staff, fork lift truck drivers, crane drivers and others exposed to specific risks?

88 .Are selected staff trained in the correct use of fire appliances?

FINAL QUESTION
Are you satisfied that your organisation is as safe and healthy as you can reasonably make it, or that you know what action must be taken to achieve that state?

ACTION PLAN

1. Immediate action

2. Short-term action (14 days)

3. Medium-term action (6 months)

4. Long-term action (2 years)

Auditor_____ Date_____

- **FIG 6.2 Health and safety audit**

SAFETY SAMPLING EXERCISE

			LOCATIONS
1.	Housekeeping	Max 20	
2.	Fire protection	Max 20	
3.	Machinery safety	Max 20	
4.	Electrical safety	Max 15	
5.	Internal storage arrangements	Max 10	
6.	Use of hazardous substances	Max 10	
7.	Manual handling operations	Max 10	
8.	Personal protection	Max 10	
9.	Welfare amenities	Max 10	
10.	Lift truck operations	Max 15	
11.	Lighting	Max 10	
12.	Access and egress	Max 10	
13.	Dust and fume control	Max 10	
14.	Hand tools	Max 10	
15.	Structural safety	Max 20	
	TOTALS	**Max 200**	

6

ACTION PLAN

1. Immediate action

2. Short-term action (14 days)

3. Medium-term action (6 months)

4. Long-term action (2 years)

Sampling Officer_____ Date_____

● **FIG 6.3 Safety sampling exercise**

ACCIDENT AND INCIDENT INVESTIGATION

The term 'accident' is defined as a 'deviation from the normal, the expected or the planned, commonly resulting in injury'. An 'incident', on the other hand, is 'an undesired event which does, or could, result in loss'.

Investigation of the direct and indirect causes of both accidents and loss-producing incidents is a reactive strategy in safety management. There are very good reasons for the effective and thorough investigation of these events, in particular, namely:

- on a purely humanitarian basis, no one likes to see people killed or injured;
- the accident or incident may have resulted from a breach of statute or regulations by the organisation, the accident victim, the manufacturers and/or suppliers of articles and substances used at work, or other persons, e.g. contractors, with the possibility of civil proceedings being instituted by the injured party against the employer and other persons;
- the injury may be reportable to the enforcing authority (HSE, local authority) under the Reporting of Injuries, Diseases and Dangerous Occurrences Regulations (RIDDOR) 1995;
- the accident or incident may result in lost production;
- from a management viewpoint, a serious accident, particularly a fatal one, can have a long-term detrimental effect on the morale of the workforce and employer/employee relations;
- there may be damage to plant and equipment, resulting in the need for repair or replacement, with possible delays in replacement; and
- in most cases, there will be a need for immediate remedial action in order to prevent a recurrence of this type of accident or incident.

Apart from the obvious direct and indirect losses associated with all forms of incident, not just those resulting in injury, there are good reasons for investigating accidents to identify the direct and indirect causes and to produce strategies for preventing recurrences. Above all, the purpose of accident investigation is *not* to apportion blame or fault, although blame or fault may eventually emerge as a result of investigation.

What incidents should be investigated?

Clearly there is a case for investigating all accidents and, indeed, classes of incident known as 'near misses'. A near miss is defined as 'an unplanned and unforeseeable event that could have resulted, but did not result in human injury, property damage or other form of loss'.

However, it may be impracticable to investigate every accident, but the following factors should be considered in deciding which accidents should be investigated as a priority:

- the *type* of accident, e.g. fall from a height, chemical handling, machinery-related;
- the *form* and *severity of injury*, or the potential for serious injury and/or damage;
- whether the accident or incident indicates the continuation of a particular *trend* in the organisation's accident experience;
- the extent of involvement of *articles and substances* used at work, e.g. machinery, work equipment, hazardous substances, and the ensuing damage or loss;
- the possibility of a *breach of the criminal law*, such as the Health and Safety at Work etc. Act;
- whether the injury or occurrence is, by law, *notifiable and reportable* to the enforcing authority; and under RIDDOR; and
- whether the accident should be reported to the insurance company as it could result in a civil *claim* for negligence being made by the injured party against the organisation.

Practical accident and incident investigation

In any serious incident situation, such as a fatal accident, an accident resulting in major injury, e.g. fractures, amputations, loss of an eye, or where there has been a scheduled dangerous occurrence, such as the collapse of a crane, speed of action is essential. This is particularly the case when it comes to interviewing injured persons and witnesses.

The following procedure is recommended:

- Establish the facts as quickly and completely as possible with regard to:-
  - the general environment;
  - the particular plant, machinery, practice or system of work involved; and
  - the sequence of events leading to the accident.
- Use an instant camera to take photographs of the accident scene prior to any clearing up that may follow the accident.
- Draw sketches and take measurements with a view to producing a scale drawing of the events leading up to the accident.
- List the names of all witnesses, i.e. those who saw, heard, felt or smelt anything; interview them thoroughly in the presence of a third party if necessary, and take full statements. In certain cases, it may be necessary to formally caution witnesses, in accordance with the Police and Criminal Evidence Act, prior to their making a statement. Do *not* prompt or lead witnesses.
- Evaluate the facts, and individual witnesses' versions of same, as to accuracy, reliability and relevance.
- Endeavour to arrive at conclusions as to the *causes* of the accident on the basis of the relevant facts.

- Examine closely any contradictory evidence. Never dismiss a fact that does not fit in with the rest. *Find out more.*
- Learn fully about the system of work involved. Every accident occurs within the context of a work system. Consider the people involved in terms of their ages, training, experience and level of supervision, and the nature of the work, e.g. routine, sporadic or incidental.
- In certain cases, it may be necessary for plant and equipment, such as lifting appliances, machinery and vehicles to be examined by a specialist, e.g. a consultant engineer, with a view to that person producing a report as to the relative fitness and safety of same.
- Produce a report for the responsible manager emphasising:
  - the identified direct and indirect causes; and
  - remedies to prevent a recurrence, including any changes necessary.

In complex and serious cases, consider the establishment of an investigating committee comprising managers, supervisors, technical specialists and trade union safety representatives.

Other parties involved in investigation

There are other persons who may also wish to investigate this accident, including:

(a) appointed employee representatives;

(b) an insurance company liability surveyor, in the event of a claim being submitted by the injured party;

(c) legal representative of the injured party to establish the cause of the accident, whether there has been negligence or a breach of statutory duty on the part of the employer and further action required; and

(d) officers of the enforcing authority, to establish whether there has been a breach of current health and safety legislation (the 'relevant statutory provisions') which may require action, such as:

    (i) the service of a Prohibition Notice;

    (ii) the service of an Improvement Notice; and/or

    (iii) prosecution.

It is essential, therefore, that any accident report produced is accurate, comprehensible and identifies the causes of the accident, both direct and indirect.

The outcome of accident investigation

Whether the investigation of an accident is undertaken by an individual, e.g. health and safety specialist, manager or by a special committee, it is necessary, once the causes have been identified, to submit recommendations to management with a view to preventing a recurrence. The organisation of *feedback* on the causes of accidents is crucial in large organisations, especially those who operate more than one site or premises.

An effective investigation should result in one or more of the following recommendations being made:

- the issuing of specific instructions by management covering, for instance, systems of work, the need for more effective guarding of machinery or safe manual handling procedures;
- the establishment of a working party or committee to undertake further investigation, perhaps in conjunction with members of the safety committee and/or safety representatives;
- the preparation and issue of specific codes of practice or guidance notes dealing with the procedures necessary to minimise a particular risk, e.g. the use of a permit to work system;
- the identification of specific training needs for groups of individuals, e.g. managers, foremen, supervisors, machinery operators, drivers, and the implementation of a training programme designed to meet these needs;
- the formal analysis of the job or system in question, perhaps using job safety analysis techniques, to identify skill, knowledge and safety components of the job;
- identification of the need for further information relating to articles and substances used at work, e.g. work, equipment, chemical substances;
- identification of the need for better environmental control, e.g. noise reduction at source or improved lighting;
- general employee involvement in health and safety issues, e.g. the establishment of a health and safety committee; and
- identification of the specific responsibilities of groups with regard to safe working practices.

Above all, a system of monitoring should be implemented to ensure that the lessons which have been learned from the accident or incident are put into practice or incorporated in future systems of work, and that procedures and operating systems have been produced for all grades of staff.

PRINCIPLES OF RISK ASSESSMENT

A risk assessment is carried out to identify the risks to employees and any other persons arising out of, or in connection with work, or the conduct of an undertaking. It should identify how the risks arise and how they impact on those affected. This information is needed to make decisions on how to manage those risks so that decisions are made in an informed, rational and structured manner, and the action taken is proportionate.

What is risk assessment?

Risk assessment involves:

- the identification of hazards at work;
- the weighing up of the risks from the hazards;

- deciding how to control the risks; and
- implementing a control strategy.

A risk assessment has three purposes:

- to identify all the factors which may cause harm to employees and others (*the hazards*);
- to consider the chance of that harm actually befalling anyone in the circumstances of a particular case, and the possible consequences that could come from it (*the risks*); and
- to enable employers to plan, introduce and monitor preventive measures to ensure that the risks are adequately controlled at all times.

Without effective assessment there can seldom be effective control.

EXTRACT FROM THE HSC APPROVED CODE TO THE MANAGEMENT OF HEALTH AND SAFETY AT WORK REGULATIONS 1999

'HAZARD' AND 'RISK' – THE DISTINCTION

'Hazard'

A hazard is something with the potential to cause harm. This can include articles, substances, plant or machines, methods of work, the working environment and other aspects of work organisation.

'Risk'

A risk is the likelihood of potential harm from that hazard being realised. The *extent* of the risk will depend on:

- the likelihood of that harm occurring;
- the potential severity of that harm, i.e. of any resultant injury or adverse health effect; and
- the population which might be affected by the hazard, i.e. the number of people who might be exposed.

The purpose of the risk assessment is to help the employer or self-employed person to determine what measures should be taken to comply with the employer's or self-employed person's duties under the *relevant statutory provisions* and Part II of the Fire Precautions (Workplace) Regulations 1997 (FPWR). This covers the general duties in the HSWA and the requirements of Part II of the FPWR and the more specific duties in the various acts and regulations (including these Regulations) associated with the HSWA.

Once the measures have been determined in this way, the duty to put them into effect will be defined in the statutory provisions. For example a risk assessment on machinery would be undertaken under these Regulations, but the Provision and Use of Work Equipment Regulations (PUWER 1998) determine what precautions must be carried out. A risk assessment carried out by a self-employed person in circumstances where he or she does not employ others does not have to take into account duties arising under Part II of the FPWR.

Suitable and sufficient

A suitable and sufficient risk assessment should be made. 'Suitable and sufficient' is not defined in the Regulations. In practice it means the risk assessment should do the following:

- The risk assessment should identify the risks arising from or in connection with work; the level of detail in a risk assessment should be proportionate to the risk and the level of risk arising from the work activity should determine the degree of sophistication of the risk assessment.
- Employers and the self-employed are expected to take reasonable steps to help themselves identify risks, e.g. by looking at appropriate sources of information, such as relevant legislation, appropriate guidance, supplier manuals and man-ufacturers' instructions and reading trade press, or seeking advice from compe-tent sources. The risk assessment should include only what an employer or self-employed person could reasonably be expected to know; they would not be expected to anticipate risks that were not foreseeable.
- The risk assessment should be appropriate to the nature of the work and should identify the period of time for which it is likely to remain valid.
- For activities where the nature of the work may change fairly frequently or the workplace itself changes and develops (such as a construction site), or where workers move from site to site, the risk assessment might have to concentrate more on the broad range of risks that can be foreseen.

Risk assessment in practice

There are no fixed rules about how a risk assessment should be carried out; indeed it will depend on the nature of the work or business and the types of hazards and risks.

In some cases employers may make a first rough assessment to eliminate from consideration those risks on which no further action is needed. This should show where a fuller assessment is needed, if appropriate, using more sophisticated tech-niques.

Employers who control a number of similar workplaces containing similar activities may produce a 'model' (or 'generic') risk assessment reflecting the core hazards and risks associated with those activities. 'Model' assessments may also be developed by trade associations, employers' bodies or other organisations con-cerned with a particular activity. Such 'model' arrangements may be applied by employers or managers at each workplace, but only if they:

(a) satisfy themselves that the 'model' assessment is appropriate to their type of work; and

(b) adapt the 'model' to the detail of their own actual work situations, including any extension necessary to cover hazards and risks not referred to in the 'model'.

A risk assessment should:

- ensure the significant risks and hazards are addressed;
- ensure all aspects of the work activity are reviewed, including routine and non-routine activities;
- take account of non-routine operations, e.g. maintenance, cleaning operations, loading and unloading of vehicles, changes in production cycles, emergency response arrangements;
- take account of the management of incidents such as interruptions to the work activity, which frequently cause accidents, and consider what procedure should be followed to mitigate the effects of the incident;
- be systematic in identifying hazards and looking at risks, whether one risk assessment covers the whole of the work activity or the assessment is divided up;
- take account of the way in which the work is organised, and the effects this can have on health;
- take account of risks to the public;
- take account of the need to cover fire risks. (See the guide *Fire safety: An employer's guide*.)

Stages of the risk assessment process

- Identify the hazards
- Identify who might be harmed and how
- Evaluate the risks from the identified hazards
- Record the significant findings.

Recording

The significant findings of a risk assessment should include:

- a record of the preventive and protective measures in place to control the risks
- what further action, if any, needs to be taken to reduce risk sufficiently;
- proof that a suitable and sufficient risk assessment has been made.

Review and revision

Risk assessments should be subject to review and revision, in circumstances where, for example:

- the nature of the work changes; and/or
- developments suggest that it may no longer be valid.

Assessment under other regulations

The requirements of, for example, the COSHH Regulations, may need to be considered in conjunction with a work activity risk assessment.

[Further information on risk assessment can be found in *The Law and Practice of Risk Assessment* (Jeremy Stranks): Pearson Education]

COMPETENT PERSONS

One way of ensuring the operation of a safe system of work is by the designation and employment of specifically trained operators who appreciate the risks involved, or by the use of external specialists. This may be for undertaking certain inspections, examination and testing of work equipment, or for undertaking certain activities where there may be a high degree of foreseeable risk.

The expression competent person occurs frequently in construction safety law. For example, under the Construction (Health, Safety and Welfare) Regulations 1996 certain inspections, examinations, operations and supervisory duties must be undertaken by competent persons.

It should be noted that the term 'competent person' is not generally defined in law except in the Electricity at Work Regulations 1989 and the Pressure Systems Regulations 2000. Therefore the onus is on the employer to decide whether persons are competent to undertake these duties. An employer might do this by reference to the person's training, qualifications and experience. Broadly, a competent person should have practical and theoretical knowledge as well as sufficient experience of the particular machinery, plant or procedure involved as will enable him to identify defects or weaknesses during plant or machinery examinations, and to assess their importance in relation to the strength and function of that plant and machinery (*Brazier* v. *Skipton Rock Company Limited* (1962) 1 AER 955).

Duties and functions of competent persons

These duties and functions are outlined below.

Noise at Work Regulations 1989

A competent person must make a noise assessment which is adequate for the purpose of:

(a) identifying employees' noise exposure; and

(b) providing the employer with appropriate information as to enable him to facilitate compliance with his duties.

Pressure Systems Safety Regulations 2000

The user of an installed pressure system and the owner of a mobile pressure system shall not operate the system or allow it to be operated unless he has a written scheme for the periodic examination, by a competent person, of the following parts of the system:

(a) all protective devices;

(b) every pressure vessel and every pipeline in which (in either case) a defect may give rise to danger; and

(c) those parts of the pipework in which a defect may give rise to danger.

The user or owner shall:

(a) ensure that the scheme has been drawn up, or certified as being suitable, by a competent person;

(b) ensure that:

(i) the content of the scheme is reviewed at appropriate intervals by a competent person for the purpose of determining whether it is suitable in current conditions of use of the system; and

(ii) the content of the scheme is modified in accordance with any recommendations made by that competent person arising out of that review.

Under these regulations, the competent person:

(a) advises the user or owner on the scope of the written scheme for periodic examination;

(b) draws up or certifies the schemes of examination; and

(c) undertakes examinations under the scheme.

A competent person is defined as a competent individual person (other than an employee) or a competent body of persons corporate or unincorporated; and accordingly any reference in these regulations to a competent person performing a function includes a reference to his performing it through his employees.

Electricity at Work Regulations 1989

No person is to carry out a work activity where technical knowledge or experience is necessary to prevent danger or injury, unless he has such knowledge or is under the appropriate degree of supervision. (Whilst the term does not appear, competence is implied.)

Management of Health and Safety at Work Regulations 1999

Competent persons must be appointed by an employer to assist the employer in, firstly, complying with legal requirements and, secondly, to assist him in implementing procedures for serious and imminent danger and for danger areas, thus:

Every employer shall appoint one or more competent persons to assist him in undertaking the measures he needs to take to comply with the requirements and prohibitions imposed upon him by or under the relevant statutory provisions and by Part II of the Fire Precautions (Workplace) Regulations 1997.

A person shall be regarded as competent where he has sufficient training and experience or knowledge and other qualities to enable him properly to assist the employer in undertaking the above measures.

Every employer shall nominate a sufficient number of competent persons to implement those procedures (i.e. for serious and imminent danger and for danger areas) in so far as they relate to the evacuation from premises of persons at work in an employer's undertaking.

A person shall be regarded as competent for the above purposes where he has sufficient training and experience or knowledge and other qualities to enable him properly to implement the evacuation procedures in the above paragraph.

Construction (Design and Management) Regulations 1994

Competence must be taken into account by:

(a) a client when appointing a planning supervisor;

(b) any person when arranging for a designer to prepare a design;

(c) any person when arranging for a contractor to carry out or manage construction work.

Construction (Health, Safety and Welfare) Regulations 1996

Competent persons must be appointed by an employer for the supervision of:

(a) the installation or erection of any scaffold and any substantial addition or alteration to a scaffold;

(b) the installation or erection of any personal suspension equipment or any means of arresting falls;

(c) erection or dismantling of any buttress, temporary support or temporary structure used to support a permanent structure;

(d) demolition or dismantling of any structure, or any part of any structure, being demolition or dismantling which gives rise to a risk of danger to any person;

(e) installation, alteration or dismantling of any support for an excavation;

(f) construction, installation, alteration or dismantling of a cofferdam or caisson; and

(g) the safe transport of any person conveyed by water to or from any place of work.

Competent persons are also required to inspect places of work as specified in Schedule 7 to the regulations below.

CONSTRUCTION (HEALTH, SAFETY AND WELFARE) REGULATIONS 1996
SCHEDULE 7

Regulation 29(1)

PLACES OF WORK REQUIRING INSPECTION BY A COMPETENT PERSON

Column 1 Place of work	Column 2 Time of inspection
1. Any working platform or part thereof or any personal suspension equipment provided pursuant to paragraph (3)(b) or (c) of regulation 6	1. (i) Before being taken into use for the first time; and (ii) after any substantial addition, dismantling or other alteration; and (iii) after any event likely to have affected its strength or stability; and (iv) at regular intervals not exceeding 7 days since the last inspection.
2. Any excavation which is supported pursuant to paragraphs (1), (2) or (3) of regulation 12	2. (i) Before any person carries out work at the start of every shift; and (ii) after any event likely to have affected the strength or stability of the excavation or any part thereof; and (iii) after any accidental fall of rock or earth or other material.
3. Cofferdams and caissons	3. (i) Before any person carries out work at the start of every shift; and (ii) after any event likely to have affected the strength or stability of the cofferdam or caisson or any part thereof.

6

PREVENTION AND CONTROL STRATEGIES

A basic principle in health and safety is that of preventing exposure to a particular risk. If it is not possible to prevent exposure, then the risk must be controlled by the use of various protective measures. A 'safe place' strategy, such as machinery guarding, or the provision of local exhaust ventilation systems to prevent exposure to airborne contaminants, should always be invoked in preference to a 'safe person' strategy, such as health surveillance or the provision of information, instruction and training.

Thus, in the process of risk assessment, once the risk has been identified and assessed, employers must either prevent the risk arising or, alternatively control same. Much will depend upon the magnitude of the risk in terms of the controls applied. In certain cases, the level of competence of operators may need to be considered prior to their undertaking certain work, such as work on electrical systems.

General principles of prevention

Schedule 1 to the Management of Health and Safety at Work Regulations outlines the general principles of prevention thus:

(a) avoiding risks;

(b) evaluating the risks which cannot be avoided;

(c) combating the risks at source;

(d) adapting the work to the individual, especially as regards the design of workplaces, the choice of work equipment and the choice of working and production methods, with a view, in particular, to alleviating monotonous work and work at a predetermined work-rate and to reducing their effects on health;

(e) adapting to technical progress;

(f) replacing the dangerous by the non-dangerous or the less dangerous;

(g) developing a coherent overall prevention policy which covers technology, organisation of work, working conditions, social relationships and the influence of factors relating to the working environment;

(h) giving collective protective measures priority over individual protective measures; and

(i) giving appropriate instructions to employees.

However, in many cases, it may not be possible to prevent exposure to risks using one or more of the strategies outlined above.

Hierarchy of control

The ideal situation is that all risks should be eliminated.

In many cases, however, it is not possible to completely eliminate a risk. Some form of control must, therefore, be applied.

A typical hierarchy of control, outlining some of the more common techniques, from high risk to low risk, is indicated below.

1. Substitution

Where substances may present a serious risk to employees, substitution of a less toxic material for a more highly toxic one, is a common technique.

Typical examples are the substitution of:

(a)   toluene for benzene; and

(b)   trichloroethane 1.1.1 for carbon tetrachloride.

2. Enclosure/Containment

This means the enclosure or containment of the risk in such a way that access is denied to same. This strategy is based on the containment of an offending agent or environmental stressor, such as machinery noise, to prevent its liberation into the working environment.

Total enclosure or containment of a process may be possible by the use of bulk tanks and pipework to deliver a liquid directly into a closed production vessel. Similarly, enclosure guarding is used for large machines where guarding of individual risks may be impracticable. Enclosures may take a number of forms, such as acoustic enclosures for noisy machinery, dust enclosures and paint spray booths.

3. Isolation/Separation

The isolation or segregation of a process using potentially dangerous substances or sources of radiation may simply mean relocating it to a controlled area, thereby separating the majority of the workforce from the risk. In this case, a system for limiting access to the controlled area only to trained and competent operators is necessary. In certain cases electrical equipment may be isolated in switch rooms or at such a distance from the floor that operators cannot come into contact with it (secure by position).

4. Environmental control

The employer has a duty to ensure the provision and maintenance of a working environment that is safe and without risks to health. Risks to health and safety may be associated with extremes of temperature, poor levels of lighting and ventilation, noise and vibration, and the presence of airborne contaminants, such as dust, fumes and gases. Control of the environment may be through mechanical ventilation systems, the use of local exhaust ventilation systems, improved lighting and sound proofing.

Structural aspects must also be considered in terms of safe floors, passages and stairways, access to and egress from the workplace, and the storage and disposal of waste items. Planned preventive maintenance systems aim to ensure structural integrity of the workplace. Procedures for waste storage and disposal, particularly in the case of flammable or toxic wastes, should be well-controlled and subject to regular monitoring.

5. Machinery and plant maintenance

The provision and maintenance of appropriate guards and safety devices is essential in preventing injuries associated with operating machinery. The requirements of the Provision and Use of Work Equipment Regulations 1998 and BS EN 292 'Safeguarding of machinery' must be complied with. Electrical services should be maintained and portable electrical appliances examined and maintained according to manufacturers' instructions.

Planned preventive maintenance procedures should ensure the integrity of all machinery, plant and appliances.

6. Change of process

Improved design or process engineering can bring about changes to provide better operator protection. Employees must be encouraged to report deficiencies in machinery and plant.

7. Controlled operation/safe systems of work

This implies the need for high standards of supervision and control where, in particular, there may be a high degree of foreseeable risk. Controlled operation may take a number of forms:

- the use of mechanical or remote control handling systems, e.g. with radioactive substances;
- the operation of formally established safe systems of work and the use of trained and supervised operators with such systems;
- the use of permit to work systems, e.g. for entry into confined spaces; and
- restriction of certain activities to highly trained and supervised staff, e.g. persons working in high voltage switch rooms.

Risk assessment commonly identifies the need for formally written and adopted safe systems of work. Operators must be instructed and trained in safe systems of work, with regular supervision to ensure operation of same.

8. Competent persons and authorised persons

In certain cases, it may be necessary to designate certain people as competent persons, on the basis of their skill, knowledge and experience, to undertake certain tasks. Such persons must be appointed by law in the case of, for instance, the inspection of scaffolds, excavations, lifting appliances and tackle.

In other cases, the use of certain items of equipment may require a period of training and instruction. It may be necessary to specify that only those authorised to use the equipment, such as lift trucks, may use same. Authorised persons may also be appointed to undertake certain electrical switching operations.

9. Information, instruction and training

There are duties under the Health and Safety at Work Act and the majority of Regulations to provide information, instruction and training to employees and others, particularly with regards to the hazards that may exist and the precautions necessary. This should be linked with effective supervision to ensure the lessons learned are being put into practice.

10. Warning systems

Warning systems take many forms, from simple safety signs posted at prominent points to more sophisticated devices, such as visible and audible alarms.

11. Personal protective equipment

The provision and use of items of personal protective equipment (PPE), such as eye protection, ear protection, safety boots and shoes, should be seen as the last resort, when all other measures have failed or, perhaps, as an interim measure until some more permanent form of protection, such as local exhaust ventilation, is provided.

The sole reliance on PPE as a means of protecting people from risks is a very dangerous strategy and relies on a high degree of supervision to ensure operators use or wear the PPE all the time they are exposed to the risk.

12. Health surveillance

Health surveillance may take many forms, such as pre-employment health screening by an occupational health nurse, on-going health examinations to assess continuing fitness for work or, where people may be exposed to hazardous substances, various forms of biological monitoring, such as blood and urine tests, vision screening of display screen equipment users, etc.

The need for health surveillance may be identified from risk assessments undertaken at an earlier stage.

INFORMATION, INSTRUCTION AND TRAINING

In order to maintain a safe and healthy workplace, it is essential that employees and other persons, such as the employees of contractors, visitors and temporary workers, are adequately informed, firstly of the hazards that could arise and, secondly, the action they may need to take to protect themselves from these hazards.

Most modern health and safety legislation requires employers to provide their employees and other persons with information, instruction and training on a wide range of issues.

The duty to inform, instruct and train

The terms 'information', 'instruction' and 'training' are not defined in law. However, the giving of information implies the imparting of factual knowledge by one person to another. Instruction, on the other hand, involves actually telling people what to do and may incorporate an element of supervision to ensure they do it correctly. 'Training' is defined by the Department of Employment as 'the systematic development of attitude, knowledge and skill patterns required by an individual to perform adequately a given task or job'.

The duties on the part of employers and others to provide information, instruction and training to employees and other persons are incorporated in the Health and Safety at Work etc. Act 1974 (HSWA) and regulations made under the Act. In most cases, this duty is qualified by the term 'so far as is reasonably practicable'. However, under much recent European Union-driven legislation, such as the Management of Health and Safety at Work Regulations 1999, this duty is absolute or strict.

Health and Safety at work etc. ACT 1974

Duties of employers towards employees

Employers must provide their employees, in particular, with details of the hazards to which they could be exposed and the precautions necessary on the part of those employees to prevent exposure to same. This is an important requirement of HSWA, thus:

> Every employer shall ensure the provision of such information, instruction, training and supervision as is necessary to ensure, so far as is reasonably practicable, the health and safety at work of his employees. (Section 2(2)(c))

Duties of employers towards non-employees

The giving of information by an employer to non-employees is a matter which is not well-understood by many employers. Typical examples are where contractors' employees may be working at a second employer's premises, e.g. roofing contractors, electrical contractors, or where sub-contractors are working under the direction of a principal contractor. Section 3 of HSWA is quite explicit here.

> In such cases as may be prescribed, it shall be the duty of every employer and every self-employed person, in the prescribed circumstances and in the prescribed manner, to give to persons (not being his employees) who may be affected by the way in which he conducts his undertaking the prescribed information about such aspects of the way in which he conducts his undertaking as might affect their health or safety. (Section 3(3))

The *Swan Hunter* Case

The above requirement was classically illustrated in *R v. Swan Hunter Shipbuilders Ltd [1981] The Times*. This prosecution by the HSE followed the fire on board HMS Glasgow, which was under construction in the Swan Hunter yard. In this situation, oxygen had leaked from a hose that passed down into a compartment on the lowest deck. The last-known person to use the oxygen hose, which was the source of the leak, was an employee of Telemeter Installations Ltd. As a result of an oxygen-enriched atmosphere in confined spaces below decks, there was an explosion and eight workmen were burnt to death.

Here the main contractor, Swan Hunter Ltd, and the sub-contractor, Telemeter Installations Ltd, were fined £15,000 and £3,000 respectively for failing to inform new sub-contract labour of the hazards arising from oxygen-enriched atmospheres, particularly in poorly ventilated spaces below decks.

Duties of manufacturers, etc. of articles and substances

Section 6 of the HSWA distinguishes between the duties in respect of both articles and substances, thus:

It shall be the duty of any person who designs, manufactures, imports or supplies any *article* for use at work:

> to take such steps as are necessary to secure that there will be available in connection with the use of that article at work adequate information about the use for which it is designed and has been tested, and about any conditions necessary to ensure that, when put to that use, it will be safe and without risks to health (Section 6(1)(c))

6

It shall be the duty of any person who manufactures, imports or supplies any *substance* for use at work:

> to take such steps as are necessary to ensure that there will be available in connection with the use of that substance at work adequate information about the results of any tests which have been carried out on or in connection with the substance and about any conditions necessary to ensure that it will be safe and without risks to health when properly used. (Section 6(4)(c))

It should be noted that the term *article* is broadly defined as meaning:

(a) any plant designed for use or operation (whether exclusively or not) by persons at work, or who erect or install any article of fairground equipment; and

(b) any article designed for use as a component in any such plant or equipment.

A *substance* means any natural or artificial substance (including micro-organisms) intended for use (whether exclusively or not) by persons at work.

The Product Liability Chain

Principal duties to provide information rest with manufacturers, designers and importers of articles and substances for use at work. Secondary duties, in particular to ensure that the information is passed on, rests with wholesalers, retailers and others in the supply chain.

Duties under Regulations

Construction (Design and Management) (CDM) Regulations 1994

The CDM Regulations impose, in the main, absolute requirements and prohibitions with respect to 'construction work'. This term is very broadly defined and includes not only construction activities, such as the erection, extension or conversion of a building, but preparation for an intended structure, assembly of prefabricated elements, demolition operations and the installation of services to a structure.

There must be a formal procedure where each of those holding duties, namely clients, designers, planning supervisors, principal contractors and contractors, exchange information on a range of safety-related issues.

Provision of information by clients to planning supervisors

A client, that is, a person for whom project work is carried out, has the following duties.

Every client shall ensure that the planning supervisor for any project carried out for the client is provided (as soon as is reasonably practicable but in any event before the commencement of the work to which the information relates) with all the information mentioned in the paragraph below about the state or condition of any premises at or on which the construction work included or intended to be included in the project is or is intended to be carried out.

The information required to be provided by the above paragraph is information which is relevant to the functions of the planning supervisor under these regulations and which the client has or could ascertain by making enquiries which is reasonable for a person in his position to make (Regulation 11)

Provision of information by principal contractors to contractors

Principal contractors must provide information to other contractors.

The principal contractor appointed for any project shall ensure, so far as is reasonably practicable, that every contractor is provided with comprehensible information on the risks to the health or safety of that contractor or of any employee or other persons under the control of that contractor arising out of or in connection with the construction work.

The principal contractor shall ensure, so far as is reasonably practicable, that every contractor who is an employer provides any of his employees at work carrying out the construction work with:

(a) any information which the employer is required to provide to those employees in respect of that work by virtue of regulation 8 of the Management of Health and Safety at Work Regulations; and

(b) any health and safety training which the employer is required to provide to those employees in respect of that work by virtue of regulation 11(2)(b) of the Management of Health and Safety at Work Regulations (Regulation 17)

Control of Substances Hazardous to Health (COSHH) Regulations 2002

Where employees are liable to be exposed to a substance hazardous to health as part of their work, an employer must undertake a suitable and sufficient assessment of the risks to those employees and of the steps that need to be taken to meet the requirements of the regulations. He must then provide particular information, instruction and training as outlined below. It should be noted that this duty is extended to non-employees who are liable to be exposed.

Every employer who undertakes work which may expose an employee to a substance hazardous to health shall provide that employee with suitable and sufficient information, instruction and training.

The information, instruction and training provided under the above paragraph shall include:

(a)   details of the substances hazardous to health to which the employee is exposed including:

   – the names of those substances and the risks which they present to health;
   – any relevant occupational exposure standard, maximum exposure limit or similar occupational exposure limit;
   – access to any relevant safety data sheet; and
   – other legislative provisions which concern the hazardous properties of those substances;

(b)   the significant findings of the risk assessment;

(c)   the appropriate precautions and action to be taken by the employee in order to safeguard himself and other employees at the workplace;

(d)   the results of any monitoring of exposure in accordance with regulation 10 and, in particular, in the case of a substance hazardous to health for which a maximum exposure limit has been approved, the employee or his representatives shall be informed forthwith, if the results of such monitoring show that the maximum exposure limit has been exceeded;

(e)   the collective results of any health surveillance undertaken in accordance with regulation 11 in a form calculated to prevent those results from being identified as relating to a particular person; and

(f)   where employees are working with a Group 4 biological agent or material that may contain such an agent, the provision of written instructions and, if appropriate, the display of notices which outline the procedures for handling such an agent or material.

Every employer shall ensure that any person (whether or not his employee) who carries out work in connection with the employer's duties under these regulations has the necessary information, instruction and training. (Regulation 12)

Health and Safety (Display Screen Equipment) Regulations 1992

These regulations apply only to defined 'users' and 'operators' of display screen equipment. A *user* is an employee who habitually uses display screen equipment as a significant part of his normal work. An *operator*, on the other hand, is a self-employed person who habitually uses display screen equipment as a significant part of his normal work. The terms 'habitually', 'significant' and 'normal' are important in determining defined users and operators.

Many people use display screen equipment (DSE) as an ancillary aspect of their work. However, where people do not meet one or other of these definitions completely, the regulations do not apply.

Regulation 6 imposes an absolute duty on employers to provide training in specific situations.

Where a person:

(a) is a user on the date of coming into force of these regulations; or

(b) is an employee who does not habitually use display screen equipment as a significant part of his normal work, but is to become a user in the undertaking in which he is already employed,

their employer shall ensure that they are provided with adequate health and safety training in the use of any workstation on which they may be required to work.

Every employer shall ensure that each user at work in his undertaking is provided with adequate health and safety training whenever the organisation of any workstation in that undertaking upon which he may be required to work is substantially modified. (Regulation 6)

Training should be adapted to the requirements of the particular display screen tasks, be adapted to users' skills and capabilities and be refreshed or updated as the hardware, software, workstation, environment or job are modified.

Users must be provided with information about their daily work routine, the form of training they will receive and the measures taken by the employer to comply with his duties under the regulations, e.g. the outcome of workstation risk analysis. This duty to inform must also be read in conjunction with the general duty to provide information, which is comprehensible, and relevant under the Management of Health and Safety at Work Regulations.

Regulation 7 of the Health and Safety (Display Screen Equipment) Regulations 1992 is quite specific on this issue, thus:

> Every employer shall ensure that users at work in his undertaking are provided with adequate information about the measures taken by the employer in order to comply with their duties under regulations 4 (daily work routine of users) and 6(2) (provision of training) as relate to them and their work.

> Every employer shall ensure that users employed by him are provided with adequate information about such measures taken by him in compliance with his duties under regulations.

Health and Safety (Information for Employees) Regulations 1998

These regulations require information relating to health, safety and welfare to be furnished to employees by the means of posters or leaflets in the form approved and published for the purposes of the regulations by the HSE.

The poster or form is entitled *Health and Safety Law – What you should know* and is available from the HSE and HMSO.

Ionising Radiations Regulations 1999

The duty on the part of employers to inform, instruct and train employees and others who could be accidentally exposed to ionising radiation is a significant requirement under the regulations.

Every employer shall ensure that:

(a) those employees who engaged in work with ionising radiation receive such information, instruction and training as will enable them to conduct the work in accordance with the requirements of these regulations;

(b) adequate information is given to others who are directly concerned with the work with ionising radiation carried out by the employer to ensure their health and safety, so far as is reasonably practicable;

(c) classified persons and trainees are informed of the health hazard, if any, associated with their work, the precautions to be taken, the importance of complying with the medical and technical requirements and given appropriate training in the field of radiation protection; and

(d) those employees who are engaged in work with ionising radiation who are women are informed of the possible hazard arising from ionising radiation to the foetus in early pregnancy and the importance of informing the employer as soon as they discover they have become pregnant. (Regulation 12)

Management of Health and Safety at Work Regulations 1999

Every employer shall provide his employees with comprehensible and relevant information on:

(a) the risks to their health and safety identified by the (risk) assessment;

(b) the preventive and protective measures;

(c) the procedures referred to in regulation 8(1)(a) (emergency procedures) and the measures referred to in regulation 4(2)(a) of the Fire Precautions (Workplace) Regulations 1997;

(d) the identity of those people nominated by him in accordance with regulation 8(1)(b) (competent persons for the purpose of emergencies) and regulation 4(2)(b) of the Fire Precautions (Workplace) Regulations 1997; and

(e) the risks notified to them in accordance with regulation 11(1)(c) (shared workplaces).

Noise at Work Regulations 1989

Every employer shall, in respect of any premises under his control, provide each of his employees who is likely to be exposed to the first action level or above or to the peak action level or above with adequate information, instruction and training on:

(a) the risk of damage to that employee's hearing that such exposure may cause;

(b) what steps that employee can take to minimise the risk;

(c) the steps that the employee must take in order to obtain the personal ear protectors referred to in regulation 8; and

(d) that employee's obligations under these regulations. (Regulation 11)

Personal Protective Equipment at Work Regulations 1992

Personal protective equipment (PPE) is defined in these regulations as meaning 'all equipment (including clothing affording protection against the weather) which is intended to be worn or held by a person at work and which protects him against one or more risks to his health and safety, and any addition or accessory designed to meet this objective'.

Information and instruction provided to employees must be comprehensible to those receiving same.

Employers must provide adequate and appropriate information to enable the employees to know:

(a) the risk or risks that the personal protective equipment will avoid or limit;

(b) the purpose for which and the manner in which the PPE is to be used; and

(c) any action to be taken by the employee to ensure that the PPE remains in an efficient state, in efficient working order and in good repair.

The information and instruction provided shall not be considered adequate and appropriate unless it is comprehensible to the people to whom it is given. (Regulation 9)

Pressure Systems Safety Regulations 2000

Any person who:

(a) designs for another any pressure system or any article which is intended to be a component thereof; or

(b) supplies (whether as manufacturer, importer or in any other capacity) any pressure system or any such article,

shall provide sufficient written information concerning its design, construction, examination, operation and maintenance as may reasonably foreseeably be needed to enable the provisions of these regulations to be complied with.

The employer of a person who modifies or repairs any pressure system shall provide sufficient written information concerning the modification or repair as may reasonably foreseeably be needed to enable the provisions of these regulations to be complied with.

Provision and Use of Work Equipment Regulations 1998

Every employer shall ensure that all those who use work equipment have available to them adequate health and safety information and, where appropriate, written instructions pertaining to the use of that equipment.

Every employer shall ensure that any of his employees who supervise or manage the use of work equipment has available to him similar adequate information.

The information and instructions required by either of the above paragraphs shall include information and, where appropriate, written instructions on:

(a) the conditions in which and the methods by which the work equipment may be used;

(b) foreseeable abnormal situations and the action to be taken if such a situation were to occur; and

(c) any conclusions to be drawn from experience in using the work equipment.

Information and instructions required by this regulation shall be readily comprehensible to those concerned. (Regulation 8)

Every employer shall ensure that all those who use work equipment have received adequate training for the purposes of health and safety, including training in the methods that may be adopted when using the work equipment, any risks that such use may entail and the precautions to be taken.

Every employer shall ensure that any of his employees who supervises or manages the use of work equipment has received similar adequate training.

Safety Representatives and Safety Committees Regulations 1977

An employer shall make available to safety representatives the information within the employer's knowledge necessary for them to enable them to fulfil their functions except:

(a) any information the disclosure of which would be against the interest of national security;

(b) any information that they could not disclose without contravening a prohibition imposed by or under any enactment;

(c) any information relating specifically to an individual, unless he has consented to it being disclosed;

(d) any information the disclosure of which would, for reasons other than its effect on health, safety or welfare at work, cause substantial injury to the employer's undertaking or, where the information was supplied to them by some other person, to the undertaking of that other person; and

(e) any information obtained by the employer for the purpose of bringing, prosecuting or defending any legal proceedings. (Regulation 7)

Sources of information

Many sources of information are available to employers and employees. These can be classed as formal and informal sources.

Formal (primary) sources of information

Formal sources of information come from a variety of sources. These include:

EU Directives

These are European Community instruments of legislation and are legally binding on the governments of all Member States. On this basis they must introduce national legislation, or use administrative procedures where applicable, to implement their requirements.

Acts of Parliament (statutes)

Acts of Parliament may be innovatory, i.e. introducing new legislation, or consolidating, i.e. reinforcing, with modifications, existing law. Statutes, in most cases, empower the Minister or Secretary of State to make regulations (delegated or subordinate legislation). Typical examples are the HSWA and the Disability Discrimination Act 1995.

Regulations (statutory instruments)

Regulations are more detailed than the parent Act, which lays down the framework and objectives of the system. Specific details are incorporated in Regulations made under the Act, e.g. the Control of Substances Hazardous to Health (COSHH) Regulations, which were passed pursuant to the HSWA. Regulations are made by the appropriate Minister or Secretary of State whose powers to do so are identified in the parent Act.

Approved Codes of Practice

The HSC is empowered to approve and issue Codes of Practice for the purpose of providing guidance on health and safety duties and other matters laid down in Statute or Regulations. A Code of Practice can be drawn up by the Commission or the HSE. In every case, however, the relevant government department, or other body, must be consulted beforehand and approval of the Secretary of State must be obtained. Any Code of Practice approved in this way is an Approved Code of Practice (ACOP).

An ACOP enjoys a special status under the HSWA. Although failure to comply with any provision of the code is not in itself an offence, that failure may be taken by a court, in criminal proceedings, as proof that a person has contravened the legal requirement to which the provision relates. In such a case it will be open to that person to satisfy a court that he or she has complied with the requirement in some other way.

Examples of ACOPs are those issued with the Control of Substances Hazardous to Health (COSHH) Regulations entitled 'Control of Substances Hazardous to Health', 'Control of Carcinogenic Substances' and 'Control of Substances Hazardous to Health in Fumigation Operations'.

Case law (common law)

Case law is an important source of information. It is derived from common law because, traditionally, judges have formulated rules and principles of law as the cases occur for decision before the Courts.

What is important in a case is the *ratio decidendi* (the reason for the decision). This is binding on courts of equal rank who may be deciding the same point of law. Ratio decidendi is the application of such an established principle to the facts of a given case, for instance, negligence consists of omitting to do what a 'reasonable man' would do in order to avoid causing injury to others.

Case law is found in law reports, for example, the All England Law Reports, the Industrial Cases Reports, the Current Law Year Book and in professional journals, e.g. Law Society Gazette, Solicitors Journal. In addition, many newspapers carry daily law reports, e.g. *The Times, Financial Times, Daily Telegraph* and *Independent*.

The supreme law court is the European Court of Justice, whose decisions are carried in *The Times*.

Secondary sources of information

A wide range of non-legal sources of information are available, some of which may be quoted in legal situations, however.

HSE Series of Guidance Notes

Guidance Notes issued by the HSE have no legal status. They are issued on a purely advisory basis to provide guidance on good health and safety practices, specific hazards, etc. There are five series of Guidance Notes – General, Chemical Safety, Plant and Machinery, Medical and Environmental Hygiene.

British Standards

These are produced by the British Standards Institute. They provide sound guidance on numerous issues and are frequently referred to by enforcement officers as the correct way of complying with a legal duty.

British Standard EN 292 'Safeguarding of machinery' is commonly quoted in conjunction with the duties of employers under the Provision and Use of Work Equipment Regulations 1998 to provide and maintain safe work equipment.

Manufacturers' information and instructions

Under section 6 of the HSWA (as amended by the Consumer Protection Act 1987) manufacturers, designers, importers and installers of articles and substances used at work have a duty to provide information relating to the safe use, storage, etc. of their products. Such information may include operating instructions for machinery and plant and hazard data sheets in respect of dangerous substances. Information provided should be sufficiently comprehensive and understandable to enable a judgement to be made on their safe use at work.

Safety organisations

Safety organisations, such as the Royal Society for the Prevention of Accidents (RoSPA) and the British Safety Council, provide information in the form of magazines, booklets and videos on a wide range of health and safety-related topics.

Professional institutions and trade associations

Many professional institutions, such as the Institute of Occupational Safety and Health (IOSH), the Chartered Institute of Environmental Health (CIEH) and the British Occupational Hygiene Society (BOHS), provide information, both verbally and in written form. Similarly, a wide range of trade associations provide this information to members.

Insurance companies

Most insurance companies provide an information service to clients on a wide range of health and safety-related matters.

Department of Social Security

This Government department publishes annual statistics on claims for industrial injuries benefit and other matters.

Published information

This takes the form of text books, magazines, law reports, updating services, microfiche systems, films and videos on general and specific topics.

Existing written information

This may take the form of Statements of Health and Safety Policy, in-house Health and Safety Codes of Practice, specific policies, e.g. on the use and storage of dangerous substances, and written safe systems of work, methods and operating instructions.

Such documentation could be quoted in a court of law as an indication of the organisation's intention to regulate activities in order to ensure legal compliance. Evidence of the use of such information in staff training is essential here.

Work study techniques

Included here are the results of activity sampling, surveys, method study, work measurement and process flows.

Job descriptions

A job description should incorporate health and safety responsibilities and accountabilities. It should take account of the physical and mental requirements and limitations of certain jobs and any specific risks associated with the job. Representations from operators, supervisors and trade union safety representatives should be taken into account.

Compliance with health and safety requirements is an implied condition of every employment contract, breach of which may result in dismissal or disciplinary action by the employer.

Accident and ill-health statistics

Statistical information on past accidents and sickness may identify unsatisfactory trends in operating procedures which can be eliminated at the design stage of safe systems of work.

The use of accident statistics and rates, e.g. accident incidence rate, as a sole measure of safety performance is not recommended, however, due to the variable levels of accident reporting in work situations. Under-reporting of accidents, common in many organisations, can result in inaccurate comparisons being made between one location and another.

Task analysis

Information produced by the analysis of tasks, such as the mental and physical requirements of a task, manual operations involved, skills required, influences on behaviour, hazards specific to the task, and learning methods necessary to impart task knowledge must be taken into account.

Job safety analysis, a development of task analysis, will provide the above information prior to the development of safe systems of work.

Direct observation

This is the actual observation of work being carried out. It identifies interrelationships between operators, hazards, dangerous practices and situations and potential risk situations. It is an important source of information in ascertaining whether, for instance, formally designed safe systems of work are being operated or safety practices, imparted as part of former training activities, are being followed.

Personal experience

People have their own unique experience of specific tasks and the hazards which those tasks present. The experiences of accident victims, frequently recorded in accident reports, are an important source of information. Feedback from accidents is crucial in order to prevent repetition of these accidents.

Incident recall

This is a technique used in a damage control programme to gain information about near-miss accidents.

Product complaints

A record of all product complaints and action taken should be maintained. Such information provides useful feedback in the modification of products and in the design of new products. Product complaints may also result in action by the enforcement authorities under section 6 of the HSWA and/or civil proceedings in the event of injury, damage or loss sustained as a result of a defective product or defect in a product.

HEALTH SURVEILLANCE

Health surveillance is the regular review of the health of employees exposed to various forms of health risk, for instance from hazardous substances, manual handling operations or as a result of working on specific processes. Whilst there is a general duty on an employer under section 2(1) of the Health and Safety at Work Act 1974 'to ensure, so far as is reasonably practicable, the health, safety and welfare at work of all his employees', more specific duties to provide same can be found in Regulations made under the Act, such as the COSHH Regulations 2002 and the Management of Health and Safety at Work Regulations 1999.

Legal requirements

Regulation 11 – Control of Substances Hazardous to Health Regulations 2002

Where it is appropriate for the protection of the health of his employees who are, or are liable to be exposed to a substance hazardous to health, the employer shall ensure that such employees are under suitable health surveillance. Health surveillance shall be treated as appropriate where:

(a) the employee is exposed to one of the substances specified in Column 1 of Schedule 6 and is engaged in a process specified in Column 2 of that Schedule, and there is a reasonable likelihood that an identifiable disease or adverse health effect will result from that exposure; or

(b) the exposure of the employee to a substance hazardous to health is such that:

   (i) an identifiable disease or adverse health effect which may be related to the exposure;

   (ii) there is a reasonable likelihood that this may occur under the particular conditions of his work; and

   (iii) there are valid techniques for detecting indications of the disease or effect;

and the technique of investigation is of low risk to the employee.

Regulation 6 – Management of Health and Safety at Work Regulations 1999

Every employer shall ensure that his employees are provided with such health surveillance as is appropriate having regard to the risks to their health and safety which are identified by the (risk) assessment.

The ACOP outlines the objectives of health surveillance, thus:

'The primary benefit, and therefore the objective, of health surveillance should be to detect adverse health effects at an early stage, thereby enabling further harm to be prevented.

In addition the results of health surveillance can provide a means of:

(a) checking the effectiveness of control measures;
(b) providing feedback on the accuracy of risk assessment;
(c) identifying and protecting individuals at increased risk.

Once it is decided that health surveillance is appropriate, such health surveillance should be maintained during the employee's employment unless the risk to which the worker is exposed and associated health effects are short term. The minimum requirement for health surveillance is the keeping of an individual health record. Where it is appropriate, health surveillance may also involve one or more health surveillance procedures depending upon their suitability in the circumstances. Such procedures can include:

(a) inspection of readily detectable conditions by a responsible person acting within the limits of their training and experience;
(b) enquiries about symptoms, inspection and examination by a qualified person such as an occupational health nurse;
(c) medical surveillance, which may include clinical examination and measurements of physiological or psychological effects by an appropriately qualified practitioner;
(d) biological effect monitoring, i.e. the measurements and assessment of early biological effects such as diminished lung function in exposed workers;
(e) biological monitoring, i.e. the measurement and assessment of workplace agents or their metabolites either in tissues, secreta, excreta, expired air or any combination of these in exposed workers.

Aspects of health surveillance

Health surveillance is a very broad area and must be viewed as part of an overall occupational health strategy aimed at protecting the health of people at work. Much will depend upon the risks to which people are exposed. What is important is that the form of health surveillance undertaken should be appropriate to these risks.

Health surveillance should be carried out by suitably qualified persons, such as a doctor with, preferably, a specialist qualification in occupational medicine (occupational physician) or an occupational health nurse. It may involve the assessment of hazardous substances or their by-products in the body (by the examination of urine or blood) or of body functions (e.g. blood pressure, lung function). In some cases, clinical examinations or tests may be necessary. Where medical examinations and inspections are called for, employers must provide suitable facilities on site.

Health protection

It is important at this stage to consider not only the risks to individuals through exposure to, for instance, noise and hazardous chemical substances, but also the risk to food products. The implications of the Food Safety Act 1990 and the Food Safety (General Food Hygiene) Regulations 1994, may necessitate the operation of some form of health surveillance of food handlers.

An appropriate strategy to protect people at work from health risks should incorporate the following elements.

1. Pre-employment health screening

This form of surveillance, carried out by an occupational health nurse, can include not only an assessment of general fitness for the job but also specific aspects of that job, such as vision screening of drivers and VDU operators, initial audiometry for people who may be exposed to noise and the assessment of individual disability levels where manual handling operations are involved. Such screening may also seek information as to previous food-borne illness sustained by the applicant and the possibility that the individual may still be a carrier of that illness.

2. Primary and secondary monitoring

Primary monitoring is concerned largely with the clinical observation of sick people who may seek treatment or advice on their conditions. Such observation will identify new risks which were previously not considered.

Secondary monitoring, on the other hand, is directed at controlling the hazards to health which have already been recognised. Audiometry is a classic form of secondary monitoring whereby the hearing levels of operators are tested on, say, a six-monthly or annual basis to assess whether there has been a further deterioration in hearing due to exposure to noise. On a more general basis, the annual health screening of food handlers by an occupational health nurse is an important form of secondary monitoring.

3. Supervision of vulnerable groups

Certain groups, such as young persons, the aged, the disabled and pregnant women may be more vulnerable to health risks than others. Special attention must be given to these persons in terms of health counselling, assistance with rehabilitation and reorganisation of their jobs to remove harmful factors.

4. Avoiding potential risks

This can involve an examination of the current work layouts and arrangements, taking into account ergonomic aspects of the job and the potential for fatigue. The effects of shift working, long periods of work and the physical and mental effects of repetitive tasks would be taken into account in the assessment of health risks. On this basis shift workers and those working long shifts could well be a subject for health surveillance.

Health records

Under the COSHH Regulations, employers must ensure that a health record, containing particulars approved by the HSE, in respect of each of his employees who have been or may have been exposed to substances hazardous to health, is made and maintained. That record, or a copy thereof, must be kept available in a suitable form for at least 40 years from the date of the last entry made in it.

However, apart from the duty to maintain health records under the COSHH Regulations, there is clearly a case for maintaining health records of all employees. Not only is it possible for the occupational health practitioner to obtain useful information on unidentified health risks from accumulated records, but a health record could well be important evidence where an individual may be taking a civil action for negligence against the organisation many years later, or where the enforcement agency may be taking action under health and safety legislation.

The following records on individual employees may need to be maintained:

(a) pre-employment and subsequent health questionnaires submitted by employees;
(b) details of pre-employment and subsequent health examinations and screening tests undertaken by an occupational health nurse;
(c) relevant medical and occupational history, smoking habits, disabilities and handicaps;
(d) injuries resulting from occupational and non-occupational accidents;
(e) illness occurring at work or on the way to or from work;
(f) history of sickness absence;
(g) details of occupational diseases and conditions diagnosed;
(h) care and treatment provided;
(i) advice given, recommendations and work limitations imposed;
(j) referrals made to other medical specialists or agencies;
(k) correspondence relating to the health of individual employees;
(l) dispersal of cases following emergencies and treatment; and
(m) details of communication between occupational health practitioners and others, including written reports.

The confidentiality of health records must be maintained at all times.

Forms of health surveillance

Health surveillance is, fundamentally, the regular review of the health of persons exposed to hazardous substances or working in specified processes. It should be carried out by a suitably qualified person, such as an occupational physician or an occupational health nurse.

It may involve the assessment of hazardous substances or their by-products in the body by the examination of urine or blood or the assessment of body functions, such as blood pressure, lung function. In some cases, clinical tests or examinations may be necessary.

Where medical examinations and inspections are required by law, employers must provide suitable facilities on site.

Where health surveillance is required, records must be kept listing the personal details of the employee, e.g. name, sex, age and previous and current occupations which involved exposure to hazardous substances. The records, or copies of same, should be kept for at least 30 years after the date of the last entry.

The purpose of health surveillance

1. Objectives

The objectives of health surveillance, wherever employees may be exposed to substances hazardous to health in the course of their work, are:

(a) the protection of the health of individuals by detecting adverse changes, attributed to exposure to substances hazardous to health, at the earliest possible stage;

(b) to help in assessing the effectiveness of control measures;

(c) the collection, maintenance and use of data for the detection and evaluation of hazards to health; and

(d) to assess the immunological status of employees doing specific work with micro-organisms hazardous to health.

2. Outcome

The results of any health surveillance procedure should lead to some form of action which will be to the benefit of employees.

Health surveillance under the COSHH Regulations

1. Biological monitoring

This could entail monitoring to detect, for instance, 2,5-hexane dione in urine (for hexane exposure) or carboxyhaemoglobin in the blood of workers exposed to methylene chloride.

2. Biological effect monitoring

One example is the presence of cholinesterase in the blood of workers exposed to certain organo-phosphorous pesticides.

3. Medical surveillance

This could include, for example, the provision of lung function tests for workers exposed to substances known to cause asthma and chest X-rays of workers exposed to respirable dusts, e.g. quartz.

4. Enquiries, inspection or examination by a suitably qualified person

Typical examples include an occupational health nurse administering a questionnaire covering the symptoms of asthma or rhinitis and visual examination of the hands and arms of employees for evidence of dermatitis.

5. Inspection by a responsible person

In certain cases, it may be common practice for a manager or supervisor to undertake visual examinations of hands and arms of employees for early evidence of chrome ulceration.

Record keeping requirements

To ensure effective health surveillance, it is essential that occupational health practitioners are aware of the hazardous nature of substances, the control methods in operation, air monitoring carried out and personal protective equipment being used by employees.

A record keeping system established in connection with the COSHH Regulations should incorporate the following:

1. Information on hazardous substances
   Substances (safety data sheets)
   Processes
   Potential emission situations
2. Assessments
   Details of health risk assessment in specific areas
   Results of air monitoring surveys
3. Prevention and control measures
   Testing and planned maintenance procedures
   Elimination potential
   Substitution potential
   Storage and distribution arrangements
   Totally enclosed processes
   Local exhaust ventilation systems
4. Personal protective equipment
   Selection, testing and maintenance details
   Respiratory protective equipment maintenance
5. Individual personal health file
   Health assessment/screening details
   Information, instruction and training received
   Issue of personal protective equipment
   Details of acute exposure situations.

The benefits of health surveillance

Apart from the need to comply with legislation, such as the COSHH Regulations, the regular review of the state of health of employees ensures a fit and healthy workforce, identifies early evidence of not only occupational disease but other non-occupational diseases and conditions of which employees may be unaware at the time.

Health surveillance should be viewed as an essential element of the management process, embracing a range of procedures directed at maintaining the health of the workforce. This should result in reduced sickness absence and claims for work-related disease and ill-health.

Bibliography and further reading

Bertram, D. and Taylor, D., *The Allied Dunbar Business Tax & Law Handbook* (Pitman Publishing, London, 2003)

Health and Safety Commission, *Directors' responsibilities for health and safety* (HSE Books, Sudbury, 2002)

Health and Safety Commission, *First Aid at Work: Health and Safety (First Aid) Regulations 1981: Approved Code of Practice* (HMSO, London, 1990)

Health and Safety Commission, *General COSHH ACOP (Control of substances hazardous to health) and Carcinogens ACOP (Control of carcinogenic substances) and Biological agents ACOP (Control of biological agents): Control of Substances Hazardous to Health Regulations 2002* (HSE Books, Sudbury, 2002)

Health and Safety Commission, *Health and Safety at Work etc. Act 1974* (HMSO, London, 1975)

Health and Safety Commission, *Management of health and safety at work: Management of Health and Safety at Work Regulations 1999: Approved Code of Practice and Guidance* (HMSO, London, 2000)

Health and Safety Commission, *Managing construction for health and safety: Construction (Design and Management) Regulations 1994: Approved Code of Practice* (HMSO, London, 1995)

Health and Safety Commission, *Safety representatives and safety committees: Safety Representatives and Safety Committees Regulations 1977: Approved Code of Practice and Guidance:* (HMSO, London, 1978)

Health and Safety Commission, *Safe use of work equipment: Provision and Use of Work Equipment Regulations 1998: Approved Code of Practice and Guidance* (HMSO, London, 1998)

Health and Safety Commission, *Workplace health, safety and welfare: Workplace (Health, Safety and Welfare) Regulations 1992: Approved Code of Practice* (HMSO, London, 1992)

Health and Safety Executive, *A Guide to the Health and Safety at Work etc. Act 1974* (HMSO, London, 1990)

Health and Safety Executive, *Buying new machinery* (HSE Books, Sudbury, 1998)

Health and Safety Executive, *Consulting employees on health and safety: A guide to the law* (HSE Books, Sudbury, 1996)

Health and Safety Executive, *COSHH: a brief guide to the regulations: Control of Substances Hazardous to Health Regulations 2002* (HSE Books, Sudbury, 2003)

Health and Safety Executive, *Display screen equipment at work: Guidance on Regulations: Health and Safety (Display Screen Equipment) Regulations 1992* (HMSO, London, 1992)

Health and Safety Executive, *Everyone's guide to RIDDOR 1995: Reporting of Injuries, Diseases and Dangerous Occurrences Regulations 1995* (HSE Books, Sudbury, 1999)

Health and Safety Executive, *Five Steps to Successful Health and Safety Management: special help for directors and managers* (HSE Books, Sudbury, 1994)

Health and Safety Executive, *Health and Safety in Construction, HS(G)150* (HSE Books, Sudbury, 1996)

Health and Safety Executive, *Health and Safety Regulation: A Short Guide* (HSE Books, Sudbury, 2002)

Health and Safety Executive, *Help on work-related stress; a short guide* (HSE Books, Sudbury, 2000)

Health and Safety Executive, *Manual handling: Manual Handling Operations Regulations 1992: Guidance on Regulations* (HMSO, London, 1992)

Health and Safety Executive, *Memorandum of Guidance on the Electricity at Work Regulations: Guidance on Regulations* (HSE Books, Sudbury, 1989)

Health and Safety Executive, *Personal protective equipment at work; Guidance on Regulations; Personal Protective Equipment at Work Regulations 1992* (HMSO, London, 1992)

Health and Safety Executive, *Reducing error and influencing behaviour, HS(G)48* (HMSO, London, 1999)

Health and Safety Executive, *The complete idiot's guide to CHIP: Chemicals (Hazard Information and Packaging for Supply) Regulations 1994* (HSE Books, Sudbury, 1999)

Health and Safety Executive, *Working with ADR: an introduction to the carriage of dangerous goods by road)* (HSE books, Sudbury, 2004)

Health and Safety Executive, *Workplace health, safety and welfare: a short guide for managers* (HSE Books, Sudbury, 1999)

Home Office, *The Fire and Rescue Services Act 2004* (HMSO, London)

Iles, T., *Environment, Health and Safety Handbook* (Gee Publishing Ltd, London, 2000)

Munkman, J., *Employers' Liability at Common Law,* (Butterworths, London, 1962)

Peters, R., Gill, T., Tyler, M. and Stranks, J. *Health and Safety: Liability and Litigation.* (FT Law & Tax, 1995)

Secretary of State for Employment, *Employment Rights Act 1996*, Part V, HMSO, London, 1999

Secretary of State for Employment, *The Fire Precautions (Workplace) Regulations 1997* (HMSO, London 1997)

Secretary of State for Employment, *The Health and Safety (Miscellaneous Amendments) Regulations 2002* (HMSO, London, 2002)

Secretary of State for Work and Pensions, *The Disability Discrimination Act 1995* (HMSO, London, 1995)

Secretary of State for Work and Pensions, *The Disability Discrimination Act 1995 (Amendment) Regulations 2003* (HMSO, London, 2003)

Stranks, J., *Health and Safety at Work: Key Terms* (Butterworth Heinemann, 2002)

Stranks, J., *Health and Safety for Management* (Highfield.co.uk. Limited, 2003)

Stranks, J., *The Handbook of Health and Safety Practice* (Prentice Hall, 2003)

Tromans, S., *The Environmental Protection Act 1990: text and commentary* (Sweet & Maxwell, London, 1991)

Wolf, S. and White, A., *Environmental Law* (Cavendish Publishing Limited, London, 1995)

Index